T0329330

Understanding
Satellite Navigation

Understanding Satellite Navigation

Rajat Acharya

AMSTERDAM • BOSTON • HEIDELBERG • LONDON
NEW YORK • OXFORD • PARIS • SAN DIEGO
SAN FRANCISCO • SINGAPORE • SYDNEY • TOKYO

Academic Press is an imprint of Elsevier

Academic Press is an imprint of Elsevier
32 Jamestown Road, London NW1 7BY, UK
225 Wyman Street, Waltham, MA 02451, USA
525 B Street, Suite 1800, San Diego, CA 92101-4495, USA
The Boulevard, Langford Lane, Kidlington, Oxford OX5 1GB, UK

ISBN: 978-0-12-799949-4

British Library Cataloguing-in-Publication Data
A catalogue record for this book is available from the British Library

Library of Congress Cataloging-in-Publication Data
A catalog record for this book is available from the Library of Congress

For information on all Academic Press publications
visit our website at http://store.elsevier.com/

Typeset by TNQ Books and Journals
www.tnq.co.in

Printed and bound in the United States of America

Dedicated to the loving memory of
Chiku, Charki and Gunda

Contents

Preface

I was very much motivated to write a book on Satellite Navigation Principles when my booklet published internally in ISRO on this subject for the Satellite Navigation Course by the Centre for Space Science and Technology Education-Asia Pacific (CSSTE-AP) was much appreciated by my students. From the feedback I received from people from a wide spectrum of professions and countries and represent both commercial and strategic users, as well as researchers, it was clear that a book was needed which was acceptable equally in terms of ease of understanding and pertinent. It was obvious that a good introductory book with a selective assortment of subjects and explanations simple enough for new learners was necessary. So, this book has been written with the aim of introducing the subject to beginners, coming from different backgrounds, in a very congenial form, so that the subject can be learned from scratch.

One of the main features of this book is that it explains the basic working principles right from a foundation level with just the necessary mathematics and no complex-looking equations. This book is written keeping in mind the undergraduates and all those readers who are being introduced to the subject for the first time. It is expected that the simplistic approach developed from the first principle yet incorporating all technicalities, will not only serve them well but also be a source of enjoyable learning.

This book will cover the basic working principles of the generic satellite navigation system, instead of concentrating on any particular existing system. It will emphasize the build up of fundamental ideas for each involved process based on elementary physics added to rational common sense. Every basic principle is followed by mathematical substantiation, but using only as much mathematics as is deemed necessary for target readers. In addition, this book employs MATLAB as a visualisation tool, for every important new concept that is introduced. This will allow readers to corroborate all that they have learned through simulation, which we believe makes this book unique.

Finally, it is worth mentioning that all the relevant topics on the subject have been encompassed in a comprehensive manner. However the views and the opinions presented here are those of the author and do not necessarily reflect the views of his employer, nor the Government of India.

Rajat Acharya

Acknowledgment

I wish to extend my sincere gratitude to the University of Calcutta and especially to personalities like Prof. Asish Dasgupta, Prof. Apurba Datta and Prof. Bijoy Banerjee, who have always been my inspiration. I am also grateful to the, Space Applications Centre (SAC-ISRO) for the support it has provided me. The courtesy extended by Dr Bijoy Roy, Dr Chandrashekhar, Mr Suman Aich, and Mr Ananya Roy of SAC by agreeing to review and comment on the early versions of the manuscript are also very much appreciated. Thanks are due to Dr Suman Ganguly of CFRSI for his kind cooperation. I also acknowledge the generous gestures of Dr M R Sivaraman, Dr Kalyan Bandyopadhyay, Mr Vilas Palsule and the whole team of the CSSTE-AP. I would also like to thank the entire Elsevier team and the peer reviewers of this book. Last but not the least, I sincerely thank my wife Chandrani, my son Anubrata, and my parents, for their inspiration and for the relentless sacrifices they made during the writing of this book in order to make this effort a success.

Introduction to Navigation

1

CHAPTER OUTLINE

1.1 Introduction

Navigation is a basic need for anyone who wants to move with a purpose. Navigation is the art of moving in a suitable direction in order to arrive at a desired location. Thus, even in prehistoric times, when the most primitive form of animals started moving on earth, the art of navigation existed in its most ancient form. Even today, when humans, the most evolved species on earth, move by flying in the most technologically advanced aircraft or by driving a car, or by riding a bicycle or simply walking, with a desire to reach somewhere, we perform some sort of navigation.

You may have noticed that when we move without the aid of instruments and the route to our destination is known to us, we generally use some sort of mental

Understanding Satellite Navigation. http://dx.doi.org/10.1016/B978-0-12-799949-4.00001-4

map, which is mostly pictorial in the form of landmarks and connected paths. On this map, we identify our positions and apply our previous experience to guide us and decide the course of our movement. However, this method does not work for a new destination or for places where such landmarks are not present, which is the reason why people get lost in deserts or on the oceans. In such situations, we paper or digital maps, which give similar information. However, whether paper or digital, or as mental pictures including other geographical information, these maps are aids to navigation that enable us to locate and relate our positions with respect to our destinations and show different possible ways to reach there. The decisions we make in choosing the course of our movement by comparing our position with the available information on these maps is called navigation.

Thus, it is apparent that we first need to know our position to identify correctly where we are, and then to make an appropriate decision about where to move. *Satellite navigation* is a method that provides us with the correct position on or off the earth for this purpose. Here, signals transmitted from navigation satellites are used to derive the required set of position parameters by a navigation receiver. In turn and in conjunction with the additional information, these parameters are used to further decide the course of movement.

However, positions are not sought only for movement. Sometimes our exact position is also required to be correlated with other facts or to derive ancillary information. For example, if we know our position on the earth's surface, we can easily figure out the kind of climate we must expect. Knowing precise positions of a network of points on the earth will also let one obtain the exact shape of the earth or its derivatives, such as tectonic or other crustal movements. There are many other interesting applications of navigation, which we will discuss in Chapter 10. There, we shall come to know how this knowledge about position and its derivatives can be used for many exciting applications.

The general requirement of the estimation of position is global; for that, we need to represent positions uniquely. Positions are hence represented in terms of global standards such that positions of all the points on and near the earth can be expressed by a certain unique coordinate based on a common reference. It is like the unique identity of that position. Thus, finding the position of a person is simply a matter of determining the unique identity of the place where he or she is currently located. These coordinates are hence chosen to specify the positions in a convenient manner. In later subsections of this chapter, we will learn about reference frames and coordinate systems, which forms the basis for representing the positions. Nevertheless, the definition of these coordinates assumes the existence of certain geodetic parameters.

1.1.1 Organization of this book

The philosopher Socrates said "Know thyself." At the outset of learning navigation, we can update this to say, "Know (the position of) thyself." Thus, our entire endeavor throughout this book will be to understand the fundamentals of how modern space technology is used to fix our own position, aided by advanced techniques and effective resources. Details about existing systems currently being used for this purpose will be discussed post hoc.

However, it is also important to know how the information is organized in this book. The more logically things are developed here, the more easy it will be to understand them. Thus, it is a good idea to first have a holistic view of how the different aspects of a satellite navigation system are gradually introduced in the chapters in this book. We therefore suggest that readers continue to pursue this section describing the overall organization of this book, about which many of us have a general apathy and a tendency to want to skip this explanatory material.

The first chapter of this book is informative. We will start by introducing the term 'navigation' and getting a feel for the real development of a navigation system through a chronological description from their inception up to the current state of the art. We will first learn about the historical development of the navigation system. Whilst to some history may sound boring according to Sir Francis Bacon "Histories make men wise". We will therefore take a look at the history of satellite navigation before we gear up to understand the technological aspects of the subject. Then, before we move on to the topics of satellite-based navigation, a brief introduction to its predecessors, including other forms of navigation, should prove helpful. All of these will be covered in this chapter, and reading it, we hope, will be as interesting as the technology in subsequent chapters. Chapter 2 is also information based, primarily regarding the overall architectural segments of the whole satellite navigation system. Although we will only learn in detail about the control segment in this chapter, other elements will be discussed in the following chapters. Enjoyment of this book will intensify in Chapter 3, where we describe the space segment of the architecture. From this chapter onward, there will be frequent Matlab activities illustrating the current topic. We suggest that readers attempt these activities as they come across them, rather than leaving them to the end. Chapter 4 details the satellite signals used for navigation purposes and transmitted by satellites. Their characteristics will be described and the rationale for their use explained. Chapter 5 describes the user segment and will provide the working principles of a navigation receiver and the different aspects of it. We will explain how signals are used in receivers to derive the parameters required to fix a position. Chapter 6 explains the algorithms for the derivation of the navigation parameters i.e. position, velocity, and time, by using the measurements and estimations performed in the receivers. Receiver errors in such estimations with their sources and effects are discussed in detail in Chapter 7. Chapter 8 contains the topic of differential navigation system. It is a vast subject that could easily fill a book the same size as this or even bigger. However, we have accommodated it here into a single concise chapter of only few pages. Chapter 9 looks at special topics such as the Kalman filter and the ionosphere, both of which have large implications for navigation systems. Readers may skip reading this particular chapter if they wish, without loss of continuity. However, that would be at the cost of some very interesting material. Finally, Chapter 10 provides details of some important applications of satellite navigation.

1.2 **Navigation**

Navigation is related to the art of getting from one place to another, safely and efficiently. Although, the word 'navigation' stems from the Latin word Navigare, which means 'to sail or drive a ship,' its contemporary meaning is the art of providing position and/or direction to anyone on land or sea or in space (Elliot et al. 2001).

1.2.1 **History of navigation**

The art of navigation predated the advent of mankind. Prehistoric animals moved in search of food using their innate navigation skills. Figure 1.1, however, is only indicative.

Humans have been using different techniques of navigation from the early ages of civilization. Primitive people living in caves had to hunt deep in the forest in search of food when geographical movement was not easy and finding their way back was difficult. Thus, they made special marks on trees or erected stone pillars to create landmarks in order to find their way back home. The use of sound or smoke signals were another common means of finding their way back. We will look more at the formal classifications later, but it is worth mentioning here that these most primitive methods of navigation were of the *guidance* type.

Navigation developed at sea, with most developments in the modern navigation system occurring in the process of guiding sea vehicles. In ancient times, seafaring explorers started traveling across the oceans in search of new lands, in order to increase trade and colonize. Development in the field became necessary in order to cater to the needs of voyages and the constant effort to improve them.

FIGURE 1.1

Primitive navigation.

As mentioned before, mapping the sea along side its adjacent lands is not navigation, but deriving ones own position from it and thus deciding the direction of movement toward a certain destination is. However, both systems were developed simultaneously and are sometimes treated as the same thing. In this section, we restrict our discussion to navigation only.

The first kind of sea navigating was probably done by skirting around the coast, and thus by staying in sight of land. Pictorial maps were created during this time by sailors who would draw what they could see along the coast. Using these, they could return or retrace their course on subsequent journeys. The first known coastal and river maps were from China in around 2000 B.C. and indicated sailing directions (Wellenhof et al. 2003). However, when voyages ventured further out into the sea, the only means of navigating was by observing the position of the sun and stars. This kind of navigation is termed celestial navigation. Some experienced sailors could also navigate by understanding the winds or determining the depth of the seabed, from which they could estimate their distance from the land. This was probably the earliest form of bathymetric navigation.

Written records of celestial navigation date back to the third century B.C. Some of these accounts are available in Homer's epic, *The Odyssey* (History of Navigation, 2007). The astrolabe, which measures the elevation of the sun and the stars, as shown in Figure 1.2(a) became the main instrument for positioning and was apparently used even before 600 BC (Kayton, 1989). Heron and Vitruvious gave a detailed description of the odometer, an instrument to measure distance (Wellenhof et al. 2003). During this time, Greek and Egyptian sailors started using the polar stars and constellations to navigate because they did not disappear below the horizon throughout the night. However, movement using polestars needed to be corrected with time as the stars change their positions because of the wobbling of the earth on its axis. Measurements of the instruments were aided by nautical charts. Ptolemy produced the first world map, which remained in use for many years during sea voyages. Textual descriptions for sailing directions have been in use in one form or another since then.

The middle ages in navigation were marked by the discovery of lodestone. With this, navigation became easier for sailors, who started to use it for its magnetic properties. Comparing it with detailed maps of the period, they could find their way easily even with unfavorable sky conditions allowing sailors to navigate even with limited visibility. The first true mariner's compass was invented in Europe toward the beginning of the thirteenth century A.D. Thus, when Christopher Columbus set out on his transatlantic voyage in 1492, he had only a compass, a few dated measuring instruments, a method to correct for the altitude of Polaris, and some rudimentary nautical charts as tools for navigation.

From the middle of the sixteenth century, navigation saw a rapid development in related technology when a number of instruments and methods were invented. This was when the Europeans started to settle colonies in different countries, and they used sea routes to navigate to these new lands. The improvement of navigational techniques became mandatory and the mathematical approach toward navigation made it a scientific discipline. By the seventeenth century, the quadrant had become one of the dominant instruments. Magnetic variations were studied and the magnetic dip, the angular inclination of the geomagnetic lines of force at a location, was discovered, which gave enormous support to position finding. The defined nautical mile could also now be measured with much accuracy. By the middle of the eighteenth century, the invention of instruments such as the sextant and the chronometer, marked the onset of modern times in navigation. The sextant, as shown in Figure 1.2(b), could measure the elevation of the sun, moon or a star by aligning its reflection from a semi-reflecting surface with the visible horizon seen directly through it. In the process, its own relative position could be estimated when the position of the star or the sun was known.

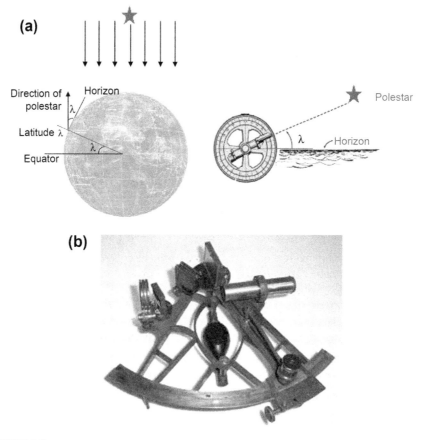

FIGURE 1.2

(a) Working principle of Astrolabe, (b) Sextant.

Time always remained an important parameter when finding a position. The sundial was a primitive clock, the oldest of which was found in Egypt at Karnak. It was also used in ancient Greece and China (Kayton, 1989). Later, the pendulum clock was invented and was used to keep time on land but was not suitable for marine platforms. Thus, for the seagoing vehicles and mariners, the best devices were still water and sand clocks. This put a serious constraint on the accurate determination of longitude, which needs a precise knowledge of time. Around the middle of the eighteenth century, a huge sum of prize money was offered to anyone who could provide a precise method of measuring longitude. In 1759, John Harrison invented a clock that was accurate within a few seconds over a period of around 6 months. Captain Cook used the Harrison Clock for his expedition to the Antarctic. Another remarkable event that took place in the determination of longitude was the landmark decision of adopting the prime meridian (0° longitude) in 1884. It remains the basis of positioning even today.

In the last decade of the nineteenth century, radio communications started in the form of wireless telegraphs. For sea goers, signals started being sent to ships not only in the form of messages, but to allow navigators to correct their chronometers. Radio communication between ships also helped sailors to make navigational decisions.

Radio-based navigation systems advanced rapidly during World War II. By this time, the quartz clock became available and microwaves were used extensively with navigational devices and British physicist Robert Watson first demonstrated the Radio Detection and Ranging system (RADAR) as a warning system against air attacks. This technology was readily implemented in ships as a navigation aid. Soon after this, Alfred Loomis suggested a radio-based electronic navigation system, which was later developed into the Long Range Navigation (LORAN) system.

Radio navigation was ushered into a new era in October 1957, when the former Soviet Union (USSR) launched the world's first artificial satellite, Sputnik. Scientists used the Doppler shift of the Sputnik's signal to obtain the satellite's position and velocity. Subsequently, a series of satellite-based navigation programs had been undertaken and established by both the United States and the USSR. During this time, satellite constellations for navigation based on both Doppler and ranging came into use. The TRANSIT satellites that started operating in 1964 were Doppler-based navigation systems, whereas the SECOR system was based on ranging. These were followed by the Russian TSIKADA and American TIMATION systems. TIMATION was planned for time transfer by sending precise clocks into space. The results of these precursor programs formed the basis of today's Global Positioning System (GPS) of NAVSTAR (Parkinson, 1996) and GLONASS. Currently, many countries and groups of nations use their own satellite-based navigation system: the Galileo system in the European Union, COMPASS in China, and IRNSS in India are some of them. We shall learn about the basic working principles of these systems in later chapters.

1.2.2 **Types of navigation**

A modern navigation system is typically a radio navigation system that is non-autonomous in nature. It means, the system operates only when an appropriate external signal is received by a receiver (Wellenhof et al., 2003). It provides position, velocity, and time (PVT) in a three-dimensional system. However, there are certain systems which only give the path that one should follow to reach the destination. Based on the nature of parameters and how they are derived, modern navigation systems can be divided into three broad types.

1.2.2.1 *Guidance*

Guidance is a type of navigation that provides only a course to a destination for the user, but with no information about its exact position. Thus, the user only knows the exact route that should be followed to lead him or her to his destination, with no knowledge of his or her present position.

Guidance is the oldest type of navigation. The movements of early travelers finding their way to their destination by observing the rising and setting of the sun and the moon and orientation using constellations were navigation of the guidance type. In modern times, when you follow the markers in a big airport directing you to reach your designated terminal, or when you decide your route on the basis of displayed signs on a highway, this is navigation of the guidance type. Thus, we frequently utilize guidance navigation throughout our lives sometimes without even realizing it.

Some modern radio navigation systems are in this group, such as the instrument landing system (ILS) and microwave landing system, which are used for aircraft.

1.2.2.2 *Dead reckoning*

Sometimes it is difficult to use guidance navigation, especially for long-range movement. In such cases, it is more convenient to know one's current position rather than be guided from origin to destination. But how do we update our position with time? Position at the current instant can be determined from the positions at any prior time using the value of time elapsed since then and some simple dynamic parameters obtained as position derivatives. Thus, current positions of any moving entity can be deduced in relation to any prior position, or even with respect to the point of its origin of movement. One of the easiest methods of doing this is by using the basic principles of Newton's laws of dynamics. From these laws, we derive the current position and velocity of the body as

$$v_k = v_{k-1} + f_{k-1} \cdot (t_k - t_{k-1}) \tag{1.1}$$

$$S_k = S_{k-1} + v_{k-1} \cdot (t_k - t_{k-1}) + \tfrac{1}{2} f_{k-1} \cdot (t_k - t_{k-1})^2 \tag{1.2}$$

where S_k and S_{k-1} are the positions at time t_k and its previous instant t_{k-1}, respectively. v_k and v_{k-1} are the corresponding velocities, and f_{k-1} is the acceleration at instant t_{k-1}.

Therefore, the position S_k and velocity v_k at any instant t_k may be derived from those at its previous instant t_{k-1}, just by knowing the acceleration, f_{k-1}, and from their previous values S_{k-1} and v_{k-1}, respectively. Similarly, the values of S_{k-1} and v_{k-1} can be derived from parameters at an instant even before them. Therefore, extending this logic backward, we can say that if we know the position S_0 at any starting point with a standstill condition, i.e. ($v_0 = 0$) at any time instant t_0, we can find out its position and velocity at any later time t_k just by measuring acceleration f at the initial instant and all intermediate instants. Because the present position is deduced from an initial *standstill* condition (i.e. the 'dead' condition of the body) from which we start reckoning (i.e. calculating the position), this kind of navigation is called dead reckoning and the term is also sometimes said to be derived from the word 'deduced' (Meloney, 1985).

There are six degrees of freedom for any massive rigid body. Degrees of freedom are the set of independent dimensions of motion of the rigid body that completely specify the movement and orientation of the body in space. They indicate the independent directions in which a body can exhibit linear or rotational motion without affecting its similar movement in any other direction. In accounting for such motions, the three directions of translational motion defining the position of the body are considered, along with the three orthogonal rotational directions of the body that give its orientation. Thus, in addition to position, dead reckoning can also be used to find the orientation of the body by measuring the angular acceleration of the body about these three rotational axes.

Navigational systems based on these inertial properties of a dynamic system belong to this category, such as the inertial navigation system (INS). Most commercial aircraft use this as their primary navigational system.

1.2.2.3 Piloting
In the piloting or pilotage type of navigation, the user derives navigational parameters (PVT) which are updated each time. New measurements are performed over the update interval, which leads to new positions for every update.

Satellite-based navigation, which is the subject of this book, belongs to this category. Other systems of this kind are hyperbolic terrestrial radio positioning systems such as LORAN etc. (LORAN, 2001; Loran, 2011)

1.3 **Referencing a position**

We fix our positions using different navigation systems. But the question is, *with respect to what?* A more fundamental question that may arise is, *do we always need to represent our position with respect to something?* If yes, then what should that 'something' be?

To answer these questions, let us start by using a simple analogy. How do we typically communicate our positions in everyday life? Verbally, we actually tell our location to someone in a fashion such as, "I am on Parallel Boulevard, about a 100 yards right from the old lighthouse," or "I am at Copley Plaza, about 50 m south of the city library," or "I have come across the airport by half a kilometer due east." Notice the common features we use in these statements. In all cases, we refer to our position in terms of distance with respect to some specific reference landmarks such as the lighthouse, the city library, or the airport. Furthermore, we mention the distance from them in definite directions. We also assume that the person to whom the position is being described already knows these landmarks used as references. The description is useless if he or she is new to these places and does not know where the references are. Thus, what we deduce from these are that we need a fixed (or apparently fixed) and defined reference to describe our position, and a distance from or direction away from these reference points. Fixed and known references, defined directions, and definite distances are thus the elements required to describe positions.

The reference used should be universally accepted and understood and should be convenient for referring to any position of interest. We have talked about specific landmarks as references in our example. These references are local and cannot be used to describe positions of any location across the globe. Therefore, to perform pragmatic position fixing, what should be the nature of these references? The obvious answer is that the reference point itself must be located with approximate equal nearness to all points whose positions are to be described so that any point in question may be represented with equal convenience.

Then, once the reference is set, the next requirement is to represent the distances of our position in the best possible way. From any pragmatic reference point O, as in Figure 1.3, there will be a shortest radial range, moving along which the point in question, P, may be reached. The distance is shown as R. However, in our three-dimensional space, this range may be in any arbitrary direction from the reference, O. Describing any arbitrary direction is impossible unless we make use of some predefined standard

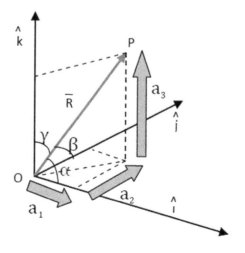

FIGURE 1.3

Orthogonal bases in a three-dimensional space.

direction for specifying it. In a three-dimensional space, there can be a *maximum* of three mutually orthogonal directions. We may fix three such directions in space, each referred to as an axis. The direction of radial range may be described by the angle it makes with such directions. The angles are represented by α, β and γ which are such that the relation $\cos^2 \alpha + \cos^2 \beta + \cos^2 \gamma = 1$ is always maintained. The same point P may also be reached by moving vectorial distances along these axes. These paths are nothing but projections of the radial vector distance R on the defined axes shown as a_1, a_2, and a_3 in Figure 1.3. Thus, we can express vector $\underline{\mathbf{R}}$ as

$$\begin{aligned} \mathbf{R} &= a_1 \widehat{i} + a_2 \widehat{j} + a_3 \widehat{k} \\ &= R \left(\cos\alpha \, \widehat{i} + \cos\beta \, \widehat{j} + \cos\gamma \, \widehat{k} \right) \end{aligned} \tag{1.3}$$

where $\widehat{i}$, $\widehat{j}$, and $\widehat{k}$ are the unit vectors along these three axes and form the basis.

In general, we need to move three orthogonal vectorial distances along the axes to move effectively from the reference point to reach any other point of choice. Any mutually orthogonal vectors can be used to represent the axes for movement, but depending on the need, we use some predefined fixed directions of these vectors because it serves no purpose if these three directions are always chosen arbitrarily.

So, represent the position of a point, we first need to fix a reference point. Then, with respect to this reference, the positions of all other points in question may be described in terms of distances along three fixed, predefined orthogonal axes.

1.3.1 Reference frame

Reference frame: Any arbitrary reference point and associated definition of three orthogonal axes, with respect to which the position of all other points may be defined, constitutes a reference frame. It is typically defined by specifying the position of the reference point and direction of the axes. The reference point is specified by attaching it to any physical system and is referred to as the origin of the frame.

Accordingly, there may be two types of reference frames, described below.

Inertial: An inertial frame of reference is one that is not itself accelerating, and hence one in which the laws of inertia are valid.

Non-inertial: A non-inertial frame of reference is one that is itself accelerating, and hence the laws of inertia are not valid.

We mentioned that the positions of other points are defined in terms of distances along the defined directions from this reference point. The unit vectors along these three orthogonal vectors thus form the basis for describing distances with respect to the reference. There can be different orthogonal sets of basis vectors, and each such set constitutes a coordinate system.

Coordinate System: A defined set of three orthogonal basis vectors associated with each axis of a reference frame, in which the position of any point in space may be described in terms of distances along the axes from the origin.

Different types of coordinate systems, such as Cartesian, spherical, and cylindrical, are used for different purposes. However, for navigation, geodetic reference frames with cartesian or spherical coordinate systems are typically used, as mentioned previously.

1.3.1.1 Heliocentric reference frame

Helios was the Greek sun god whose name was later Latinized as *Helius* to represent the sun. Thus, from the name, it is evident that heliocentric reference frames are those in which the origin of the reference frame is fixed at the center of the sun. These references are used to represent the positions of the celestial bodies or the positional elements in the solar system. However, it is not suitable for representations of positions over the earth or near it.

1.3.1.2 Geocentric reference frames

We have seen that the location of points in three-dimensional space are most conveniently described by coordinates with an origin around the points. Therefore, to suitably represent positions on the earth and around it, the chosen reference frames are typically geocentric. The origin coincides with the center of the earth and the axes align with the earth's conventional axes or planes. Geocentric reference frames can be naturally divided into different classes, as described subsequently.

To represent positions on the earth and its surroundings, a geocentric reference frame may be defined with Cartesian coordinates. But how are the axes of this frame defined? For an inertial system, the axes should not linearly accelerate or rotate, because rotation is always accompanied by acceleration.

We know that the earth is spinning about its own axis and is revolving around the sun as well. Thus, what appears to us to be fixed and stationary on the earth when we look at it standing on the earth, is not actually so. We can find that everything on the earth is rotating with it, if we look at it from space. Thus, no frame fixed with the earth can be stationary. Furthermore, as the earth revolves round the sun, so too is everything that appears to be fixed upon it.

Then what can we use as a stationary reference frame? Honestly speaking, nothing has yet been found that can be treated as absolutely stationary. No reference may be truly considered inertial. Therefore, we use relative stationarity, or those references that are approximately stationary, compared with the motion of the earth. The distant stars can be used for this purpose. When we look toward the sky from the earth, the distant stars appear to surround us in all directions. These stars apparently remain fixed at their positions throughout the year, whatever the position of the earth around the sun. These stars are at such a great distance that the range of the earth's movement is proportionately negligible compared to it. These distant stars may be assumed to form another hollow sphere of infinite radius concentric with the earth, called the celestial sphere. The deviations of these stars, as observed over the entire movement of the earth, are trivial, and hence can be considered stationary. However, the sun appears to move in this geocentric frame on an elliptical path called the ecliptic. The direction toward the distant stars on the celestial sphere from the geocenter through the point where the ecliptic crosses the equatorial plane with the northward motion of the sun is called the direction of the vernal equinox or the first point of Aries. The angular distances of the distant stars as observed from the earth are reckoned with respect to this fixed direction that marks the beginning of the Aries constellation (Figure 1.4).

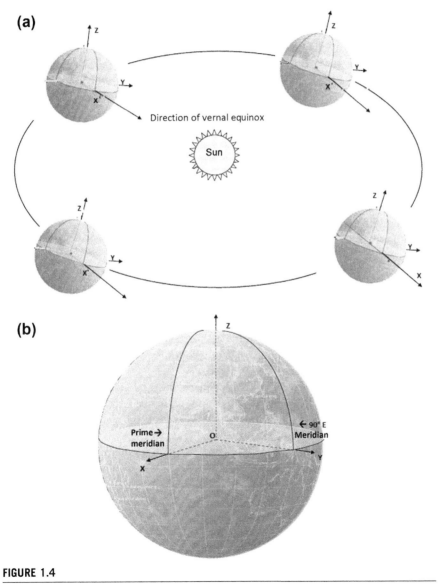

(a)

Direction of vernal equinox

Sun

(b)

Prime→ meridian

← 90° E Meridian

FIGURE 1.4

(a) ECI reference frame, (b) Earth-centered, earth-fixed reference frame.

1.3.1.2.1 Earth-centered inertial

The earth-centered inertial (ECI) is a geocentric reference frame that does not rotate with the earth, and hence, the axes retain their orientation relative to these fixed distant stars.

In the ECI frame, as illustrated in Figure 1.4(a), the origin of the frame is at the center of the earth, and the three axes of right-handed orthogonal frame are attached to it. The X and Y axes of the Cartesian coordinates remain on the equatorial plane of the earth mutually perpendicular to each other. The X axis remains always directed toward the first point of Aries, whatever the position of the earth. The Z axis remains aligned with the mean rotational axis of the earth and points toward the North Pole. Because of this, although it is geocentric, the reference frame does not rotate as the earth rotates about its axis. Thus, the coordinate system has a fixed orientation in space, and hence may be called an inertial system as long as other perturbations of the axes are considered trivial. However, the perturbations cannot be totally neglected for precise applications. To alleviate this problem, the directions are defined at a fixed epoch, i.e. the direction that these axes made on January 1, 2000, is taken as the standard (Kaplan et al. 2006). Satellites revolving around the earth experience gravitational pull effective from the point of its mean center but are not affected by the rotation of the earth. Therefore, to represent the positions of satellites revolving around the earth, ECI acts as a suitable reference point. Because the earth rotates whereas the frame does not, the position of the points fixed on the earth's surface changes with time in this reference frame.

1.3.1.2.2 Earth-centered, earth-fixed

We saw in the last section that in an ECI frame the position of locations fixed on the earth's surface changes with time. This is inconvenient for conventional positioning uses. Thus, if we define a geocentric reference frame in which the axes are fixed with the earth and rotate with it, the coordinate of any position on the earth remains fixed over time. Hence, this problem can be avoided.

Earth-centered, earth-fixed (ECEF) is a geocentric reference system in which the axes are attached to the solid body of the earth and rotate with it. This is shown in Figure 1.4(b). In this kind of system, the origin of the frame is again at the center of the earth and a right-handed frame of axes is attached to it to. The distances are primarily represented in Cartesian coordinates, XYZ. The X and Y axes remain perpendicular to each other. The X axis is fixed along the prime meridian (i.e. 0° longitude) and the Y axis is hence accordingly placed on the equatorial plane along 90° E longitude. The Z axis remains pointing toward the North Pole. But here, the axes of the coordinate system are fixed with the earth itself and are rotating with it, hence the name. As the earth rotates about its axis, the coordinate system also rotates with it, rendering the frame non-inertial. It is suitable for indicating positions located on the earth's surface and consequently moving with the earth.

1.3.1.2.3 Geographic and geodetic coordinates

In describing these reference frames, we have mentioned nothing about the shape of the earth. You may argue that one should not be concerned with the earth's shape while defining the frames. It suffices to consider the origin at the center of the earth and align the axes aligned correctly. But then, how can one place the origin of the frame at the earth's center unless you know where the center is? Obviously, we do not have access to the center of the earth and what is available to us are only measurements of the earth's. Therefore, from these measurements we can fix the frame correctly and also generate a regular surface on which we can represent the position coordinates on the surface of the earth.

First, assuming simplistically that the earth is a perfect sphere, the ECEF frame may be considered with spherical coordinates for positioning. This constitutes the geographic coordinate system (Heiskanen and Moritz, 1967). Here, the orthogonal coordinates are the radial distance of the point from the reference point at the center of the earth: i.e. the radius (R); the angular distance of the point at the origin from the predefined equatorial plane, called the latitude (λ); and the angular separation of the meridian plane of the point from the meridian plane of the prime meridian, called the longitude (φ). This exempts the use of large Cartesian coordinate values for positions on the earth's surface, except for the radius, which is approximately a constant.

However, the earth is not a perfect sphere. The true surface of the earth is too irregular to be represented by any geometric shape. Therefore, this topographic irregularity may be regarded as the variation over a smooth surface. This smooth surface is known as the geoid and is a model of the earth that has a constant gravity potential. Thus, it forms an equal gravito-potential surface such that its gradient at any point represents the gravitational force at that point in a direction perpendicular to the surface. In other words, the geoid is always perpendicular to the direction of the acceleration to the gravity, the direction pointed to by the free vertical plumb lines at any point. Considering that the water surface always remains at the same potential, the mean sea level is used as a physical reference to obtain a geoid shape. It is extended to the land surface, below, above, or on its normal topology to get a complete geoid shape (Ewing and Mitchell, 1970).

The geoid is also not a defined geometric shape. Therefore, the surface needs to be approximated to a regular geometric shape, fitted in the best possible way to this geoid, so that any position may be sufficiently estimated and represented by easy mathematical calculations pertaining to that shape.

The regular shape that best fits the geoid is the ellipsoid of revolution. This is obtained when an ellipse is rotated about its minor axis. Thus, a reference ellipsoid is chosen that represents the earth's geoid shape in the closest manner. This ellipsoid is typically described by its semi-major axis (equatorial radius) 'a' and flattening 'f' = (a − b)/a, where b is the semi-minor axis (polar radius). The reference ellipsoid must have a definite orientation with respect to the true earth shape to represent it with a minimum squared error. Fixing the orientations in a three-dimensional space is equivalent to fixing the positions of a few selected points on the earth's surface

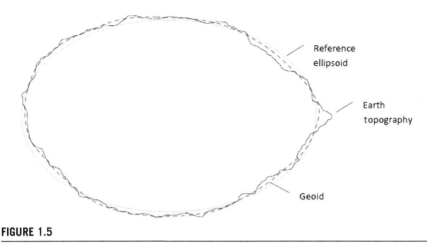

FIGURE 1.5

Ellipsoid reference.

with respect to the right-handed ECEF coordinate frame positioned at the center of this ellipsoid model of the earth. Conceptual descriptions of these surfaces are illustrated in Figure 1.5.

Realizing the ellipsoidal model in practice requires positioning a large number of points on the earth's surface to be measured including their gravitometric parameters. Hence, it is much more practical to define the ellipsoid that fits best to smaller regions in question. This has, therefore, been mostly done on a regional basis.

To determine the proper shape of the earth and the associated parameters, the measurements performed are called the geodetic measurements, a wide and complex scientific discipline. The basic parameters determined from these measurements are called geodetic datum.

The formal meaning of datum is something that is used as a basis for calculating or measuring. In the case of geodesy, these are the data defining the dimensions of the earth. For this, geodesists generate data, local or global, constituting the parameters of this reference ellipsoid, which the empirical estimations of as the best geometric fit for the earth's surface.

Before the satellite geodesy era, the coordinate systems associated with a geodetic datum attempted to be geocentric, but all parameters were obtained by empirically fitting the local measurements with an ellipsoid surface. These were regional best fits to the geoids within their areas of validity, minimizing the deflections of the vertical over these areas. The origin being computed on a local basis differed from the geocenter by hundreds of meters, owing to regional deviations. Some important regional geodetic datums are:

- Everest Datum (ED 50)
- North American Datum (NAD 83)
- Ordnance Survey of Great Britain (OSGB 36)

With the advent of space-based measurements using satellites, a specific geocenter could be defined because the satellites orbit about the natural geocenter. This point becomes the obvious choice of origin of a coordinate system because the satellite positions in space are themselves computed with respect to this point.

World Geodetic System 1984 (WGS 84): The WGS 84 is composed of a set of global models and definitions with the following features:

- An ECEF reference frame
- An ellipsoid as a model of the earth's shape
- A consistent set of fundamental constants
- A position referred to as WGS84-XYZ or WGS84-LLA (latitude, longitude, and altitude)
- An ellipsoid semi-major axis taken as 6378137.0 m, and semi-minor as 6356752.3142 m

The height of the geoid measured normally above the ellipsoid is called the geoid height. Similarly, the height of the terrain measured above the geoid is called the orthometric height, and is expressed as the height in meters above mean sea level.

Detailed information on this subtopic can be obtained from references such as Leick (1995), Larson (1996), Torge (2001), El-Rabbani (2006), and Kaplan et al. (2006).

1.3.1.3 Local reference frames

A local reference frame is constituted when the reference point (i.e. the origin of the reference frame) is located at a point local to the observer, generally on or near to the surface of the earth instead of at the center of the earth. Because these frames are used to specify positions of objects with respect to a local observer, they are so named. Typically, the origin of such frames is chosen as the position of a receiver on the earth's surface.

1.3.1.3.1 East-North-Up

One of the most useful local coordinate systems is the East-North-Up (ENU) system. It is a local coordinate system in which the E axis of a right-handed system is directed toward the local east, the N axis is along the local north, and the U axis is vertically up. These axes are suitable for representing the position of objects above the earth's surface, such as a flying aircraft or visible satellite. By local east, we mean that the axis originating from the reference point on the earth's surface is contained on the latitudinal plane and is tangential to it at that point. Similarly, by local north, we mean that the axis is directed toward the north at that point and is tangential to the meridian plane at that local point of reference. Thus, the E and N axes both lie on the local horizontal surface. The description is illustrated in Figure 1.6.

1.3.1.4 Conversions between coordinate systems

Conversion of the coordinates from one system to another is essential for navigational purposes. Navigational estimations are typically (although not essentially)

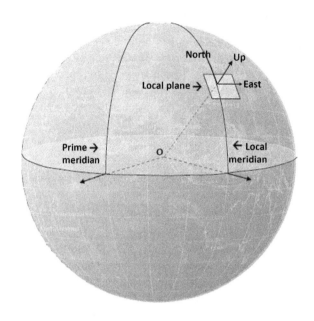

FIGURE 1.6

Local ENU coordinates.

done in ECEF Cartesian coordinates, whereas application requirements may use other systems. Therefore, the estimated coordinates are converted to geographical, geodetic, or other coordinates, as necessary. However, it may become necessary to convert them into local coordinates. Here, we discuss only the conversion between the Cartesian ECEF and ENU coordinates with spherical earth. For conversion from ECI to ECEF geodetic coordinates, see Figure 1.7 and refer to focus 3.1 (Figure 1.7).

1.3.1.4.1 Geodetic Cartesian to geodetic latitude–longitude
Refer to Figure 1.7. The Cartesian coordinates x, y, z in ECEF can be expressed in terms of the geodetic latitude λ, longitude φ, and height h as

$$
\begin{aligned}
x &= (N+h)\cos\lambda \, \cos\varphi \\
y &= (N+h)\cos\lambda \, \sin\varphi \\
z &= (N+h)\sin\lambda - \varepsilon^2 N \sin\lambda
\end{aligned}
\tag{1.4}
$$

where N is the radius of curvature in the prime vertical and is given by

$$
\begin{aligned}
N^2 &= a^4 / \left(a^2 \cos^2\lambda + b^2 \sin^2\lambda\right) \\
h &= p/\cos\lambda - N
\end{aligned}
\tag{1.5}
$$

(a)

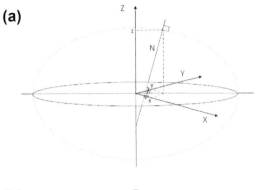

(b)

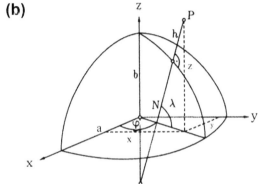

FIGURE 1.7

(a) and (b) Two different views for conversion of ellipsoid to ECEF Cartesian.

Similarly, the conversion from the Cartesian coordinates to the geodetic coordinates may be done using the following equations:

$$\varphi = \tan^{-1}(y/x) \quad \text{when } x > 0$$
$$= \pi + \tan^{-1}(y/x) \quad \text{when } x < 0,\ y > 0$$
$$= -\pi + \tan^{-1}(y/x) \quad \text{when } x < 0,\ y < 0$$
$$\lambda = \tan^{-1}\{(z + e^2(a^2/b)\sin^3 u)/(p - e^2 a \cos^3 u)\} \qquad (1.6)$$

$$\text{where } p = \sqrt{a^2 + b^2} \text{ and}$$
$$u = \tan^{-1}\{(z/p)(a/b)\}$$

1.3.1.4.2 Cartesian ECEF to ENU

To convert from conventional Cartesian ECEF to ENU coordinates with a spherical model of the earth, we need to conduct some coordinate transformation exercises. Let a point in space be represented as x, y, and z in the ECEF and E, N, and U in the ENU coordinate system with the origin located at latitude λ and longitude φ for a spherical earth of radius R.

To obtain a relationship between the two, first let us rotate the ECEF axes about the Z axis by an amount φ to an intermediate coordinate system $X'Y'Z'$ that has an earth-centered origin but with the x axis directed along the φ instead of the prime meridian. Thus, in this coordinate, the coordinates x', y', and z' may be represented as

$$\begin{aligned} x' &= x\cos\varphi + y\sin\varphi \\ y' &= -x\sin\varphi + y\cos\varphi \\ z' &= z \end{aligned}$$

(1.7)

Now, rotate the axes about the Y' axis by an amount of λ to another frame $X''Y''Z''$. This turns the $X'Z'$ plane in such a way that the new X'' axis directs vertically up at the selected point. In this frame, the x'', y'', and z'' become

$$\begin{aligned} x'' &= x'\cos\lambda + z'\sin\lambda \\ y'' &= y' \\ z'' &= -x'\sin\lambda + z'\cos\lambda \end{aligned}$$

(1.8)

This coordinate system is aligned with the ENU system with the X'' axis aligned along Up direction, Y'' along East, and Z'' along local North of the ENU origin. The only difference between them is that the origin of this transformed frame is still is at the center of the earth. So, upon radial translation of the origin to the point on the earth surface in question we get the required set of ENU coordinates as

$$\begin{aligned} U &= x'' - R \\ E &= y'' \\ N &= z'' \end{aligned}$$

(1.9)

Combining the whole, we get

$$\begin{aligned} E &= -x\sin\varphi + y\cos\varphi \\ N &= -x\cos\varphi\sin\lambda - y\sin\varphi\sin\lambda + z\cos\lambda \\ U &= x\cos\varphi\cos\lambda + y\sin\varphi\cos\lambda + z\sin\lambda - R \end{aligned}$$

(1.10)

To validate this relation, let us see what coordinate value we get for the origin of the ENU system. Obviously, it should give [0, 0, 0] in the ENU system. Its coordinate in ECEF is given by [$x = R\cos\lambda\cos\varphi$, $y = R\cos\lambda\sin\varphi$, $z = R\sin\lambda$]. Thus, when we put these values of x, y, and z in these expressions for E, N, and U, we get

$$\begin{aligned} E &= -R\cos\lambda\cos\varphi\sin\varphi + R\cos\lambda\sin\varphi\cos\varphi = 0 \\ N &= -R\cos\lambda\cos\varphi\cos\varphi\sin\lambda - R\cos\lambda\sin\varphi\sin\varphi\sin\lambda + R\sin\lambda\cos\lambda = 0 \\ U &= R\cos\lambda\cos\varphi\cos\varphi\cos\lambda + R\cos\lambda\sin\varphi\sin\varphi\cos\lambda + R\sin\lambda\sin\lambda - R = 0 \end{aligned}$$

(1.11)

This is exactly as what we expected. Focus 1.1 describes the equivalence of the range vectors in these two systems.

We have seen that in a local frame, the local horizontal plane contains the E and N axes. So, when the local frame is used with spherical coordinates, the coordinates become the radial range from the origin, the angle with the local horizontal plane, and the angle from a predefined direction—say, local north. These parameters are

---

FOCUS 1.1 COORDINATE CONVERSION

Considering the conversion formula from ECEF to local ENU, let us see how the radial range of a point in ENU is related to the range in an ECEF frame. To obtain this, let us start with the range expression in the ENU frame. If the coordinates of the point in this frame are [e, n, u], the range is given by

$$R_L^2 = \left(e^2 + n^2 + u^2\right)$$

Substituting the e, n, and u values in terms of the ECEF coordinates,

$$R_L^2 = x^2 + y^2 + z^2 + R_0^2$$
$$-2xy\sin\varphi\cos\varphi + 2xy\cos\varphi\sin\varphi\sin^2\lambda + 2xy\cos\varphi\sin\varphi\cos^2\lambda$$
$$-2xz\cos\varphi\sin\lambda\cos\lambda - 2yz\sin\varphi\sin\lambda\cos\lambda$$
$$+2xz\cos\varphi\sin\lambda\cos\lambda + 2yz\sin\varphi\sin\lambda\cos\lambda$$
$$-2R_0x\cos\varphi\cos\lambda - 2yR_0\sin\varphi\cos\lambda - 2zR_0\sin\lambda$$

where x, y, and z are the coordinates of the same point in ECEF and R_0 is the radius of the earth, which is also the radial distance of the ENU origin in ECEF. λ and φ are the latitude and the longitude of the ENU origin, respectively.

However, as the coordinate of the origin of ENU in ECEF is [$x_0 = R\cos\lambda\cos\varphi$, $y_0 = R\cos\lambda\sin\varphi$, $z_0 = R\sin\lambda$], this becomes equal to

$$R_L^2 = x^2 + y^2 + z^2 + R^2 - 2x\,x_0 - 2y\,y_0 - 2z\,z_0$$
$$= R^2 + R_0^2 - 2R \cdot R_0$$
$$= (\underline{\mathbf{R}} - \underline{\mathbf{R}}_0)^2$$

Therefore, the radial distance of any arbitrary point measured in ENU is the difference of the vectorial distance of the point and the vector representing the ENU origin, both from the geocentric EFEC origin. This is what we expected from our knowledge of vectors. This relation will be used later in Chapter 8, where we shall be deriving the error parameters of a satellite navigation system.

---

nothing but the range, R, elevation, Ele, and azimuth, Azi. Therefore, the conversion is given as

$$[E \ N \ U] = [R\cos(Ele)\sin(Azi), \ R\cos(Ele)\cos(Azi), \ R\sin(Ele)] \qquad (1.12)$$

1.4 Radio navigation system

Radio signals can traverse a long range within which they can be received by suitable receivers, and the information within can be used for different purposes, including navigation. Radio navigation has advantages in that the signals necessary to derive positions can be communicated under all weather and visibility conditions, and a lot of information required for both basic navigation as well as ancillary uses can be transferred with it reliably and with high accuracy. However, all of these advantages come with the price of more complex instruments.

Here, we shall briefly look at the working principles behind each of type of piloting: guidance system and dead reckoning system. We will also introduce other types of navigation system so that readers have a complete knowledge of the various types available. Hence, in all of these discussions, we purposefully avoid any mathematical descriptions, to make it as simple as possible. However, the related concepts of the basic foundation will be mentioned so that interested and enthusiastic readers may relate them with further reading.

1.4.1 **Piloting system**

Hyperbolic terrestrial radio navigation is a *piloting* system, based on the principle of finding the position from the time difference of arrivals of signals to a receiver from a pair of radio transmitters located at known positions. These two transmitting stations can be joined by a straight line called the principle line.

A constant time difference between signals from the two stations, A and B, as shown in Figure 1.8, represents a constant path difference between stations at the

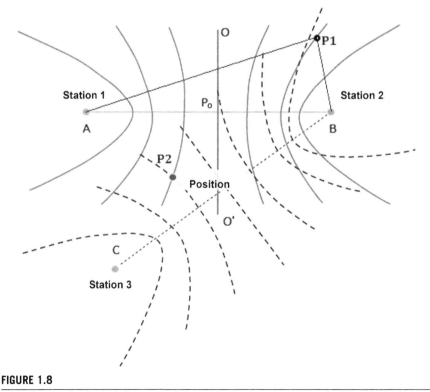

FIGURE 1.8

Hyperbolic lines of position.

receiver locations, because the radio waves may be assumed to move with a constant velocity equal to that of light, represented by 'c.' Let R_A and R_B be the ranges from stations A and B, respectively. Therefore, if two signals transmitted simultaneously from these two different stations are received by a receiver located at the point at a time difference of Δt, the range difference of these two stations from the receiver is $(R_A - R_B) = \delta = c\Delta t$.

If the difference in distance is zero (i.e. the ranges are equal), it will be equidistant from both A and B, as is point P_0 in the figure. All such possible points will construct a straight line normal to the principle line and pass through its midpoint, as shown as OO' in the figure.

If the range difference is positive, with range R_A more than range R_B, it should have proximity toward B, as is point P1 in the figure. Then, the locus of all such points with the same difference may be sought. One such point lies on the principle line with closeness toward B about the principal line. However, this time, the locus cannot be a straight line, for which the difference in ranges needs to be quadratic in nature and not constant. It also cannot have a bend toward A. Thus, it can be a curved line that bends toward B. For different values of the positive range difference, it will form different curved lines bent toward B. For negative range differences, a similar set of curves will be obtained bending toward A. The geometrical shape of these lines is called a hyperbola, each of which represents the locus of points with a definite path difference, also known as hyperbolic line of position (LOP).

If the positions of the two synchronized stations are known, the locus of the possible positions of the receiver can be determined from the geometry. However, the exact location of the receiver has yet to be fixed. In two-dimensional space, the position of the receiver can be explicitly obtained by generating a similar locus from another pair of stations. The other pair will also generate a separate LOP. The crossover point of the two loci gives the actual position, as shown for point P2 in the figure (LORAN, 2001; Loran, 2011).

1.4.2 Guidance system

An instrument landing system is a *guidance* type of navigation that provides an instrument-based technique for guiding an aircraft to approach and land on a runway. It uses a combination of radio signals to enable a safe landing even during challenging conditions such as low visibility.

The ILS provides the aircraft with a recommended path it should follow so that it maintains its horizontal position at the center of the runway and the vertical position most appropriate for a smooth landing. Thus, an ILS consists of two independent subsystems. The first, which provides lateral guidance, restricting the aircraft approaching a runway to shift laterally from the recommended path, is called the localizer. The second, which gives vertical guidance, and hence restricts vertical deviation of the aircraft from the recommended path of descent, is called the glide slope or glide path. Guidance is provided by transmitting a pair of amplitude-modulated signals from two spatially separated transmitters. A similar pair of signals is available in both lateral and vertical directions. Figure 1.9(a) and (b) illustrates the condition. This signal is received by ILS receivers in the aircraft and is

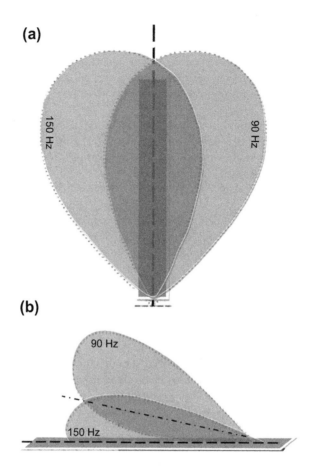

FIGURE 1.9

Radiation pattern of ILS for (a) aerial view of localizer, (b) lateral view of glide scope.

interpreted to obtain guidance information. This is done by comparing the modulation depth of the modulating signals within a particular pair through differencing. The pair of signals is spatially separated in such a way that exact cancellations of the modulating signal happen only along the recommended path of the movement of the aircraft. Once the aircraft deviates from this path, one of the components exceeds the other and there appears a nontrivial resultant value of the difference signal. Then, the aircraft positions are adjusted accordingly to bring it back to the positions of exact cancellation (Parkinson et al. 1996; Kayton, 1989; Meloney, 1985).

1.4.3 Dead reckoning system

An INS is a type of dead reckoning system. It is used by most aircraft for navigation en route and consists of inertial navigation instruments. The instrument that

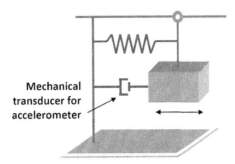

FIGURE 1.10

Schematic of the mechanical equivalent of an accelerometer.

measures translational acceleration is called the accelerometer. It is typically a piezoelectric transducer whose mechanical equivalence is shown in Figure 1.10. The other instrument, measuring the rotational acceleration, is called the gyroscope. At any instant, these instruments provide the dynamic parameters of the vehicle. These instruments give the linear and angular accelerations from which, in conjunction with the position derived at any definite time, the subsequent positions of the vehicle may be derived, as described in Section 1.2.2. An INS can provide continuous and reliable position determination but errors continue to accumulate over time owing to the integration algorithm used (Grewal et al. 2001, Figure 1.10).

Conceptual questions

1. What features should a reference frame contain that can be used to conveniently represent the positions of objects on the ground, as seen from an aircraft?
2. When do the ECI and the ECEF coordinates axes coincide? Do they coincide again after every interval of a solar day (i.e. after every 24 h) from that instant? Give reasons for your answer.
3. Is the phase information relevant for signals provided by the glide scope and the localizer of an ILS to keep the aircraft in fixed locus?
4. How many reference transmitting satellites are required to employ hyperbolic navigation using signals from space? Give reasons for your answer.

References

Elliot, J., Knight, A., Cowley, C. (Eds.), 2001. Oxford Dictionary and Thesaurus. Oxford University Press, New York.

El-Rabbani, A., 2006. Introduction to GPs, second ed. Artech House, Boston, MA, USA.

Ewing, C.E., Mitchell, M.M., 1970. Introduction to Geodesy. American Elsevier Publishing Company, New York, USA.

Grewal, M.S., Weill, L., Andrews, A.P., 2001. Global Positioning Systems, Inertial Navigation and Integration. John Wiley and Sons, New York, USA.

Heiskanen, W.A., Moritz, H., 1967. Physical Geodesy. W.H. Freeman and Company, San Fransisco, USA.

History of Navigation, 2007. Wikipaedia. http://en.wikipedia.org/wiki/History_of_navigation (accessed 19.01.14).

Kaplan, E.D., Leva, J.L., Milbet, D., Pavloff, M.S., 2006. Fundamentals of satellite navigation. In: Kaplan, E.D., Hegarty, C.J. (Eds.), Understanding GPS Principles and Applications, second ed. Artech House, Boston, MA, USA.

Kayton, M., 1989. Navigation − Land, Sea and Space. IEEE Press.

Larson, K.M., 1996. Geodesy. In: Parkinson, B.W., Spilker Jr., J.J. (Eds.), Global Positioning Systems, Theory and Applications, vol. II. AIAA, Washington, DC, USA.

Leick, A., 1995. GPS Satellite Surveying, second ed. John Wiley and Sons, New York, USA.

Loran, 2011. Encyclopaedia Britannica. http://www.britannica.com/EBchecked/topic/347964/loran (accessed 27.12.13).

LORAN, 2001. Wikipaedia. http://en.wikipedia.org/wiki/LORAN (accessed 21.01.14).

Meloney, E.S., 1985. Dutton's Navigation and Piloting. Naval Institute Press, Armapolis, Merryland.

Parkinson, B.W., 1996. Introduction and heritage of NAVSTAR, the global positioning system. In: Parkinson, B.W., Spilker Jr., J.J. (Eds.), Global Positioning Systems, Theory and Applications, vol. I. AIAA, Washington, DC, USA.

Parkinson, B.W., O'Connor, M.L., Fitzgibbin, K.T., 1996. Aircraft automatic approach and landing using GPS. In: Parkinson, B.W., Spilker Jr., J.J. (Eds.), Global Positioning Systems, Theory and Applications, vol. I. AIAA, Washington, DC, USA.

Torge, W., 2001. Geodesy, third ed. Wlater de Gruyter, New York, USA.

Wellenhof, B.H., Legat, K., Wieser, M., 2003. Navigation: Principles of Positioning and Guidance. Springer-Verlag Wien, New York, USA.

Satellite Navigation

From general navigation, which we have discussed in Chapter 1, we now move on to our specific topic of *satellite navigation* and introduce its basic concepts. Our approach in this book is to explain the fundamental principles of the generic satellite navigation system, instead of concentrating on any particular existing system. This chapter starts by formally defining satellite navigation and describing it as a service. It discusses the important aspects of a navigational service and its formal categories. Then we look at the different architectural components of a typical satellite navigation system elaborating in particular on the control segment.

2.1 Satellite navigation

2.1.1 Definition

Any type of Navigation involving satellite that sends reference information to users from which either the navigational parameters, such as position, velocity, and time (PVT), can be directly derived or that aids in improving the derivations, is called satellite navigation.

The advantage of using satellites for the purpose of navigation is that the satellites have wide spatial coverage that make the service available to many users at a time. Satellites limitations are the propagation impairments that the navigation signals experience while passing through the medium and also the resulting weak signals at the receivers.

Satellite navigation systems provide navigation services to their valid users by offering them enough information such that with appropriate resources, they can derive or improve the estimate of their own position and time at all locations within the extent of the service area of the system, at all times.

Satellite navigation is basically a piloting system in which the positional parameters are derived afresh every time. Unlike the other piloting system, LORAN, which works on the basis of 'time difference of arrival' of the signals from two sources, the positions are mainly derived on the principle of 'time of arrival' of the signal. The satellite acts as the reference in this case, whose positions are precisely known as priori. The derived positions are represented in a suitable global reference frame.

2.1.2 Navigation service

We shall start this subsection with the general definition of the service. However, we shall only define the term to such an extent that we can look at it from the perspective of satellite navigation as a service. Then, we shall quickly move on to its technical aspects. Briefness is necessary because the term 'service' has enormous economic implications, and explaining this subject is not only extremely difficult but beyond the natural capacity of the author of this book. Thankfully, that does not affect our objective, which is to understand the technical foundations of satellite navigation and not the implications of using it as a service.

Defining the term 'service' is not simple. It is one of those words for which it is easy to understand the idea yet difficult to create a crisp, clear definition that covers all of its aspects. Although, for our purposes, a lay person's understanding of the concept is perfectly adequate. In most simplistic words, service is defined as an intangible economic activity that delivers commercial value to its users by facilitating outcomes that the users of the service want to achieve or experience, without exercising ownership of the service (Service (Economics), 2003).

So, services are like utilities such as delivering water to our houses through a pipe or providing power or internet connectivity through a cable. Road, rail, and air companies that offer a means of commuting to their users, are types transport service. Tutors provide services pertaining to education, whereas physicians provide medical services.

Services are provided via an interface: power cables have terminals, whereas telephone connectivity has terminal equipment.

On the other hand, what do users consume from the service? They interact with these offerings of the services, typically running on knowledge, skill, or technology, and consume the effect of the transactions. They can experience, collect, add, update, find, view, and compute on that. In totality, they can execute a process on the offerings to fulfill the need for which they opted for the service.

During this process, all that users care about is what they are receiving, not about what happened in order for them to receive the service. Users are, however, typically expect a certain quality of service to be delivered to them. The quality of the service is important to the users and specifies whether the service can satisfy their requirements. However, users take responsible for the process and how and where it is applied, and can also use it as a platform that can be used with many other applications.

This is a very brief and simplified description of what a service is, but it is sufficient for our needs at this point. Satellite navigation is also a service, and as a service, it delivers position and time estimates to users through radio waves transmitted from satellites as signals. These signals, which carry enough information so that the user's position may be derived is the service that it provides. Users interact with these signals using their receivers, which act as the interface and perform the process of estimation to obtain their positions and, thus, provide the service value.

2.1.3 Service parameters

The parameters that define the quality of the navigation service are its accuracy, availability, continuity, and integrity (Colney et al., 2006). These service parameters are further defined below:

2.1.3.1 Accuracy

Accuracy is the measure of correctness of the position and other estimates carried out by the user with respect to their true position. In plain words, it describes how close to the true values the estimations are, using the service. Accuracy depends on many technical factors, such as the signal quality, correctness of navigational data and propagation effects, satellite geometry, and receiver goodness etc. We shall discuss these in more technical detail in Chapter 7. These factors, determining the accuracy and intrinsic service quality, are required to be assessed to provide a quantitative figure of expected quality of the service offered.

It is also important to discuss the difference between the terms 'accuracy' and 'precision'. Although we sometimes use them interchangeably these two terms are entirely different in meaning. Accuracy is the conformity to truth, with the technical definition being the degree of closeness of the measure or estimate of a quantity to its actual value. The navigational service's position accuracy is given by the difference between the true positions of the user to the estimated position. The smaller the difference, the better the accuracy. For any ensemble process, it is typically represented by the average absolute error of estimation.

On the other hand, precision is the measure of repeatability of a value, and it represents how well you adhere to the same value of your estimates for different tests under the same conditions. This is defined by the deviation of the estimates from the most expected value over the ensemble. It, thus, provides an idea of the distribution of the estimates about the mean and is represented by the standard deviation of the estimates, which is also the standard deviation of the estimation errors.

A measurement system can either be accurate but not precise or precise but not accurate; it can also be neither or both. For example, a measurement carrying a constant bias, along with a large random error like a Gaussian white noise is neither accurate nor precise. If the samples are obtained through averaging of several measurements, the random noise, but not the bias, is removed, and then the measurement is precise but still not accurate. Now, if the bias is removed and only the noise remains, it becomes accurate but not precise. If, on removing the bias, we get averaged samples, the measurement becomes both accurate and precise (Box 2.1).

To explain this, refer to Figure 2.1 which consists of a true position T of a body and its estimates represented by scattered points around it. In estimating the position, there is a radial error that exists, which is randomly scattered. Accuracy describes the closeness of the points to the true position, T. Points that are located closer to T are considered more accurate. The accuracy of all the estimates as a whole is the mean radial distances of the points from T.

Considering the many points representing the estimations under the same condition, precision is the width of the distribution of points. The ensemble of all points is considered precise when they all have the same radial error. So, when the estimates are precise, although not necessarily accurate, the points form a well-defined geometry, with its radial distance from the true position defining the accuracy. The more these points are scattered, the lower the precision of the estimate. Similarly, a large cluster of points scattered widely around T is accurate but not precise. However, it is not possible to reliably achieve accuracy in individual measurements without precision.

2.1.3.2 Integrity

Integrity is the capability of the system to indicate whether the service is able to continue with the promised performance in terms of quality. Because any service

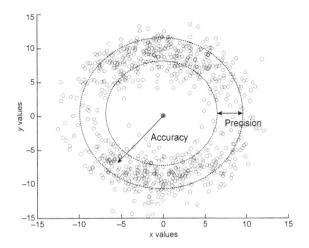

FIGURE 2.1

Difference between accuracy and precision.

is adhered to with some minimum quantitative specification of the quality of service, it is necessary for the system to disseminate the information of whether the technical parameters are within the specified warranted limits at any *time* or not. It also includes the ability to provide timely warning of any failure to do so, the latency being of importance here. This is an important concept in navigation because the user, on the basis of the integrity values, may decide on the usability of the service.

2.1.3.3 Availability

Availability is defined by the probability that the navigation system will make the useable navigation signal resources accessible to the user within a specified coverage area with warranted quality, such that the user can find his or her position and time. So, the time that simultaneously meets the condition of signal accessibility and service quality and integrity is accounted for in calculating the availability of a

---

BOX 2.1 MATLAB EXERCISE

The MATLAB program accuracy_precision.m was run that represents the measurement of a value $x_0 = 0$ and $y_0 = 0$ where the measurements are done under two different conditions, (1) with random error only and (2) with bias and random error. The plots thus obtained will be given later. The bias values were taken as 2 and 3 for x and y, respectively, and the value of the standard deviation σ was assumed as 2 (Figure M2.1(a) and (b)).

Continued

BOX 2.1 MATLAB EXERCISE—cont'd

Run the program with different biases and for different values of σ and observe the nature of the points representing the measurements.

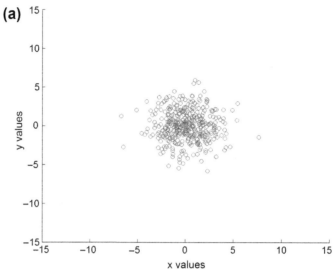

(a)

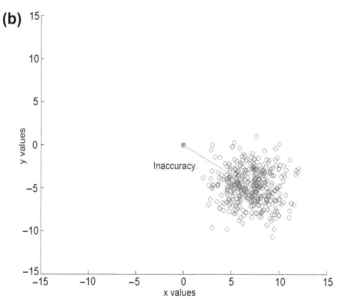

(b)

FIGURE M2.1

(a) Good accuracy, poor precision. (b) Poor accuracy, poor precision.

system. Availability is a function of both physical characteristics of the environment and the technical capabilities of the service provided (Colney et al. 2006).

2.1.3.4 Continuity

Continuity is the probability that a service, once available, will continue to be available and perform during the duration of a phase of operation. It is, thus, the ratio of the number of times that the service sustains over the total period of use once the user starts using the service/the total number of uses, measured over a finite time interval.

2.1.4 Categories of satellite navigation

The satellite navigation may be categorized on a different basis, and for most of the categories, this basis is any one of the characteristic parameters of the navigation service. Herein, we mention briefly just some prominent categories of the satellite navigation system.

2.1.4.1 Primary and differential services

Evident from the name itself, this category of navigation system is based on the elementary characteristics of the service in terms of the essential contents of its signal. A primary navigation service is the one in which the basic signal containing fundamental information, necessary and sufficient for estimation of absolute positions, time and their derivatives are offered to the users.

In contrast, a differential navigation may not provide position estimates all by itself but is the one that contains the correction information using which, the position and timing estimates obtained from primary services can be improved in terms of accuracy and reliability. Herein, the corrections are obtained relative to a reference whose position is known precisely and can only be used in addition to the signals from the primary system.

2.1.4.2 Open and restricted services

This category is based on the accessibility of the signals. A satellite navigation system may be termed an open service when its signal can be accessed by anybody, typically free of any direct charge. It is thus suitable for mass market applications and is generally popular in terms of use. Restricted service, on the other hand, is the service where access to the useable signal is restricted to a certain closed group of users. It has some additional advantages over open services, in terms of accuracy, availability, and integrity, and normally has more protection against viable threats.

2.1.4.3 Global and regional services

The navigation service can also be categorized on the basis of the area of available service. On the basis of availability, which also defines the extent of the service, it can be categorized into global or regional services. For global service, the navigation signals from the systems are available across the globe at any point of time. For regional satellite navigation services, the availability of the signals is restricted only within a limited geographical region over the earth. However, for regional services, the signals may be accessed beyond the defined area of service, but the specified quality of the service is not guaranteed there.

2.1.4.4 Standard and precise services

This division of the navigation system is based on the accuracy and precision of the navigational solutions offered by the service. Standard navigation services offer a fair value of accuracy and precision with the best possible estimation of positions and times. The performance is suitable for general navigational applications and is meant for the use of the masses. The precision services, as the name suggests, offer much higher precision, as well as accuracy, to the position and time solutions to their users and are typically meant for strategic uses. Thus, precision services are restricted, whereas the standard services are open and free. A single system can offer two different services distinguished by their accuracy (Parkinson, 1996).

2.2 Architectural components

The architectural components of a typical satellite navigation system consist of the following three segments (Spilker and Parkinson, 1996):

Space segment: This consists of a constellation of satellites orbiting the earth in space and transmitting the necessary signals, from which the user can derive his or her position (i.e. avail the service).

Control segment: Consists of resources and assets that on the ground monitor, control, and maintain the space segment satellites.

User segment: This consists of the users who interact with the signal to derive their own position, thus availing the navigation service.

The schematic of the segments and their interactions are shown in Figure 2.2.

To understand why we need such architecture, we have to start from the user segment. Although we shall discuss the user segment in greater detail in a later chapter, we need this preamble to understand the design of such architecture at this point. In navigation service, the user, as mentioned in Section 2.1, should be completely able to derive his or her position by interacting with the signal transmitted from the satellite, which is the service offered. For deriving his or her own position, the user is required to know the position of the reference, which, in this case, is the satellite, and the receiver range from it; moreover, he or she needs to receive this from more than one satellite. The range is calculated using the propagation time and the latter is obtained by comparing the signal transmission time from the satellite with the signal receiving time at the receiver. So, the satellite clock, from which the transmission time is derived, is also required to be highly stable and accurate. Even a trivial clock shift at the satellite is necessary to be corrected at the receiver during estimation.

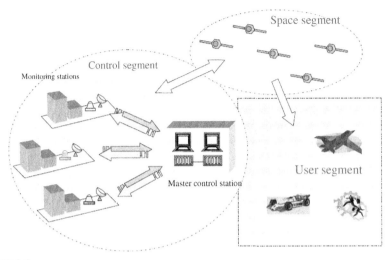

FIGURE 2.2

Segments of a satellite navigation system.

All these data, including the satellite position, transmission time of the signal, satellite clock correction factor, and much more similar information, need to be disseminated using radio aids to the user wherever he or she is within the service area. They should be transmitted in such a form that the information may be obtained from the signal or derived from it through proper processing. This dissemination of information and simultaneous satisfaction of the requirement of multiple visibilities over the whole service area around the clock are possible through the use of a constellation of satellites. This justifies the establishment of the space segment consisting of a fleet of satellites.

However, there must be some definite setup to look after these satellites. The derived position, being a sensitive function of the position of the satellite, needs to be precisely known. The positions of these satellites are hence required to be derived through proper monitoring of their dynamics. Any anomaly in the dynamic behavior of the satellite is to be corrected at the appropriate time. Similarly, a system time is required to be maintained while any aberration in the satellite clock with respect to the system time is to be estimated for corrections. This is done by the ground-based control segment. The salient features of the three segments may be listed as follows:

Control segment:
- Situated on ground.
- Multiple entities are connected through a network.
- Monitors the dynamics and behavior of the satellites in the space segments and their clocks.
- Predicts, updates, and uplinks data to be transmitted by the satellites.
- Commands maneuvers of satellites if necessary.

Space segment:
- Transmits the necessary data to the users for deriving the PVT information correctly.
- Orientation of satellites designed to serve the service area continuously.
- Design of space segment may be for serving local, regional, or global areas.
- Frequency and power levels are judiciously chosen for optimal performances.

User segment:
- Consists of the users of the service.
- User receive-only terminals to receive the transmitted data and compute their PVT.
- Receivers of different capabilities and accuracies are used.

Herein, we shall discuss the control segment while the details of the other two segments will be discussed in subsequent chapters.

2.3 Control segment

We started with the notion of understanding the satellite navigation system with a system-independent perspective, covering all theoretical possibilities. Theoretically, the individual architecture of the different segments may vary and the variations may depend on the nature of services offered and area of service. Likewise, the control segment architecture is also determined by various factors. However, describing the architectural possibilities in a wide spectrum at this point will not be helpful in understanding but will invite unnecessary complexity. So, instead of encompassing all such variants, we shall only concentrate on the commonality of the architectural features and discuss only these features in detail. It would be rather helpful if the reader, after completing the book, revisits this chapter once again to find any alternative possibilities, which, I am sure, he or she will be capable of doing at that point. It is also suggested that he or she compares it with the architecture of the wide area differential system that we shall read about in Chapter 8.

The control system consists of resources and assets of the satellite navigation system situated on the ground used to monitor, control, and update the satellites in the space segments. It is distributed over the entire service area and carries the necessary intelligence to perform the processes that help in the functioning of the system in an appropriate manner. Proper functioning of the system has the implicit meaning that the service requirements must be fulfilled effectively in the entirety.

First, let us find out what elements need to be present for the proper functioning of the service. The offerings of the service are the information-laden signal, and this must be available to users. This, in turn, requires proper dynamic behavior of the satellite and proper generation of the signal. Therefore, a major objective of the control segment is to monitor the satellites in the space segment, for their proper dynamic behavior, including their positions in orbit, proper functioning, health, and the necessary station keeping. Any inconsistencies therein should be handled such that they do not affect the service. Second, the mandatory signal elements must be made available to the satellites to be successively transmitted to the user through the signals. This segment is also responsible for the important job of system timekeeping for which it maintains the clock resources on the ground. Finally, it performs an overall management of ground assets.

The control segment architecture has the key driving requirements, such as the following:

- Proper provision of the control resources.
- Adequate redundancy of the resources commensurate with the system requirements, avoiding any irrelevant duplication.
- Appropriate location of ground elements to meet the overall objective.
- Generation of independent timescale and achieve time synchronization across the system.

Architecturally, it is not a single entity but an ensemble of distributed resources, dispersed over the total service area of the system. The individual units that compose the control segment are the monitoring stations (MS), master control station (MCS), ground antenna system, and associated processing resources (Francisco, 1996).

2.3.1 Monitoring station

The monitoring stations are the components of the ground segment used for the purpose of continuously tracking the navigation satellites and their signals. To accomplish this end, these stations are distributed over a diversified geographical region. Therefore, we shall discuss two aspects of monitoring stations, their receiver characteristics and their geographical locations for effectively performing this job.

2.3.1.1 Receiver characteristics

The task of tracking the navigation satellites is accomplished by continuously receiving the signal and data it sends. From the received signal, the range measurements are also done. These form the basis from which the performance of the satellites can be analyzed. The navigation signal derivatives and the data, along with the local ground meteorological information, are communicated by the MS to the MCS for analysis. The main informational details derived from the analysis of the data collected at these tracking stations and required for the control of the system are the future positions of the satellites and the clock corrections.

Because the measured ranges are affected by the propagation impairments, such as the ionospheric delays, the measurements from these receivers must be adequate to estimate them and remove the same. Dual-frequency receivers data of the MS are used for doing so. However, the corrections are actually done after the data is transferred to the MCS. Furthermore, these channels are required to be calibrated to remove any interchannel bias. Likewise, all possible corrections to the signals and their derivatives are done. We shall read about these errors in detail in Chapter 7. However, that will be in connection to the signals received at the user receiver.

The receiver, in general, tracks the received signal for obtaining the characteristics of its individual components. For monitoring receivers, because the measurements of the signal are necessary not only in a continuous manner but in such a way that any anomaly in the data is significantly manifested, tracking is required to be performed even if the signals are not consistent. Therefore, the monitoring stations are required to be equipped with receivers that can continue tracking the satellite behavior even when the signals are not in conformance with the one they are designed for. The receivers are expected to be adaptive enough to be able to accommodate and follow such aberrations or abnormal data from satellites to the maximum extent possible. Moreover, the number of active channels should not limit the tracking of any visible satellite. To provide the necessary redundancy, the monitoring stations with duplicate or triplicate receivers may be installed to give the system enough robustness against the loss of track by any of the stations.

Another mandatory requirement of the receivers at these control stations is that they must have a stable clock. So, an atomic frequency standard with a precision counter is necessary to be running in these receivers. Besides all such reference stations are required to be synchronized, which may be achieved if they get triggered by a common external source. This is done by the MCS.

In addition to the satellite ranging data and station data, the local meteorological data are also required to be collected at these sites to remove any signal ranging error due to the effects of tropospheric elements. For this, the receiver is collocated with meteorological sensors. The parameters of interest include the surface pressure, surface temperature, and relative humidity. The schematic of monitoring station instruments is shown in Figure 2.3.

2.3.1.2 Suitable positions

Because the monitoring station has to track the satellites around-the-clock, these stations are required to be distributed in such positions that the combined system does not lose the visibility of any of these satellites. That is why the selection of the station locations across the whole service region is to be done in a manner so that all the satellites are visible to some of the monitoring stations at all times throughout the period of operation.

To serve the objective of deriving future satellite positions, their current positions are required to be estimated. This is done by measuring the satellite ranges by more than one monitoring station. The position estimation solutions are better when the monitoring stations are maximally dispersed. So these stations are most often distributed across the globe for a global system, and for regional systems, they are evenly distributed across service area. The station locations must enable them to range simultaneously from multiple stations at any instant of time and for each of the satellites.

To maximize the period that the monitoring station sees a particular satellite, their locations must be such that the satellite visibility is not restricted even for low elevation of the satellites. Accordingly, the locations of the monitoring stations may be selected where there is clearance up to this lower elevation angle. A suitable elevation cutoff is kept to avoid large propagation through troposphere and

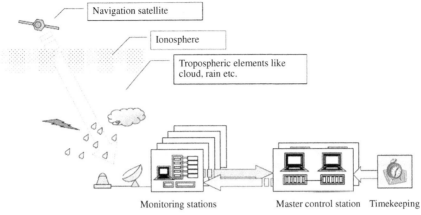

FIGURE 2.3

Schematic of monitoring station instruments.

multipath. However, performance monitoring using the contained data can still be extended to even lower elevations. So, locations over the oceanic islands are suitable, particularly for global services, because they do not have the issue of blockage of lower angles and the horizon-to-horizon visibility extent is large.

Finite inclination of the satellite orbits allows the satellite to traverse equally to both hemispheres. For the purpose of even vision of the satellites during their excursion on both the hemispheres, equatorial locations are suitable for global operations. This allows viewing of the satellite up to the extreme ranges of the satellites in both the hemispheres. With a large part of the equator being on the oceans, it facilitates both of the previously described criteria. However, this may not be possible for regional services where the monitoring stations are required to be restricted within a limited geographical boundary, typically, but not necessarily, within the service area.

The precise location of the MS receiver may be identified through surveying, on which the accuracy of the satellite position estimation is dependent. Moreover, noise survey is also needed to eliminate the influence of any external noise on the measurement system of these stations.

Individual monitoring stations, located across the service area, must be connected among themselves and with the central resource of the MCS to form a network. There must be secure communication channels connecting these units with proper redundancy, thus providing the required digital communication backbone to the system with the highest order of availability. This ensures timely and efficient exchange of data. High-speed optical or microwave data links are preferred options for establishing such networks. However, VSAT communication systems may also be used as standby.

These monitoring stations, although geographically separated, operate under the control of the MCS. It is from the MCS that the tracking orders and configuration commands come. It is necessary to synchronize the clocks of the monitoring stations and may be done through common view algorithms that exploit the available space and time geometry. Readers will find details of this common view technique in Chapter 10.

However, there can be exceptions to this and the satellites may be required to be monitored with monitoring stations distributed over a limited geographical region. In such cases, tracking of the satellites and all associated activities are required to be established when the satellites are within the visibility zone of these stations.

2.3.2 Ground antenna

The ground antenna is required for communication of data and command between the satellite and the control segment. It should bear a full duplex channel through which the satellite commands and navigation data are uploaded to the satellites, while the telemetry, containing the status information, and the satellite response to the ground commands and other related data are received from it.

Because this is entirely meant for communication, suitable lower bands and larger antenna sizes result in large transmission power flux density. It provides a large signal/noise ratio for reliable communication even at lower elevations, where the noise is naturally large.

Because of the similar criterion of visibility of the satellites like the monitoring stations, the locations of these antennas are also required to be globally distributed or across the regional area of service. Data upload can take place without any problem at any selected time. Moreover, these antennas also need to be always connected to the MCS and controlled by it in the same way as the monitoring stations. So, despite the ground antennas being a separate architectural entity of the control segment, they can be physically located, along with the monitoring stations, because the location criteria are the same for both. Over and above, collocation of these two resources helps in managing them efficiently.

The ground antenna can also support additional ranging in this communication band if necessary using two-way techniques.

2.3.3 Master control station

The MCS is a central resource consisting of the computational and decision-making facilities for the proper management of the system. It is the site of the management of the space resources executed through the proper analyses of data. It analyzes the health and performance of the individual satellites and makes a decision with conse-quent issuance of appropriate commands for necessary actions. In addition, it also generates the input to the navigation message and manages the proper functioning of the different units of the control segment, including the monitoring station and the ground antenna. Furthermore, the MCS also keeps the system clock ensemble from which it derives the system time using a definite algorithm. So, the MCS is responsible for all aspects of constellation command and control, including routine satellite and payload monitoring and satellite maintenance, in addition to any anomaly resolution. In addition to its scheduled job of navigation message generation and upload, it also performs the management of signal performance and timely detection of failures.

As discussed in the previous section, the MCS is connected with the monitoring stations that track the individual satellites and dispense them with the data. These data, received in almost real time from the different monitoring stations, along with all the derivatives obtained from them, form the basis of all further decisions taken at the MCS.

After the master station accesses the regular and the critical data, all of them are continually processed and analyzed. The first activity that is required to be done on the data is to process the incoming data and identify any inconsistency.

2.3.3.1 Data processing

Navigation data processing is the core processing part that takes place in the ground segment of any navigation system. The data measured by the distributed monitoring stations are required to be passed on to the central facility of the MCS, where all the resources for the said processing are available. The combined data are then simultaneously processed, and the processed products are then used for generating the navigation message parameters. These messages are subsequently uploaded to the satellites and disseminated to the users at an appropriate time through the navigation signal. The centralized processing architecture not only makes the best processing resources available to the maximum extent of data but also makes the distribution of the processed products through various distributed units easier.

The raw tracking data collected at the monitoring station include the range measurements from the input to the processing. The total process to convert these new data to a packaged database in a specific message format consists of the following (Francisco, 1996; Dorsey et al. 2006).

1. *Measured data preprocessing*
2. *System state estimation*
3. *Navigation data frame generation*

2.3.3.1.1 Measured data preprocessing

The preprocessing of the data consists of correcting the measurement errors that remain in the measured values obtained by the monitoring stations. It includes ionospheric error correction, tropospheric error correction, and other propagation error corrections, required to remove the excessive range estimates of the satellites. The ranges of the satellites measured at each of the monitoring stations in two different frequencies are available to the MCS. From this, the ionospheric delay for the range measurements may be derived with precision. The derived delay is then used for the correction of the ranges. The meteorological measurements coming from these stations serve as additional input. They are used to correct the range errors that occur as the satellite signal passes through the tropospheric regions of the atmosphere. This may be done using the standard models available for the purpose. Effects of the earth's rotation during the propagation time also need correction. Hence, the earth's orientation relative to the inertial scale also forms an important parameter for consideration. Thereafter, relativistic corrections and overall smoothing of measured data are also performed on these measurements. The last of these is done using the carrier-aided smoothing. The monitoring stations being situated at the locations surveyed for noise and multipath, are unlikely to experience these impairments. These are, hence, not serious matters for concern. In addition, it also derives the range rate errors. The range errors and the range rate errors thus derived are used by the MCS for necessary corrections to meet its different objectives.

2.3.3.1.2 System state estimation

A planet in deep space follows a fixed path around its parent body in accordance with Newton's laws of motion. Its orbit is Keplerian or perfectly elliptical, and its position can be predicted exactly for any given future instant of time. A navigation satellite moves around the earth under the earth's gravitational field in accordance with the same laws. We shall learn about their motion in our next chapter in much more detail.

The navigation satellites, however, operating at an altitude of a medium earth orbit or even a geostationary earth orbit, are subject to external forces that produce orbital irregularities or perturbations. To improve the accuracy of the system, these perturbations must be accurately predicted such that the satellite positions are determined exactly at any instant of time.

Consequently, measurements done at the monitoring stations are analyzed and estimation of different necessary parameters are done using a Kalman filter. The Kalman filter is an advanced estimation algorithm that is used in the whole process from position estimation to ephemeris generation. The required parameters are considered as filter states and are estimated in the presence of noise, for which the Kalman filter provides excellent performance.

The current positions of the satellites are definitely one of the required states because, from this information, the trajectory will be derived and the future ephemeris of the satellites will be predicted. The measured range being a function of satellite and monitoring station positions, the range residuals from the values derived from surveyed position and approximate satellite position are attributed to the position error of the latter and timing inaccuracy based on their relative confidence. From this, the true positions and satellite clock aberrations are simultaneously estimated by the filter. The other parameters simultaneously derived are the clock drifts and sometimes the propagation delays. A consolidated filter using the inputs from all the monitoring stations is used to give a better solution. The total states are the solution of one large filter implementation.

Once the precise positions of the satellites and their temporal variation up to the current instant are derived, the trajectories of the satellites, including the perturbation effects, are determined. Subsequently, the computing stations, having known the up-to-date trajectory of the satellite, compute an orbit that best fits the same. The orbital information is then extrapolated to give satellite positions for regular future time intervals (e.g. every minute for the following few hours). The geometry thus formed is used, and the corresponding orbital parameters are derived from it along with the expected perturbations. Weighted least square fit may be used to evaluate numerical value of these Keplerian parameters. The filter also simultaneously finds the clock aberrations of each of the satellites from the system time.

However, there are many other parameters that need to be estimated. Because the monitoring stations are located at well surveyed positions and fixed, their locations are known a priori. But, few other monitoring station parameters, like their relative clock offset and drift with respect to the system time, are required to be established using the filter. Furthermore, the wet tropospheric correction values are also required to be estimated. The ionospheric corrections are already done using dual-frequency measurements and dry tropospheric errors from the meteorological parameters during the preprocessing. Hence, these parameters need not be derived again.

The recursive estimates of the Kalman filter are comparatively efficient and provide numerical robustness to the systems (Francisco, 1996; Yunck, 1996). The state errors are to be kept at minimum to a submetric level to support performance compliance. Errors add up when unmodeled components get included in the measurements or when the true temporal dynamics violate the assumed ones. So, to reduce these effects, any measurement blunders are blocked by setting the range of possible values and measurement variances. Periodic regeneration of output is performed whenever a force event is introduced that invalidates the model or when the residual for the KF indicates large nonlinearity.

2.3.3.1.3 Navigation data generation

The derived parameters are used as input to the navigation data. To permit continued services, many advanced data sets may be generated a priori. Because the raw measurements are used for predicting the future orbits of satellite, the quality improves with an increase in measurement accuracy and numbers of monitoring stations. Typically, each set of the derived ephemeris is valid for a few hours, with the quality of prediction deteriorating with time. This is called the aging effect of the ephemeris. For this reason, it could be better if new sets of ephemeris would have been available at lower intervals. But, the practical constraint is that new sets were required to be estimated more frequently, increasing the computation load of the ground system. A greater concern is that a large volume of data would be required to be uploaded to the satellite, which is preferably kept modest for various reasons.

The MCS needs to estimate parameters with stringent consistency to generate a high-quality navigation message with integrity. The message generation facility has the importance of reducing the computational load on the satellite. However, some functionalities may be housed in the satellites as the space technology evolves and advances through time.

The messages generated at the MCS are uploaded to the satellites for broadcasting to the user. To do this, the total navigation data sets are generated and arranged for regular uploading. Data sets are generated at periodic intervals to befit the accuracy requirements. The same ephemeris data are used over a certain period of time while updating of the current values occurs at every definite interval.

2.3.3.2 Telemetry and telecommand

The master station also requires the generation of a command for configuration of monitoring stations or even the satellites. These commands should reach the respective units for execution. However, the data transfer, command generation, and procedures should be highly secured and must be encrypted for the reason of security and authenticity, with the protocol being preestablished.

Analyzing the telemetry signal and generating telecommand are the responsibility of the MCS. The telemetry signals are continuously received at the ground antenna system. These signals are then required to be transmitted to the MCS through secured channels. The collocation of the ground antenna with the monitoring stations gives the added advantage of transmitting this telemetry information through the same channel that delivers the ranging data to the MCS. The information thus received at the MCS may be used to generate activities, as described later.

When the information contained within the telemetry data is analyzed at the MCS, the analysis result, consisting of the operable range of parameters measured, state values concerning satellite operations, the operation of equipment, and verification of execution of commands, is used to take decisions on subsequent necessary actions. These decisions taken lead to the issuing of the telecommands, which are then required to be transmitted back through the command channels.

The MCS may also need to perform planned and unplanned station-keeping maneuvers. Communicating control signals from the ground to the satellite to initiate maneuvers of the satellite dynamic parameters and to change the state or mode of operation of equipment is an important task to be handled by the MCS. Maneuvers require imparting of unaccounted forces from within the satellite, which are not included in the models of normal satellite dynamics, for which commands are also additionally generated at appropriate times.

The communication links thus require high availability and reliability in addition to security. Reliability and security may be achieved simply through successive repeat and encryption of the data and command. On the other hand, the onboard command handling facility and computing power enable information processing and execution of the command. However, the onboard processor can take some decisions by itself. It reduces the load of the control segment. But, hierarchically, these controls must be much simpler and should not affect the system operation.

2.3.4 Navigational timekeeping

2.3.4.1 International timekeeping

The Bureau International de Poids et de Mesures (BIPM), located in Paris, France, has the prime responsibility of global timekeeping. They compute and distribute the International Atomic Time (TAI) timescale, which is obtained by properly using and controlling data from more than 200 clocks distributed around the world (Lombardi, 1999; Petit, 2003).

However, TAI is an atomic clock time and has no relation with the solar day observed due to the spinning of the earth, to which our conventional perception of time is related. So, they do not take into account the factors like the deceleration of earth rotation for correction and strictly adhere to the atomic definition of time. This makes TAI unsuitable for public time coordination.

To overcome this aspect, BIPM generates another timescale, called Universal Time Coordinated (UTC). This time is equivalent to TAI in terms of time progression. However, it accommodates corrections due to earth rotation factors by keeping provision of the so-called 'leap seconds,' which are accordingly added to the TAI as necessary to ensure that the sun crosses the Greenwich meridian precisely at 1200 h noon UTC with a precision of within 0.9 s when averaged over a year.

So, UTC (and TAI) is a 'paper' timescale and is the reference timescale for the worldwide time coordination, providing the basis for the standard time in different countries and for different services.

2.3.4.2 System timekeeping

In navigation, time is the most important factor to be used for providing the services and to be disseminated to users. The necessary estimations are dependent on it. The control segment should therefore also perform as the system timekeeping facility to generate, maintain, and distribute the reference system time. A smaller ensemble of atomic clocks is used to keep the system time, and it may consist of highly stable active hydrogen maser (AHM) and cesium atomic clock, with appropriate time and frequency measurement equipment. The AHM clocks exhibit the best short-term stability, whereas cesium clocks exhibit good long-term stability. Hence, the combination of the maser and cesium clocks provides excellent short- and long-term stabilities. Appropriate algorithms are used for combining the individual time of the clocks in the assembly and hence may generate the best system time.

So, in addition to the physical generation, for realization of clock timings, the employed algorithms may assimilate individual measurements of atomic clocks through estimation of weight factors. Weights are derived using some statistical techniques based on the standard deviation of variations of the individual clocks. The clocks may be weighted based on their stabilities, and the ensemble clock (paper clock) thus becomes a weighted average of all the clocks, ensuring high stability and accuracy of the resulting timescale. The master control center thus derives the system time. This system time is atomic and not related to the geodynamics. Hence, the difference between the system time and the UTC is always monitored, and a correspondence with UTC is maintained and provided to the users for correcting and converting the system time to UTC, as required.

For the sake of completeness of our discussion on clocks and timing, it is recommended that, before concluding the chapter, we briefly mention the basic physics of an atomic clock. This will help us in understanding how it acquires such enhanced accuracy and what makes even these atomic clocks drift. An atomic clock is a combination of an arrangement of atomic emission and electronics to probe the same. It works on the principle of *tuning* an ordinary oscillator by a precise reference. The reference is an emission due to atomic electron transition from an excited state to a lower state of energy. When enclosed in some definite magnetic field, a particular atomic energy state of some specific atoms shows a hyperfine spectral splitting. If the emission from the ordinary oscillator is used to excite the electrons, the latter will absorb the incident energy and will jump from the lower to the higher of the states obtained as a result of splitting. These excited electrons on transiting back to the normal level emit the radiation of same energy proportional to its frequency. If the frequency of the incident energy derived from the ordinary oscillator used for excitation is gradually changed, it will show a resonating effect when the frequency of the oscillator is exactly equal to the frequency corresponding to the energy difference of the transition states. If δE is the energy difference, then the frequency of the illuminating signal at which this occurs will be ν, where $\delta E = h\nu$ and h is the Plank's constant. The emission may be observed by some well-responsive sensors to identify the maximum intensity for resonance. Alternatively, the condition of the maximum count of excited atoms may be identified for resonance. Once it is identified, the oscillator is fixed to the corresponding frequency that created resonance. Thus, the ordinary oscillator may be tuned precisely to the frequency, ν.

The activity of searching for the resonance frequency for transition is called probing. The hardware used to probe the atoms consists of highly sensitive electronics with adequate protection from the external magnetic fields. The resonating condition is more accurately identified when the interaction time of the probe with the atoms is higher. The atoms thus need to be sloth, and the condition may be created by retarding the atoms and reducing their dynamicity with lasers, a technique known as laser cooling.

For an atom whose split levels of energy corresponds to a frequency of ν, the time taken for ν numbers of oscillations to occur in such an emission is considered as 1 s. For different atomic sources, δE values are different. Consequently, ν is also different, making the numbers of oscillations that defines 1 s different for different atoms. The standard is defined for Cesium atoms, for which the frequency is 9,192,631,770 Hz.

A lucid elaboration of time and frequency measurements can be found in Lombardi (1999) and Jespersen and Fitz-Randolph (1999). A schematic figure of the arrangement of an atomic clock is shown in Figure 2.4.

Conceptual questions

1. What could be the advantages and disadvantages of eliminating the control segment by adding all its functionalities in the space segment?
2. What methods can you devise to identify whether an anomaly detected in the measurements by monitoring stations is actual or measurement defects?
3. What are the alternatives to correcting the effects in the measurements when the clocks of the monitoring stations are not synchronized?

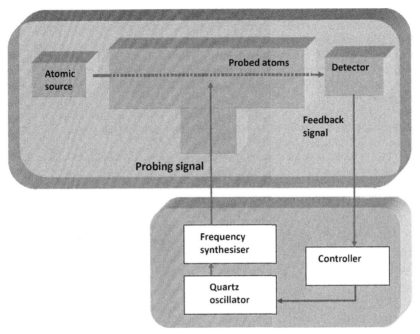

FIGURE 2.4

Schematic of an atomic clock.

4. Assuming that the physical package performs identically, what are the features of the constituent atomic resonance cavity that the accuracy of the clock is dependent on? How is the reference set with respect to which the accuracy of the clock can be defined?

References

Conley, R., Cosentino, R., Hegarty, C.J., Leva, J.L., de Haag, M.U., Van Dyke, K., 2006. Performance of standalone GPS. In: Kaplan, E.D., Hegarty, C.J. (Eds.), Understanding GPS Principles and Applications, second ed. Artech House, Boston, MA, USA.

Dorsey, A.J., Marquis, W.A., Fyfe, P.M., Kaplan, E.D., Weiderholt, L.F., 2006. GPS system segments. In: Kaplan, E.D., Hegarty, C.J. (Eds.), Understanding GPS Principles and Applications, second ed. Artech House, Boston, MA, USA.

Francisco, S.G., 1996. GPS operational control segment. In: Parkinson, B.W., Spilker Jr, J.J. (Eds.), Global Positioning Systems, Theory and Applications, vol. I. AIAA, Washington DC, USA.

Jespersen, J., Fitz-Randolph, J., 1999. From Sundials to Atomic Clocks: Understanding Time and Frequency, second ed. New York, USA.

Lombardi, M.A., 1999. Time measurement and frequency measurement. In: Webster, J.G. (Ed.), The Measurement, Instrumentation and Sensors Handbook. CRC Press, Florida, USA.

Parkinson, B.W., 1996. Introduction and heritage of NAVSTAR, the global positioning system. In: Parkinson, B.W., Spilker Jr, J.J. (Eds.), Global Positioning Systems, Theory and Applications, vol. I. AIAA, Washington DC, USA.

Petit, G., Jiang, Z., 2008. Precise point positioning for TAI computation. Int. J. Navig. Obs., Article ID 562878 2008. http://dx.doi.org/10.1155/2008/562878, 2007.

Service (Economics), 2003. Wikipedia. http://en.wikipedia.org/wiki/Service_(economics) (accessed 21.01.14.).

Spilker Jr, J.J., Parkinson, B.W., 1996. Overview of GPS operation and design. In: Parkinson, B.W., Spilker Jr, J.J. (Eds.), Global Positioning Systems, Theory and Applications, vol. I. AIAA, Washington DC, USA.

Yunck, T.P., 1996. Orbit determination. In: Parkinson, B.W., Spilker Jr, J.J. (Eds.), Global Positioning Systems, Theory and Applications, vol. I. AIAA, Washington DC, USA.

Satellites in Orbit

3

CHAPTER OUTLINE

This chapter is dedicated to detailing the space segment of a satellite navigation system. We learned in Chapter 2 that the space segment consists of satellites in their respective orbits that transmit navigation data to users. In this chapter, we shall learn about the important orbital parameters of these reference satellites and how to estimate their position using these parameters, along with some important aspects related to it. Finding satellite positions is also known as reference positioning, a mandatory task for final user position fixing. Here, we shall start our discussion with the Kepler's laws that defines the dynamics of satellites moving around the earth. A few relations will be subsequently derived that will help us to obtain insight into the dynamics of satellites moving around the earth and will also assist in finally deriving the satellite's position. Continuing, we shall also learn about the different factors that perturb the satellite's motion. Then, a few important orbits will be discussed, along with the rationale for selecting parameters for different navigation purposes.

3.1 Kepler's laws and orbital dynamics

In Section 1.3, we learned that to define the position of a point, all we need is a predefined reference and the distances of the point from this reference along three definite axes. In a satellite navigation system, the position is described with respect to the geocentric reference frame, and the satellites act as the secondary reference points defined therein. Thus, to fix the position of any point in a designated coordinate system, the position of the satellites in that frame and the radial distances of the point from these satellites are needed. Therefore, these two parameters are extremely important and need to be readily available for the purpose of user positioning.

To know the position of the satellites, one must understand the dynamics of the satellites in orbit, and for that, it is necessary to know Kepler's laws. Johannes Kepler (1546–1630) was a German mathematician who examined observations about the planets moving around the sun, made by his predecessor, Tycho Brahe (1546–1601), and formulated simple but useful laws. These laws are equally true for satellites moving around the earth or for any other celestial bodies. Kepler's laws state that (Feynman et al. 1992):

1. Every planet revolves around the sun in an elliptical orbit with the sun at one of its foci.
2. A line joining a planet and the sun sweeps out equal areas in equal intervals of time.
3. The square of the orbital period of a planet is directly proportional to the cube of the semi-major axis of its orbit.

Why are these laws valid for any two celestial bodies? It is because these laws are basically derivatives of the fundamental gravitational laws, which are universal.

Our objective is not to prove Kepler's laws. Rather, we shall see how, by using these laws, we can conveniently find the position of the satellites. For this, we require some elementary knowledge of geometry, and will need to recall the law of gravitation.

3.1.1 Ellipse

We start with the geometry of ellipse, which, according to Kepler's first law, is the shape of the orbit. An ellipse is a variation of a circle in which the radii along two orthogonal directions are different. Thus, it is like a circle compressed along one direction, as in Figure 3.1. The greatest diameter, A_1A_2, is called the major axis, whereas the shortest diameter, B_1B_2, is called the minor axis. These two axes normally intersect at the center of the ellipse, O. There are two points, F_1 and F_2, equidistant from the center on the major axis, called the focus of the ellipse. Their significance follows from the standard definition of the ellipse, which states that an ellipse is formed by the locus of a point that maintains a constant sum of distance from two fixed points on the major axis. These two fixed points are the foci, F_1 and F_2. A simpler way to say the same thing is that if you take any point on an ellipse and find that the sum of its distances from the two foci is C, for any other point on the ellipse, the sum will also be C. Therefore, referring to Figure 3.1, $F_1P_1 + P_1F_2 = F_1P_2 + P_2F_2$. The greatest radius, $OA_1 = OA_2 =$ "a," is called the semi-major axis, whereas the shortest radius, $OB_1 = OB_2 =$ "b," is the semi-minor axis.

3.1.1.1 Eccentricity (ε)

The ellipse has two-dimensional geometry; hence, its shape can be represented by two independent parameters. Typically, semi-major axis "a" and semi-minor axis "b" are used for this purpose. But here, we shall use the semi-major axis and the eccentricity. Eccentricity, ε, has a definite relation with "a" and "b," given by:

$$\varepsilon = \sqrt{1 - \left(\frac{b}{a}\right)^2} \tag{3.1a}$$

or:

$$b = a\sqrt{1 - \varepsilon^2} \tag{3.1b}$$

Therefore, it relates between the two parameters, 'a' and 'b,' one of which can be derived from the other. The distances of each of focus from the center (i.e. OF_1 and OF_2) of the ellipse are equal and are given by:

$$f = a\,\varepsilon \tag{3.2}$$

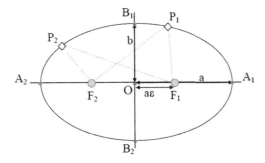

FIGURE 3.1

Geometry of an ellipse.

Thus, eccentricity indicates how far the focus is away from the ellipse's center, much like the way it is used in literary terms. Zero eccentricity refers to the state in which the semi-major and semi-minor axes are equal with the foci coinciding with the center, and consequently the ellipse is converted into a circle. The larger the value of ε is, the further is the focus from the center with a larger difference between "a" and "b." The geometry tends toward a straight line as ε approaches unity.

Taking any focus-say, F_1 in Figure 3.1—the distance F_1A_1 is the shortest range of the ellipse from the point. A_1 is called the point of perigee, or simply perigee, for the focal point F_1. Similarly, F_1A_2 is the longest range with A_2 known as apogee. These perigee and apogee distances may be obtained by subtracting and adding the focal length, respectively, to the length of the semi-major axis. Therefore, we get

$$F_1A_1 = a(1 - \varepsilon) \tag{3.3a}$$

$$F_1A_2 = a(1 + \varepsilon) \tag{3.3b}$$

and

$$F_1A_1/F_1A_1 = (1 - \varepsilon)/(1 + \varepsilon) \tag{3.3c}$$

Extending the idea for F_2, we get, $F_2A_2 = a(1 - \varepsilon)$ and $F_2A_1 = a(1 + \varepsilon)$. Thus, the sum of the distance of A_1 from F_1 and F_2 is $a(1 - \varepsilon) + a(1 + \varepsilon) = 2a$. Similarly, the sum of the distance of A_2 from F_1 and F_2 is 2a. With A_1 and A_2 as points on the ellipse, this corroborates the fact of the equal sum of the distances from the foci to the point on the ellipse. Extending this, we can conclude that the sum remains 2a for all other points on the ellipse.

3.1.2 Elliptical orbit

Kepler's law states that the orbit of the satellites is elliptic. However, this includes the possibility that the orbit is circular, because a circle is a special case of ellipse with zero eccentricity.

Before proceeding further, we need to discuss how the shape of the elliptical orbit of a satellite is realized and maintained in space. This can be obtained by considering that in such a two-body problem, the forces acting on the orbiting satellite are gravitational, attracting the satellite toward the earth, and centrifugal, owing to its velocity directed radially outward. Furthermore, if no external forces are acting and the earth is relatively fixed, then by virtue of a law of physics, a dynamic variable of the satellite called angular momentum remains unchanged. This is called the conservation of angular momentum. Angular momentum is given by

$$L = mv \times r \tag{3.4a}$$

Here, v is the linear velocity of the satellite with mass "m," and "r" is its radial distance from the center of the earth. By virtue of Kepler's law, the center of the earth is the focus of the orbit of the satellite. The cross-product $v \times r$ multiplies "r" with the component of v perpendicular to r only. Thus,

$$L = mv_\theta r \tag{3.4b}$$

where v_θ is the cross-radial component of the velocity of the satellite. For circular orbits, v_θ is equal to the total velocity, v. If the satellite mass m is fixed, the term $v_\theta r$ is a constant.

You may have recognized that this is synonymous with Kepler's second law. From Figure 3.2, we can see that the area swept by the satellite at the center of the earth is

$$A = \frac{1}{2} \int_s r \times ds \qquad (3.5a)$$

where r is the radial distance of the satellite from the center and s is the linear path traversed by the satellite. Thus, the area traversed per unit time by the satellite is

$$dA/dt = d/dt\left\{\frac{1}{2} \int_s r \times ds\right\} \qquad (3.5b)$$

Assuming "r" to be invariant over the infinitesimal time dt, this can be written as (Roychoudhury, 1971)

$$\begin{aligned} dA/dt &= \frac{1}{2} r \times ds/dt \\ &= \frac{1}{2} r \times v \\ &= \frac{1}{2} r v_\theta \end{aligned} \qquad (3.6)$$

Because we have seen that the angular momentum, $r v_\theta$, is a constant, half of it must also be so. This makes dA/dt a constant. Thus, Kepler's second law is an alternate way of stating the basic physical law of conservation of angular momentum for the planets and satellites.

To understand the formation of elliptical orbit, let us start with the hypothetical condition that the satellite is at an infinite distance from the earth. At this distance, it will remain out of the influence of the earth, with no gravitational force acting upon it. Therefore, with no mutual force existing, the potential energy, PE, of the earth—satellite system under this condition is zero.

However, as the satellite comes within a finite range, owing to the earth's gravitational pull, the satellite starts moving toward the earth. This force is given by

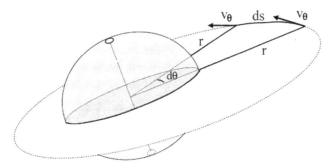

FIGURE 3.2

Area swept by satellite.

$F = -\mu/r^2\,\hat{r}$, where $\hat{r}$ is the unit vector directed radially outward from the center of the earth and $\mu = GM$, where G is the gravitational constant of the earth and M is its mass. r is the distance of the satellite from the center. Because most orbital parameters are independent of mass, in the expressions here and in all subsequent derivations, we shall consider unit mass. We will mention at the appropriate place where the mass of the satellite is to be multiplied for pragmatic cases. As the satellite moves under the influence of the earth's gravitational force, work is done on the satellite by the force; this work amounts to

$$W(r) = \int_{\infty}^{r} -\frac{\mu}{r^2}\cdot dr$$

$$= \mu/r\Big|_{\infty}^{r}$$

$$= \mu/r$$

(3.7a)

where the satellite is at any finite distance r from the center of the earth. The potential energy of the system, which by definition, is the work done "by the body against the force," is thus $-W$. This was initially zero, but now it turns negative and is equal to

$$PE(r) = -W$$
$$= -\mu/r$$

(3.7b)

The negative value indicates that effective work has been done on the body along the force from the condition of no mutual influence in which there was infinite separation. Hence, the potential energy of this two-body system has been lost to a final negative value and the satellite has come under the influence of the gravitational force. Therefore, it will eventually fall to the earth unless some opposite force acts upon it. Such a force can be imparted to the satellite if it revolves round the earth. The cross-radial motion during the revolution creates a centrifugal force opposite the gravity. The force amounts to v_θ^2/r at a range r from the earth's center, where v_θ is the cross-radial component of the linear velocity of the satellite. For circular orbits, this velocity is always equal to the total velocity, v. For other shapes of orbit, v_θ is only a component of v while a finite radial component, v_r, exists. This velocity v required for the condition of force balance adds some positive kinetic energy (KE) to the satellite.

Now, consider the left of the Figure 3.3. For a satellite at a distance "a" from the earth's center, if the velocity v is always cross-radial in direction and "just appropriate" in magnitude such that the resultant centrifugal acceleration v^2/a balances the gravitational acceleration μ/a^2, then in absence of any net acceleration on the satellite, the radius is maintained at the constant value "a." Therefore, we obtain circular orbits with a constant radius. The required condition is thus

$$v^2/a = \mu/a^2$$

$$v = \sqrt{\frac{\mu}{a}}$$

(3.8a)

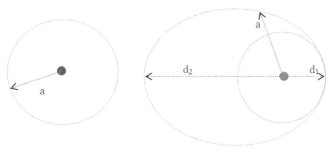

FIGURE 3.3

Circular and elliptical orbit.

Replacing the velocity with the constant angular momentum for unit mass, $L = va$, we get

$$L^2 = \mu a$$

or,

$$L^2/\mu = a \tag{3.8b}$$

Here, the term μ is a physical constant and we have no hold over it. However, for the term L, although it remains constant for a closed system, we can adjust its value for a satellite by imparting definite cross-radial velocity at a definite radius. Thus, the value of L determines its equilibrium radius "a" for a circular orbit. Remember that this L is the angular momentum of the satellite per unit mass. Therefore, when the satellite mass is "m," the total angular momentum must be divided by "m" to get the corresponding value of "L." Thus, to provide a definite circular shape to an orbit of any arbitrary radius "a," the satellite at distance "a" from the earth's center should be given and maintained with a cross-radial velocity of $v = \sqrt{\frac{\mu}{a}}$, as in Eqn (3.8a). The corresponding L thus produced satisfies Eqn (3.8b).

Under such a condition, the KE of the body is $KE = \frac{1}{2} v^2$ and the potential energy is $PE = -\mu/d$. Putting $d = a$ and replacing the value of v in the expression for KE and adding, we get the total energy TE as

$$
\begin{aligned}
TE &= -\mu/a + \tfrac{1}{2}v^2 \\
&= -\mu/a + \tfrac{1}{2}\mu/a \\
&= -\tfrac{1}{2}\mu/a \\
&= -\tfrac{1}{2}\mu^2/L^2
\end{aligned} \tag{3.9}
$$

Here, we have used the fact that under balanced conditions, $a = L^2/\mu$. This equation states that the total energy of the body is determined by the value of L or the radius "a." It increases toward zero as "a" increases.

Because the effective energy remains negative, the earth−satellite system remains coupled under the influence of the earth's gravitational field. Only when some excess positive energy will be acquired by the satellite will the total energy again be zero and the satellite attain the condition it had at infinity. Energy can be added by adding motion to the satellite. Only then will it be free from the influence of the earth even at finite range r. For this, the requirement is that the energy of a satellite in circular motion at a radius "a" must be increased by an additional amount of ½ μ/a, to make the total energy zero. Thus, if the incremental energy is expressed as ΔE,

$$\Delta E \geq \tfrac{1}{2}\,\mu/a \tag{3.10}$$

If excess energy is provided to this satellite in circular orbit with radius "a" through the additional cross-radial velocity, so that its final value is v, the excess energy added is given by

$$\Delta E = \text{Final energy} - \text{Initial engery}$$

$$= \left(\tfrac{1}{2}v^2 - \mu/a\right) - \left(-\tfrac{1}{2}\,\mu/a\right)$$

$$= \left(\tfrac{1}{2}v^2 - \tfrac{1}{2}\,\mu/a\right) \tag{3.11a}$$

This excess energy allows the satellite to deviate from its original orbit. To break free from the Earth's influence, this should be more than the amount ½ μ/a. Thus, the excess energy above the balanced condition required to achieve the condition of no gravitational influence is given by

$$\tfrac{1}{2}v^2 - \tfrac{1}{2}\,\mu/a > \tfrac{1}{2}\,\mu/a \tag{3.11b}$$

Conversely, for the satellite to remain under the influence of the earth and move in a closed orbit (Maral and Bousquet, 2006),

$$\tfrac{1}{2}v^2 - \tfrac{1}{2}\,\mu/a < \tfrac{1}{2}\,\mu/a$$

$$\text{or,} \quad v^2 < 2\mu/a$$

$$\text{or,} \quad L^2/\mu < 2a \tag{3.11c}$$

Equations (3.11c) could also be reached by considering the final energy to be less than zero. The detailed derivation is just for the purposes of understanding. Therefore, consolidating the facts we have discussed so far, the value of the angular momentum per unit mass of a satellite, L, at distance "a" from the earth's center must be such that $L^2/\mu a = 1$ to maintain a circular orbit of radius "a." Furthermore, if the L value is increased, it remains bound under the earth's influence until $L^2/\mu a = 2$ when it breaks free. But what happens to the orbit shape when the value of this term is between 2 and 1, or even less than 1? That is what we are going to learn next.

A variation of the circular orbit results is an elliptical orbit. To the right of Figure 3.3, if the orbiting satellite at A_1 at a distance d_1 from the earth has a velocity v_1 that is essentially cross-radial and is more than that required for circular orbit,

$$v_1 > \sqrt{\frac{\mu}{d_1}}$$

$$\text{or,}\quad L^2/\mu > d_1$$

$$\text{or,}\quad L^2/(\mu\, d_1) > 1 \qquad (3.12a)$$

At the same time, the value of L is well within the limit to keep the satellite bound. Hence,

$$L^2/(\mu\, d_1) < 2 \qquad (3.12b)$$

We call the factor L^2/μ, as the radius "h" of the equivalent circular orbit or the equivalent circular radius for a given L; then, Eqns (3.12a) and (3.12b), respectively, turn into

$$h/d_1 > 1 \quad \text{and} \qquad (3.12c)$$

$$h/d_1 < 2 \qquad (3.12d)$$

To represent the amount by which the expression on the left-hand side is greater than 1, we add a small term to the right of it. Let this term be called ε' for now, to distinguish it from the conventional expression of ε representing the eccentricity of the ellipse. Thus, we say that

$$L^2/(\mu\, d_1) = 1 + \varepsilon' \qquad (3.13a)$$

$$\text{or,}\quad h/d_1 = 1 + \varepsilon' \qquad (3.13b)$$

As $v_1 > \sqrt{\frac{\mu}{d_1}}$, there will be an excess centrifugal force over the gravitational pull, and consequently, some excess KE of the satellite over that required for a circular orbit at radius "d_1." Because of this excess centrifugal force, the satellite will initially gain radial acceleration and will move away from the circular path, gaining outward radial velocity, v_r. The radial distance r will gradually increase, and this will consequently decrease the cross-radial velocity, v_θ, to conserve L. However, the gravitational pull on the satellite will also lessen. The effective radial acceleration of the satellite at any distance r thus will be

$$f_r = v_\theta^2/r - \mu/r^2 \qquad (3.14a)$$

Replacing the term v_θ by an equivalent expression with constant L

$$f_r = L^2/r^3 - \mu/r^2 \qquad (3.14b)$$

Because L and μ are finite constants and the first term falls off faster than the second with r, there must be a finite r at which this force becomes zero. Beyond this point, the net force changes direction and remains attractive toward the earth. Therefore, the condition at the radius "r_0" where the forces balanced becomes

$$L^2/\mu r_0 = 1 \qquad (3.14c)$$

$$\text{or,}\quad r_0 = h$$

Using this equation, we see that the balance of centrifugal force and gravitational force occurs when the radius is equal to the equivalent circular radius h for the given L. However, because of positive radial acceleration up to this point, there will be a finite radial velocity that will keep the satellite moving in a radial direction even at the point where the forces balance.

Replacing the L^2/μ term from Eqn (3.13a), we find that the forces balance under the condition

$$r_0/d_1 = 1 + \varepsilon' \tag{3.15}$$

The force reverses when "r" exceeds beyond that. Using Eqn (3.12c and d), we can say that this reversal must occur between $d_1 < r_0 < 2d_1$ for the satellite to remain bound. It can easily be shown that at these two threshold cases, the corresponding ε' values are 0 and 1, respectively.

The orbital conditions are illustrated in Figure 3.4. With excess initial centrifugal force, as the satellite moves radially outward, it gains radial velocity up to a point at which there is a force balance with no excess centrifugal force. Thus, even at this point, the satellite does not stop its radial excursion owing to inertia. The radial velocity it has gained lets it continue to move radially outward. However, as it does so, exceeding the radius beyond this point, the gravitational pull starts to override the centrifugal force and it experiences an effective inward force and becomes decelerated. As a result, its residual radial velocity gradually reduces to zero. Therefore, the total velocity becomes effectively cross-radial again.

The satellite has now reached its furthest point A_2 at a distance d_2. That this point will be formed directly opposite A_1 is evident from the geometry of the ellipse, because at any other point except this, there is a radial component of the velocity, as shown in Figure 3.5. From point A_2, with the effective inward force acting, it starts receding toward A_1 with gradually increasing radial velocity. The cross radial velocity keeps the angular excursion increasing with the continued motion. As the radial distance again falls below r_0, the effective force again reverses direction. The satellite returns to the same condition at A_1, once again following an elliptical shape of orbit. This process continues.

The radial velocity at any intermediate points between A_1 and A_2 may be obtained by using Newtonian dynamics. The radial velocity $v_r\,(r)$ when the satellite

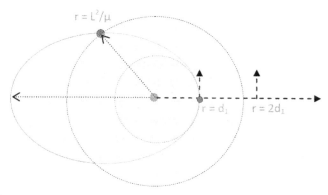

FIGURE 3.4

Orbits determined by L^2/μ.

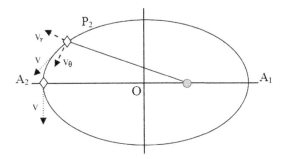

FIGURE 3.5

Velocity components of a satellite in an elliptical orbit.

is at a radial distance r can be derived from the effective radial acceleration as (Strelkov, 1978)

$$f_r = v_\theta^2/r - \mu/r^2$$

$$dv_r/dr \, dr/dt = v_\theta^2/r - \mu/r^2$$

$$v_r dv_r = \left(v_\theta^2/r - \mu/r^2\right) dr$$

$$v_r^2/2 = \int \left(v_\theta^2/r - \mu/r^2\right) dr \tag{3.16}$$

where we have taken the position of the satellite perigee as the initial condition when the radial velocity was zero. Replacing the expression for v_θ by L/r and integrating between $r = d_1$ to $r = r$, we get

$$v_r^2 = -L^2\left(1/r^2 - 1/d_1^2\right) + 2\mu\left(1/r - 1/d_1\right) \tag{3.17}$$

Using the force balancing condition, $r = L^2/\mu$, we can see that at the point of force balance, the radial velocity does not become zero. Rather, it becomes

$$v_r^2 = \mu^2/L^2 + L^2/d_1^2 - 2\mu/d_1 \tag{3.18}$$

For the satellite to remain held by the earth's gravity, the gravitational force must supersede on the outward centrifugal force and consequently turn the radial velocity back toward the earth. For that, the velocity must come down to zero at a finite value of $r \geq d_1$. To achieve the condition of $v_r = 0$, the criterion that must be fulfilled is

$$L^2\left(1/r^2 - 1/d_1^2\right) = 2\mu\left(1/r - 1/d_1\right) \tag{3.19a}$$

This relates radius "r" to known parameters L, μ, and d_1 when $v_r = 0$. One obvious solution to this equation for zero radial velocity is $r = d_1$, which is the condition with which we started. The other solution is $r = d_2$ such that

$$L^2/\mu = 2/\left(1/d_2 + 1/d_1\right)$$

$$\text{or,} \quad 2/h = \left(1/d_1 + 1/d_2\right) \tag{3.19b}$$

Points, d_1 and d_2 in an ellipse with zero radial velocity are its apogee and perigee, and the sum of these two lengths is equal to twice the semi-major axis, a. Similar to that, the sum of the inverses of these two lengths is twice the inverse of length h, which is the equivalent circular radius. Using Eqn (3.19b), we get

$$2d_1/h = (1 + d_1/d_2) \qquad (3.19c)$$

Using our definition of ε' and replacing its value from Eqn (3.13a), we get

$$2/(1 + \varepsilon') = 1 + (d_1/d_2)$$

$$\text{or,} \quad d_1/d_2 = 2/(1 + \varepsilon') - 1$$

$$= (1 - \varepsilon')/(1 + \varepsilon') \qquad (3.19d)$$

The satellite, according to Kepler's law, moves in an orbit of closed elliptical geometry with two radii of zero radial velocities as the perigee and the apogee. Thus, the shorter of these two values, i.e. d_1, is the perigee and $d_2 = (1 + \varepsilon')/(1 - \varepsilon')d_1$ is the apogee of the effective orbit of the satellite. But look at the ratio! This ratio of d_2/d_1 is the ratio of the apogee to the perigee of an ellipse. As per the geometry, this should be equal to $(1 + \varepsilon)/(1 - \varepsilon)$, where ε is the eccentricity of the ellipse, as in Eqn. (3.3c). Thus, comparing the two, we can say that what we were calling ε' is nothing but the eccentricity ε of the elliptical orbit the satellite makes. Therefore, ε' and ε are identical (i.e. $\varepsilon' = \varepsilon$). So, using the expressions, of Eqn. (3.13b), we get

$$d_1 = h/(1 + \varepsilon) \qquad (3.20a)$$

Again from Eqn. (3.3a), $d_1 = (1 - \varepsilon)a$. Thus, comparing the two, the semi-major axis of the elliptical orbit may be related to the equivalent circular orbit h for the given L by the relation

$$h/(1 + \varepsilon) = (1 - \varepsilon)a$$

$$\text{or,} \quad h = (1 - \varepsilon^2)a \qquad (3.20b)$$

Hence, h, although greater than perigee length d_1, is always lesser than a. Thus, it is evident from Eqn. (3.13a) and from the previous observation that eccentricity ε of the elliptical shape of the orbit is determined by the value of L and the distance d_1 at which the velocity is fully cross-radial.

We have thus already derived the expression for radial velocity v_r and determined the distances of the apogee and perigee using the same. At these two points, the radial velocities are zero and only the cross-radial component of the velocity v_θ exists. Now, let us find a more convenient expression for the velocity and energy of the satellite at these two opposite points, A_1 and A_2, (i.e. velocities at the perigee and the apogee). This can be done by considering that the total angular momentum and the total energy of the satellite are conserved. To do that, let us recall that for elliptical orbits, the distance is $d_1 = a(1 - \varepsilon)$ at perigee A_1, and the distance is $d_2 = a(1 + \varepsilon)$ at apogee A_2, where a is the length of the semi-major axis.

Moreover, let the velocities at these two points be v_1 and v_2. Now, because the angular momentum is conserved,

$$v_1 d_1 = v_2 d_2$$

$$\text{or,} \quad v_2/v_1 = d_1/d_2$$

$$\text{or,} \quad v_2/v_1 = a(1 - \varepsilon)/a(1 + \varepsilon) = (1 - \varepsilon)/(1 + \varepsilon) \qquad (3.21)$$

Thus, the ratio of the velocities at these two points is also a function of the eccentricity ε. Like the way the radius swings between $(1 + \varepsilon)$ and $(1 - \varepsilon)$ portions of the semi-major axis "a," the velocity also varies between the same factors of a velocity.

We have seen already that the total energy of the satellite in orbit is the sum of the potential energy as a result of the gravitational force and the KE, entirely due to the linear velocity. Thus, at the perigee and apogee points, the total energy is, respectively

$$TE_1 = -\mu/a(1 - \varepsilon) + \tfrac{1}{2}v_1^2 \quad \text{and} \qquad (3.22a)$$

$$TE_2 = -\mu/a(1 + \varepsilon) + \tfrac{1}{2}v_2^2 \qquad (3.22b)$$

Here, the potential energy of the satellite at any distance r from the earth is taken as $PE = -\mu/r$ and the KE is taken as $KE = \tfrac{1}{2} v^2$, where v is the corresponding velocity of the satellite.

Because the total energy is conserved, these two values may be equated to obtain

$$-\mu/a(1 - \varepsilon) + \tfrac{1}{2}v_1^2 = -\mu/a(1 + \varepsilon) + \tfrac{1}{2}v_2^2$$

$$\text{or,} \quad -(\mu/a)\{1/(1 - \varepsilon) - 1/(1 + \varepsilon)\} = \tfrac{1}{2} v2^2 - \tfrac{1}{2}v_1^2$$

$$\text{or,} \quad (\mu/a)\{2\varepsilon/(1 - \varepsilon^2)\} = \tfrac{1}{2}v_1^2\left\{1 - (v_2/v_1)^2\right\}$$

$$\text{or,} \quad (\mu/a)\{2\varepsilon/(1 - \varepsilon^2)\} = \tfrac{1}{2}v_1^2\left\{1 - (1 - \varepsilon)^2 / (1 + \varepsilon)^2\right\}$$

$$\text{or,} \quad (\mu/a)\{2\varepsilon/(1 - \varepsilon^2)\} = v_1^2\{2\varepsilon/(1 + \varepsilon^2)\}$$

$$\text{or,} \quad v_1^2 = \mu/a(1 + \varepsilon)/(1 - \varepsilon) \quad \text{or,} \quad v_1 = \sqrt{\frac{\mu}{a}}\sqrt{\frac{1 + \varepsilon}{1 - \varepsilon}} \qquad (3.23)$$

Similarly, from this, the expression for v_2 is

$$v_2 = \{(1 - \varepsilon)/(1 + \varepsilon)\}v_1$$

$$= \sqrt{\frac{\mu}{a}}\sqrt{\frac{1 - \varepsilon}{1 + \varepsilon}} \qquad (3.24)$$

Therefore, the velocity of the satellite at the perigee should be $\sqrt{\frac{(1+\varepsilon)}{(1-\varepsilon)}}$ times greater than that required to keep it in a circular orbit, whereas the velocity v_2 at the apogee must be lesser by a factor of $\sqrt{\frac{(1-\varepsilon)}{(1+\varepsilon)}}$ than the same velocity. Therefore, the elliptical shape may be obtained by imparting velocity v_1 at perigee A_1 or by giving velocity v_2 at apogee A_2. This shape remains almost unchanged in space unless there factors dissipate its energy. These factors are discussed in Section 3.3.

Thus, we can conclude that whenever a cross-radial velocity is imparted to a satellite at radius "d" such that the corresponding angular momentum is L and the corresponding centrifugal acceleration overbalances or underbalances the gravitational force, the satellite executes an elliptical orbital motion with eccentricity ε such that $r_0/d = (1 = \varepsilon)$ or $(1 - \varepsilon)$, respectively, where r_0 is the radius of the equivalent circular orbit corresponding to L.

The total energy of the satellite may be obtained by estimating the energy at any specific point, say A_1, because it is conserved. The total energy at A_1 is

$$
\begin{aligned}
E_{A1} &= -\mu/d_1 + \tfrac{1}{2}\, v_1^2 \\
&= -\mu/\{a(1-\varepsilon)\} + \tfrac{1}{2}\,(\mu/a)\{(1+\varepsilon)/(1-\varepsilon)\} \\
&= -\tfrac{1}{2}\,(\mu/a)\{2/(1-\varepsilon) - (1+\varepsilon)/(1-\varepsilon)\} \\
&= -\tfrac{1}{2}\,(\mu/a)
\end{aligned}
\tag{3.25}
$$

This expression depends only on its semi-major axis "a" and is independent of the eccentricity, ε. This is also equal to the energy that the satellite would have had in a circular orbit of radius "a." Let us see it in a different way.

The excess KE at the perigee of length $a(1 - \varepsilon)$ over the requisite value for circular orbit of radius a is

$$
\begin{aligned}
\delta E_k &= \tfrac{1}{2}\,(\mu/a)[\{(1+\varepsilon)/(1-\varepsilon)\} - 1] \\
&= \tfrac{1}{2}\,(\mu/a)[2\varepsilon/(1-\varepsilon)] \\
&= (\mu/a)\,\varepsilon/(1-\varepsilon)
\end{aligned}
\tag{3.26}
$$

The excess potential energy is

$$
\begin{aligned}
\delta E_p &= -\mu/a(1-\varepsilon) + \mu/a \\
&= -\mu/a\{1/(1-\varepsilon) - 1\} \\
&= -(\mu/a)\,\varepsilon/(1-\varepsilon)
\end{aligned}
\tag{3.27}
$$

Therefore, the kinetic and the potential energy deviate equally and opposite, and hence the excess energies cancel out each other. The total energy remains the same as if it had been in a circular orbit of radius "a" as obtained in Eqn. 3.9. However, this magnitude of derivation is only for the specific position of the satellite at the perigee. For all other locations, the deviation amounts differ, although they always remain equal and opposite.

Equation (3.25) also shows that reduced energy leads to a reduction in a, but the orbiting velocity increases with the decreasing radius. Thus, although it looks anomalous, reduced energy leads to an increase in the velocity of the satellites with a reduction in the radius.

Once the shape of the elliptical orbit is defined by "a" and "b" or "ε," we may be interested in defining the position of the satellite in this orbit. This may be done by using the angle obtained at the focus or center of the ellipse by the satellite. To do so, we define three angular positions, as discussed subsequently (Maral and Bousquet, 2006; Pratt et al. 1986).

3.1.2.1 True anomaly (v)

True anomaly of a satellite is the angle subtend by the satellite at the focus of its elliptical orbit at any instance between the direction of the perigee and the direction of the satellite, counted positively in the direction of the movement of the satellite (Figure 3.6).

If v is the angle of true anomaly, the expression for range "r" of any point on the ellipse as a function of v becomes

$$r = a(1 - \varepsilon^2)/(1 + \varepsilon \cos v) \qquad (3.28a)$$

You can validate this by comparing the ranges for $v = 0°$ and $v = 180°$, and also by putting $\varepsilon = 0$ for a circular orbit. The variations of the orbital shape factors and corresponding parameters of the satellite with true anomaly of the orbit has been elaborated in the Matlab exercise given in Box 3.1.

3.1.2.2 Eccentric anomaly (E)

Now, let us consider a Cartesian coordinate with an origin O at the center of the ellipse with eccentricity ε and semi-major axis "a." The x axis is along the direction of perigee on the semi-major axis and the y axis is along the semi-minor axis, as in Figure 3.6. The points on the ellipse maintain a relation between its two coordinates, x and y, which is the equation of the ellipse and is given by

$$x^2/a^2 + y^2/b^2 = 1 \qquad (3.28b)$$

To understand the eccentric anomaly, let us consider a circle concentric with the ellipse and with radius "a," as in the semi-major axis of the ellipse. The circle is called the principle circle for the given ellipse. Now, we assume a projection of satellite S at S′ located on the principle circle and with the same x coordinate value as the actual satellite S located on the ellipse.

The angular argument of S′ at the center with the positive x axis, which is also the direction of the perigee, is called the eccentric anomaly, E. It is also measured positively

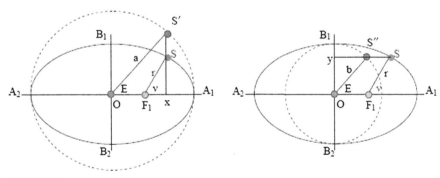

FIGURE 3.6

True anomaly and eccentric anomaly.

BOX 3.1 MATLAB EXERCISE

The MATLAB program orbit.m was run to generate the following variations of the radial range, radial and cross-radial velocity; angular momentum; and kinetic and potential energy with true anomaly. For a given initial perigee length of 7000 km, if the value of L exceeds that required for circular orbit by a factor of 1.25, the following shown in Figure M3.1 plots are generated.

Run the program for different ranges and factors. See what happens when the factor is more than 1.3. You approximately find the ratio of the ranges at the perigee and apogee from the results you obtain upon running the program. Find "a" and compare the expression for ε using it.

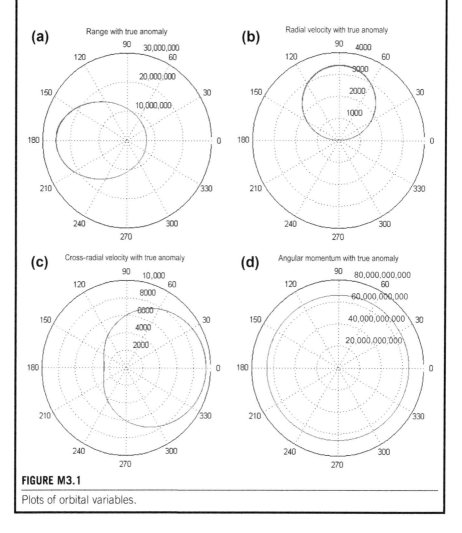

FIGURE M3.1

Plots of orbital variables.

in the direction of the movement of the satellite. Similarly, if we consider a similar concentric circle with radius "b" and consider the projection of S on it at S'' such that they have the same "y" coordinate, S'' will also subtend the same angle at the center.

So, x and y may also be represented as

$$x = a \cos E$$
$$y = b \sin E \qquad (3.29)$$
$$= a\sqrt{1 - \varepsilon^2} \sin E$$

Figure 3.6 shows the relationship between true anomaly v and eccentric anomaly E. Because the x projection remains the same for both cases, the relation can be expressed as

$$a \cos E = a \varepsilon + r \cos v \qquad (3.30a)$$

Thus, dividing the whole equation by a and replacing the expression for r, we get

$$\cos E = \varepsilon + \left(1 - \varepsilon^2\right)\cos v / (1 + \varepsilon \cos v)$$
$$\text{or,} \quad \cos E = (\varepsilon + \cos v)/(1 + \varepsilon \cos v) \qquad (3.30b)$$

Conversely,

$$(\cos E - \varepsilon)/(1 - \varepsilon \cos E) = \cos v \qquad (3.30c)$$

Therefore, one can derive the value of either the v or E from knowledge of the other. The radial range r can be expressed in terms of E as

$$r = a(1 - \varepsilon \cos E) \qquad (3.31)$$

Another important thing is that because S' is the image of S, the time S requires to traverse its complete elliptical orbit is equal to the time required by S' to complete its circular orbit.

3.1.2.3 Mean anomaly (M)

From the third law of Kepler, it is evident that the time periods of satellites moving around the earth depend solely on their semi-major axis, irrespective of their eccentricities. A satellite S_1 with any arbitrary eccentricity "ε" with a semi-major axis "a" will have the same period of revolution about the earth as that of satellite S_2 in circular orbit with radius "a" and zero eccentricity. Thus, in Figure 3.7, satellite S_1 and S_2 have the same period because they have the same semi-major axis $OA_1 = a$. If they start at the same time T_0 from point A_1, they will traverse their respective orbits and will complete a full rotation and come back to the same point at the same time instant T_1. It does not matter whether the focus is off-center, as for S_1, or at the center, as for S_2, even though satellite motion is governed by the force at the focus.

This is because satellite S_1 moves faster than S_2 when it is around its perigee, and is nearer to the earth at "F_1" than S_2, which is at distance "a." But the leverage is equally lost, as at other times when S_1 is around its apogee, A_2, the distance more

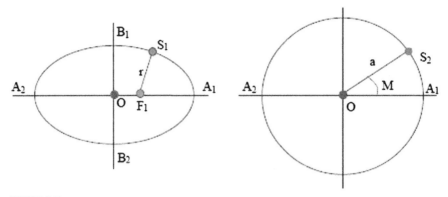

FIGURE 3.7

Mean anomaly and true anomaly.

than "a," moves slower than S_2. This is also evident from the fact that $v \times r$ of the satellite is constant, as mentioned before.

Because both satellites will traverse the same angular distance in equal time, the mean angular velocity of the satellite S_1 in elliptical orbit is equal to the constant angular velocity of the satellite S_2 in circular orbit. Thus, satellite S_2 in circular orbit may be considered to execute the mean motion of any satellite in elliptical orbit with the same length of the semi-major as far as the angular motion is concerned. However, the instantaneous angles that the two will make at any instant t at the focus will be different. It is easier to estimate the angle made by S_2 but it is difficult to do so for S_1. Can we relate these two angles? Let us attempt to do that by relating the areas they sweep; if we succeed, then from the straightforward estimation of the first we can derive that of the second.

First, let us consider that both S_1 and S_2 start from their perigee at the same time, $t = 0$, the mean anomaly, M of the satellites after any interval t is equal to the angle that S_2 makes in that time. In the same period of time, let the eccentric anomaly of S_1 be E.

To establish a relation between these angles that these two satellites sweep, let us consider that in period T, the area swept by S_1 and S_2 will be D_1 and D_2, respectively. Then

$$D_1 = \pi ab \quad \text{and}$$

$$D_2 = \pi a^2 \tag{3.32}$$

Furthermore, from Kepler's second law, individually, they will each sweep equal area in an equal interval of time. Thus, the sweeping rate of S_1 and S_2 will be, respectively,

$$\dot{D}_1 = D_1/T = \pi ab/T$$

$$\dot{D}_2 = D_2/T = \pi a^2/T \qquad (3.33)$$

Therefore, the area sweeping rates are obviously constant for each although they are different for the two satellites and the area swept is a linear function of time. In any arbitrary time interval t from $t = 0$, the area swept by S_1 and S_2 will be, respectively,

$$\Delta_1 = A_1 F_1 S_1 = (\pi a b/T)t$$

$$\Delta_2 = A_1 O S_2 = (\pi a^2/T)t$$

$$\text{So,} \quad \Delta_1/\Delta_2 = b : a \qquad (3.34)$$

We conclude from this that the ratio of the area swept by S_1 and S_2 on their respective orbits will always be in the ratio b:a.

Therefore, satellite S_2 on the equivalent circular orbit of radius "a" will have a constant angular velocity $\omega = \sqrt{\mu/a^3} = n$ and will make an angle $M = n \times t$, in time t with the reference axis along the perigee where "t" is the time elapsed after it crossed the perigee. This angle M is thus the angle subtend at the focus by the mean motion and is called the mean anomaly for the satellite in elliptical orbit.

Now, using the relation established between the corresponding areas swept, it can be easily determined that the area swept by the mean motion at the center is $\Delta_2 = \frac{1}{2} M a^2$.

Again, consider the geometry as in Figure 3.8 with satellite S on its true elliptical orbit, and also its principle circle carrying its image S'. Let us consider areas $A_1 F_1 S$ and $A_1 F_1 S'$, which are the areas swept by satellite S on the true elliptical orbit and its projection S', respectively, on the reference circle, in time t at focus F_1. From the geometry, the corresponding areas may be related by

$$A_1 F_1 S = A_1 X S - X F_1 S \quad \text{and} \qquad (3.35a)$$

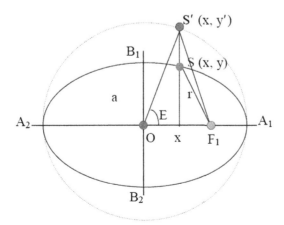

FIGURE 3.8

Relation between eccentric and mean anomaly.

$$A_1F_1S' = A_1XS' - XF_1S' \tag{3.35b}$$

Now, $A_1XS/A_1XS' = \int\limits_{X}^{A_1} y\,dx \Big/ \int\limits_{X}^{A_1} y'\,dx$

$$= \int\limits_{X}^{A_1} b\sqrt{1 - \left(\frac{x}{a}\right)^2}\,dx \Big/ \int\limits_{X}^{A_1} a\sqrt{1 - \left(\frac{x}{a}\right)^2}\,dx \tag{3.36}$$

$$= b : a$$

Here we have used the equation for the ellipse as $y^2/b^2 = 1 - x^2/a^2$, and that for the reference circle as $y^2/a^2 = 1 - x^2/a^2$. Again,

$$XF_1S/XF_1S' = XS/XS'$$
$$= y/a \sin E$$
$$= b \sin E/a \sin E$$
$$= b/a$$

So,

$$A_1F_1S/A_1F_1S' = (A_1XS - XF_1S)/(A_1XS' - XF_1S')$$
$$= (b : a) \tag{3.37a}$$

Therefore, these two areas, A_1F_1S and A_1F_1S', are also in the ratio b:a. However, A_1F_1S is the true area swept by the satellite at the focus (i.e. $A_1F_1S = \Delta_1$). So,

$$\Delta_1/A_1F_1S' = b/a$$
$$= \Delta_1/\Delta_2 \tag{3.37b}$$

We can say thus that $A_1F_1S' = \Delta_2$, the area swept by the satellite with mean motion at the center in the same time, which is again equal to $\frac{1}{2} Ma^2$. Therefore,

$$A_1F_1S' = \frac{1}{2} Ma^2 \tag{3.37c}$$

Using the equivalence in the geometry we have derived in the previous equation, we get

$$A_1F_1S' = AOS' - OF_1S'$$
$$= \frac{1}{2} Ea^2 - \frac{1}{2} a\varepsilon\, a \sin E$$

So,

$$\frac{1}{2} Ma^2 = \frac{1}{2} Ea^2 - \frac{1}{2} a\varepsilon\, a \sin E$$
$$M = E - \varepsilon \sin E \tag{3.38}$$

This is also known as Kepler's equation, from which we can observe that at $E = 0$ and $E = \pi$, the eccentric anomaly and the mean anomaly becomes the same.

Otherwise, M trails the E in the first two quadrants, whereas it leads the E in the next two. In general, we derive the value of E from M using iterative techniques.

Thus, these three parameters are equivalent because they can be transformed from one to the other. Moreover, knowing the semi-major axis, we will know the angular sweep rate of the mean motion of the satellite. Using time elapsed from the instant of crossing the perigee we can derive the mean anomaly, M. From this, we can obtain the eccentric anomaly, E, using the previous equation with the known value of eccentricity. Again, from the relation between the eccentric anomaly and true anomaly, the latter may be derived conveniently. Therefore, knowing the semi-major axis, the eccentricity, and the time from periapsis, we can derive the true anomaly, and hence the position of the satellite in the orbit. These three parameters are required to explicitly define the position of the satellite in its orbit.

3.2 Orbital orientation relative to earth

The discussion up to this point has defined only the shape of the orbit and the position of the satellite with respect to the direction of the perigee. Once the position of the satellite in orbit is known, it is necessary to find the orientation of the orbit with respect to the earth. This is because, ultimately, we need the satellite position in earth's reference. This may be done by fixing the orbit orientation in the earth-centered inertial (ECI) frame, which is described in Chapter 1. Therefore, we consider the ECI axes on the erect earth with the origin coinciding with the earth's center. As per the definition of the ECI, the origin moves with the earth, but the orientation of the axes remains fixed in space and does not change with the earth's rotation or its movement in space.

3.2.1 Orientation parameters

We shall define the parameters that fix the orbit in space and the location of the satellite in the orbit with respect to the ECI (Maral and Bousquet, 2006, Pratt et al. 1986). These orbital orientation parameters are depicted in Figure 3.9. To understand the definitions of these parameters, it is first necessary to appreciate that because the earth's center is at the focus of the orbit, it remains simultaneously on the orbital and equatorial (XY) planes. The orbital plane is the infinite plane passing through the geocenter and containing the orbit of the satellite. To define the orientation of the plane with respect to the equatorial plane, we need to define two parameters: inclination, i and right ascension of the ascending node (RAAN), denoted by Ω.

3.2.1.1 Inclination (i)

The plane of the satellite orbit may make some definite angle with the equatorial plane of the earth (i.e. the XY plane of the coordinate system). This angle is called the inclination. We know that the focus of the orbit coincides with the center of the earth on the equatorial plane. So, if we draw a vector perpendicular to the orbital

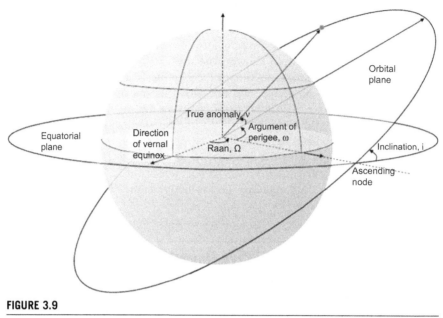

FIGURE 3.9

Orbital orientation parameters.

plane (orbital plane vector) at the focus, the acute angle it makes with the positive z axis, which is perpendicular to the equatorial plane, is the inclination.

3.2.1.2 Right ascension of the ascending node (Ω)

For the same inclination, there may be different azimuthal orientations of the orbit. This is also evident from the fact that the normal on the orbital plane at the focus can make the same angle with the z axis (i.e. have same inclination) for many different relative orientations with the equatorial plane. Thus, the true orientation of the plane is specified by the parameter, right ascension of the ascending node (RAAN).

The intersection of the orbit and the equatorial plane is a straight line passing through the earth's center (i.e. the origin of the considered ECI coordinate system). This line simultaneously remains on the orbital and equatorial plane and cuts the true orbit at two points: one where the satellite moves from the southern hemisphere to the northern hemisphere, and the other where it moves from the north to south. The first point is called the ascending node, and the second is called the descending node. The RAAN is the angle subtend at the center of the earth between the direction of the ascending node and the positive x axis of the ECI coordinate, measured positively. Defining otherwise, RAAN is the ECI longitude of the satellites ascending node. This fixes the azimuthal orientation of the orbit.

These two parameters fix the orbital plane with respect to the equatorial plane. However, the orientation of the actual orbit on this plane has yet to be fixed. This is done by defining the parameter argument of perigee.

3.2.1.3 Argument of perigee (ω)

For a definite inclination and the RAAN, the plane of the orbit with respect to the earth is fixed, but not the exact orbit itself on the orbital plane. On this defined orbital plane, different orientations of the major axis lead to distinctly separate orbits. For an orbital plane with a given "i" and "Ω," one orbit may have a semi-major axis perpendicular to the nodal line, whereas another may have the axis with a different angle made with this line at the focus, giving rise to two different orbits.

The argument of perigee is the angle made at the focus, measured positively between the direction of the ascending node and the direction of the perigee of the orbit.

It can thus be concluded that knowledge of five parameters (a, ε, i, Ω, and ω) completely defines the orbit of the satellite in space. Furthermore, the position of the satellite in this orbit, with respect to the perigee, can be defined by v, which is obtained from M via E. Again, M can be derived from semi-major axis "a" and time.

The sum of ω + v is the angle made by the satellite with the direction of the ascending node at the earth's center, measured positively from the nodal line on the orbital plane. This angle is called the nodal angular elongation or argument of latitude, "u" (Maral and Bousquet, 2006). This is useful in the case of a circular orbit where the perigee is undefined.

Once it is defined, the orientation of the orbit about the earth remains fixed with reference to space, i.e. with respect to the ECI frame, but it always changes with respect to the earth-centered earth-fixed (ECEF) as the earth spins. Thus, the coordinates in ECI are mostly converted to those in the ECEF frame for convenience, for which the angular motion of the earth is required to be considered.

The shape and the orientations of a satellite orbit for different orbital parameters is derived in the Matlab exercise of Box 3.2. Focus 3.1 describes the procedure for deriving the position coordinates.

BOX 3.2 MATLAB EXERCISE

Refer to Figure M3.2a and b. The MATLAB program sat_pos.m was run to generate the coordinates in an ECI frame for orbits with two different sets of parameters. The first orbit has eccentricity $\varepsilon = 0$ and inclination i = 55*. The second orbit has $\varepsilon = 0.4$ and i* = 35 and $\omega = 10$*. The RAAN in both cases was about 32*. These coordinates were then plotted with a globe for referencing. However, the orbital range is scaled for better representation and the following plots were generated.

Note how the eccentric anomaly is determined in the program using the nonlinear least-squares method. This can be replaced by the iterative Newton–Raphson method.

Continued

BOX 3.2 MATLAB EXERCISE—cont'd

Run the program for different eccentricities, RAAN, and ranges and factors. See what happens when the parameters are changed. Because of scaling, sometimes the orbit may not be well represented compared with the globe.

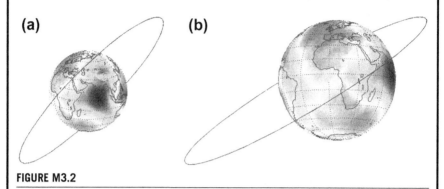

(a)   **(b)**

FIGURE M3.2

Different orbital shapes.

FOCUS 3.1 SATELLITE COORDINATE DERIVATION

We have already seen how, from the given semi-major axis and time from the crossing of the perigee, we can derive the true anomaly, v, through an estimation of the mean anomaly, M, and the eccentric anomaly, E. Given v and five other parameters, we now attempt to find the coordinate of the satellite in ECI. (Spilker, 1996a) Assume a Cartesian coordinate with an origin at the focus, x axis on the orbital plane along the perigee, and y axis also on an orbital plane. The Cartesian coordinates of the satellite on these axes are:

$$x = r * \cos v = a\left\{1 - \varepsilon^2\right\}/\{1 - \varepsilon * \cos v\} * \cos v$$

$$y = r * \sin v = a\left\{1 - \varepsilon^2\right\}/\{1 - \varepsilon * \cos v\} * \sin v$$

These coordinates are transformed into another reference with the x axis directed along the ascending node, and the y axis on the orbital plane and perpendicular to the x axis. Because ω is the angle between the new x axis and the former (i.e. the perigial line), in the new frame the coordinates become:

$$x_1 = x \cos \omega - y \sin \omega = r * \{\cos v * \cos \omega - \sin v * \sin \omega\} = r * \cos(v + \omega) = r * \cos(u)$$

$$y_1 = x \sin \omega + y \cos \omega = r * \{\sin v * \cos \omega + \cos v * \sin \omega\} = r * \sin(v + \omega) = r * \sin(u)$$

We must appreciate that the sum $(v + \omega)$ is the nodal angular elongation, "u." Effectively, we have just decomposed the range vector components over this angle, as evident from this expression.

Consider another reference frame with an x axis along the ascending node, a y axis perpendicular to it but on the equatorial plane, and a z axis perpendicular to it along the North Pole. This is only a rotation of the reference frame about the x axis by the amount of inclination angle "i." The coordinates in this frame become:

$$x2 = r * \cos(u)$$

$$y2 = r * \sin(u) * \cos(i)$$

$$z2 = r * \sin(u) * \sin(i)$$

Effectively, we have just decomposed the y1 vector into its component on the equatorial plane and normal to it.

These coordinates convert to coordinates in the ECI frame, which is nothing but rotation about the z axis by angle Ω. The vectors become:

$$xi = r\{\cos(u) * \cos\Omega - \sin(u) * \cos(i) * \sin\Omega\}$$
$$yi = r\{\cos(u) * \sin\Omega + \sin(u) * \cos(i) * \cos\Omega\}$$
$$zi = r\{\sin(u) * \sin(i)\}$$

Using the intrinsic relation, the expressions become:

$$xi = a\{(1-\varepsilon2)/(1-\varepsilon * \cos v)\}\{\cos(v+\omega) * \cos\Omega - \sin(v+\omega) * \cos(i) * \sin\Omega\}$$
$$yi = a\{(1-\varepsilon2)/(1-\varepsilon * \cos v)\}\{\cos(v+\omega) * \sin\Omega + \sin(v+\omega) * \cos(i) * \cos\Omega\}$$
$$zi = a\{(1-\varepsilon2)/(1-\varepsilon * \cos v)\}\{\sin(v+\omega) * \sin(i)\}$$

For further converting it to ECEF, another rotational transformation of the coordinated about the current z axis is required by an angle equal to the current ECI longitude of the ECEF X axis.

3.3 **Perturbation of satellite orbits**

The six parameters are sufficient to define the position of the satellite definitely in ECI. However, this is true only when the intrinsic assumptions taken are valid. The assumptions taken are:

a. The earth is spherical with homogeneous density.
b. No force is exerted on the system that is external to the earth and the satellite in question.
c. The mass of satellite is small compared with the mass of the earth.

In actuality, these assumptions are not perfectly correct. The deviations lead to the perturbations of the satellite motion. The most important of these factors are listed below.

3.3.1 **Perturbation factors**
3.3.1.1 *Nonhomogeneity of the earth*

In all of the previous discussions, it has been assumed that the earth is perfectly spherical with uniform density. However, its shape is actually nearer to an oblate spheroid with varying density. Thus, the center of mass of the earth is not exactly at the geometric center of its shape. Satellites revolving around the earth experience different gravitational forces and hence adjust accordingly to different effective distances from the true location of the earth's center of mass. For example, considering the center of mass of the earth is skewed toward the denser region, each time a satellite passes across the effectively denser portions of the earth, it is nearer to the center of mass, and hence experiences greater gravity pull. It consequently reduces range and speeds up, deviating from the designated orbit. This affects the shape of the orbit and the direction of its axes, although to a small extent.

3.3.1.2 External forces

There are many celestial bodies present, nearer to the earth and the satellite. Among them, the ones that most affect the motion of satellites the most are the moon and the sun. The sun, mainly due to its massiveness; and the moon, owing to its proximity, exert significant gravity pull and perturb satellites from their designated course. Moreover, they result in a change in inclination angles and RAAN of the satellite orbital plane. The perigial direction is also deviated as a result. However, the perturbation is expectedly small and depends on the relative location of the satellite and these perturbing elements.

3.3.1.3 Aerodynamic drag

Satellites experience some opposing forces as they traverse the atmosphere, however thin it is. The drag is significant at low altitudes around 200—400 km and negligibly low at altitudes of about 3000 km. The effect of atmospheric friction is a constant loss of energy and hence a decrease in the semi-major axis of the orbit. The circular orbit may retain its shape, but the elliptical orbit loses eccentricity and tends to become circular.

3.3.1.4 Radiation pressure

Solar radiation incident on a satellite exerts pressure on it. Satellite solar panels kept open perpendicular to the solar flux to obtain maximum power also receive maximum solar pressure. This pressure forces the satellite against its motion as it moves toward the sun during its orbit, while it adds to the movement when it recedes from the sun. Effectually, it modifies the orbital parameters of the satellite.

Out of all these factors, the non-homogeneity of the earth and the external gravitational forces perturb the satellite orbit in such a way that the total energy of the body is conserved, whereas others dissipate energy from it.

3.3.2 Implications for the system

Because of these perturbations, the trajectory of satellites is not a closed ellipse with a fixed orientation in space, but an open curve that continuously evolves in time in both shape and orientation. These perturbations may be modeled to the extent of first-order variation. Because the effectual result of these perturbations is variation in orbital parameters, additional derivatives of the orbital parameters such as da/dt, dε/dt, di/dt, dΩ/dt, and dω/dt etc become necessary to describe the position of the satellites correctly and more accurately. Because the final position estimation of the user is sensitive to the satellite's position, it is important to consider these parameters for their precision estimation.

From previous discussions, we know that the purely elliptical Keplerian orbit is precise only for a simple two-body problem under ideal conditions. Perturbations lead to a modified elliptical orbit with correction terms to account for these variations. Hence, the typical parameters transmitted in a satellite navigation system to make satellite positions available to users is more than just the six Keplerian parameters, so that the orbits estimated from these parameters are precise, at least to the first order of the perturbations.

Keplerian orbital parameters for the future along with the perturbation terms, constituting the ephemeris, are predicted by the ground monitoring facilities. They are derived from the range measurements using least-square curve fitting of the orbit done by a Kalman filter, which we learned about in Chapter 2.

3.4 **Different types of orbit**

From previous discussions, we know that there are a few parameters that define the shape and orientation of the orbits. For example, the semi-major axis determines the period and the range, the eccentricity, defines the shape of the orbit, the inclination and the RAAN, fixing the orientation. Satellite orbits may be categorized based on these individual parameters.

The most common categories of satellite orbits are made based on their radius, or, more generally, on their semi-major axis when the orbit is not circular. Primarily, there are three such types of orbits: low earth (LEO), medium earth (MEO), and geosynchronous (GSO). These orbits are described below.

Satellites in LEOs have heights ranging from 500 to about 1500 km above the earth's surface. Because of their nearness to the earth, satellites in these orbits have small periods of about 1.5–2.0 h. These orbits are used by satellites mainly for remote sensing purposes, because being nearer to the earth, they have better resolution of the images they take, and also receive the maximum reflected power from the earth owing to their proximity. However, because these satellites need to move through large resistive drag, their average lifetime is much lower than others, typically ranging from 2 to 5 years. It is also evident that because the range is smaller, the orbiting speed is fast, even more than that of the earth's spin, irrespective of the inclination. Thus, the visibility time of these satellites from any definite point on the earth is considerably small, typically about 15–20 min per pass.

Medium earth orbit satellites have a radius from around 8000 km to 25,000 km with associated periods ranging from 2.0 to 12.0 h with a typical visibility time of a few hours per pass. These satellites are generally used for navigation purposes and mobile satellite services. A portion of the range for these orbits and at some height beyond it is not generally used for satellites because there is an inhospitable environment at these heights for satellites because of the abundant presence of a belt of high energy-charged plasma particles, called as the Van Allen belt (Tascoine, 1994).

Satellites in GSO are placed at a definite distance from the earth, in circular orbits, such that the angular velocity of the satellites is equal to the angular velocity of the earth. Thus, the satellite revolves in space once in about 24 h, like the earth. Consequently, the satellite maintains almost the same longitude in which it has been placed. Geosynchronous satellites may have different inclinations depending on the application requirement. The more the inclination angle is, the more it sweeps across the latitudes on both sides of the equator, making a figure-of-eight as the ground trace owing to the difference in the east–west rotational velocity. These satellites are at a distance of about 36,000 km from the earth's surface, and because of the large distance, they require a large power of transmission but have large footprints (i.e. area of coverage).

Geostationary satellites are in a particular type of GSO with zero inclination. Thus, these satellites are fixed on the equatorial plane and ideally remain above the same fixed point over the equator, and are therefore mostly used for fixed communication services.

Unlike circular or near-circular orbits, certain orbits are characterized by their eccentricity. The highly elliptical orbit is used for satellites meant for communication at higher latitudes. These orbits have a large eccentricity $\varepsilon \sim 0.8$ to 0.9, and hence the ratio of their perigee to the apogee distance is more than 10. Because the satellites sweep equal areas in unit time, as per Kepler's law, they traverse faster when they are nearer the earth than when they are farther from it. So, when the satellites in these orbits are at the apogee, they move the slowest and thus are visible for a longer time. Such orbits are hence used at higher latitudes with apogee directed appropriately to get the visibility of satellites over a large interval.

The repetition period of satellites in different orbits depends on their range, and the number of times they reappear at the same location over the earth changes accordingly. The plot in Figure 3.10 represents the variation of these two parameters with the range.

If we drop a normal from the satellite, the point where it intersects the earth's surface will constantly change if there is a relative motion between the earth and the satellite. The locus of this point will form the ground trace of the satellite. The formation of the ground trace of a satellite in GSO orbit is described in Box 3.3.

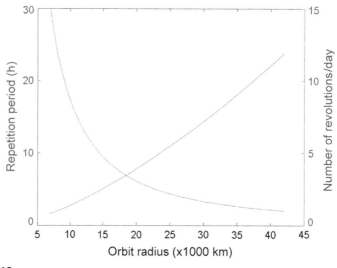

FIGURE 3.10

Variation of repetition period with range and number of revolutions per day.

BOX 3.3 MATLAB EXERCISE

The MATLAB program ground_trace.m was run to generate the coordinates of a circular orbit at a GSO orbit with a definite given inclination of $i = 35°$ and equator crossing node at 74°E. This is shown in Figure M3.3. Note the formation of the figure-of-eight. The next plot is obtained for $\varepsilon = 0.07$, $\omega = 15°$, $i = 55°$, and node at 274°E.

Observe the variation in the shape and orientation of the trace. This results from the eccentric nature of the orbit; it moves faster at one half and slower at the other half, which gives this particular shape.

Run the program for a different nodal point, eccentricity, and inclination, and observe the variations.

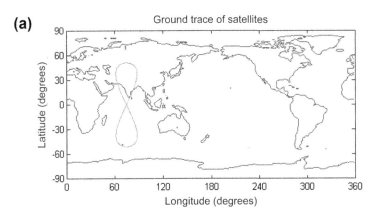

(a)

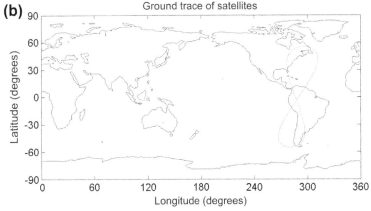

(b)

FIGURE M3.3

Ground trace of satellites.

3.5 **Selection of orbital parameters**

In choosing different orbital parameters during the design of the system, different issues come into consideration. Not only does technical feasibility matter in the process; one also has to be careful about the practicality of implementation. Here, we shall outline only a few theoretical aspects that determine the choice of parameters. However, this provides only a preliminary concept for constellation design. For a detailed designing approach, many other factors need to be considered (Spilker, 1996b); interested readers may refer to Walker (1984) and Rider (1986).

The earth rotates with a fixed angular velocity about its axis, whereas satellites also rotate about the center of the earth. From the discussion in Section 3.1 we know that the velocity of satellites in an ECI frame depends on the distance from the earth's center (i.e. its range). The nearer the satellite is, the faster it moves. Satellite velocities reduce as they move away from the earth's surface, and at about 42,000 km from the center, they have an angular velocity equal to the angular velocity of the earth; thus, satellites in this orbit remain quasi static over one particular longitude. If the inclination is zero (i.e. for GEO satellites), the satellite remains fixed on the longitude at the equatorial plane and is always visible on the equator. If it has an inclined orbit (i.e. GSO), it moves up and down along the longitude, making a figure-of-eight.

In satellite navigation, satellite orbits are chosen depending on the visibility criterion of the satellites (Dorsey et al. 2006), which again depends on the service we need. For a regional service, in which the satellites need to serve only a particular section of the earth, and furthermore, if the region is nearer to the equator with a limited latitudinal extent, it is convenient if the satellites always remain fixed on different longitudes across the service area. Under such conditions, it is likely that all satellites will be visible over the whole service area all of the time.

As we mentioned earlier, and as will be explained later in Chapter 7, the accuracy of estimation of the position depend on the user–satellite geometry, which is determined by the relative position of the satellite with respect to the users. Therefore, satellites need to be dispersed the most to obtain the best possible accuracy of positions. To achieve this, all satellites cannot be kept in GEO; some also need to be on the GSO, swinging across the latitudes and giving wide inter-satellite separation.

The inclination requirement for the GSO is based on the latitudinal extent that satellites need to serve. For the satellites with inclinations $i°$, the loci of the sub-satellite points will form a great circle on the earth, making angle $i°$ with the equatorial plane. This means that the satellite can scale up to a maximum latitude of $i°$ north and south. Users located up to this latitude will see the satellites overhead. For those located above it, satellites will always be seen southward, which obviously will affect the accuracy of the position estimates.

The same design will not be effective for a regional service at higher latitudes. If we use the same GEO–GSO combination even in this case, the satellites will mostly be visible southward for places in the northern hemisphere and the northern sky will remain devoid of them. The opposite will happen for the southern hemisphere. Moreover, because of the equator-ward excursion of the satellite, it will be out of visibility more often than not. Because visibility remains a major criterion, highly elliptical orbits may be used for such cases. The apogee of different satellites may be kept configured above the service area by properly choosing the RAAN, inclination, and argument of perigee. The range and visibility may be adjusted by appropriately designing the semi-major axis and the eccentricity. Then, the satellites will move slowly over the region and faster when it goes antipodal, giving a large proportion of the orbital period to the visible time span. However, there can be other appropriate designs. If we go still higher in latitude, the range of satellite visibility extends across the longitudinal hemispheres. Thus, visibility conditions are not simple, and similar straightforward conclusions cannot be drawn.

When it comes to global services, the whole globe is required to be covered by satellites. At the same time, at least seven to nine satellites must remain visible from any point on the earth. There are two options. As before, satellites can be put into GEO−GSO orbit. However, at GEO−GSO orbit, the geometry of the satellite constellation relative to a finite location on the earth's surface only has north−south excursion. Sometimes it creates difficulties in achieving an adequate dispersion of satellites, and thus better accuracy of position estimates. Moreover, the power required for transmission from this orbit is relatively large.

The second option is to keep satellites at a lower orbit. Here, with a few more satellites in the constellation, one can provide complete coverage with the required numbers of visible satellites. The number is obtained from the visibility requirement. Let us perform some simple calculations to verify this. Although four satellites are enough for position and time estimations, they do not give enough options to users from which to choose the best four. Because position estimation is sensitive to user−satellite relative geometry, the more satellites the user sees in the sky, the more options he has to choose from for a suitable combination of satellites to obtain better accuracy. Here, we shall do an approximate workout to find the number of satellites required to fulfill these criteria.

For simplicity, let us assume that the earth is spherical and a user receiver placed on the earth surface can see the sky down and up to the horizon (i.e. the lowest look angle has zero elevation). This is shown in Figure 3.11. It is not a pragmatic assumption, because it is almost impossible to have an unobstructed vision up to this lower elevation. However, this makes our calculations simple by easing the geometry without affecting the result much.

With these considerations, the geometry shown in the figure reduces to a simpler condition in which the angle θ that we get at the center of the earth of the maximum extent of visibility of the receiver, is $\cos^{-1}\{R/(R + h)\}$.

So, the part of the sky that lies within the visibility range of the receiver at point P subtends angle θ at the center of the earth. The area covered at satellite constellation height $R + h$ by this angle at the geocenter may be obtained by integrating the infinitesimal area dS at radius $(R + h)$ over an azimuth of 0 to 2π and over a polar angle of 0 to θ. The total surface S becomes

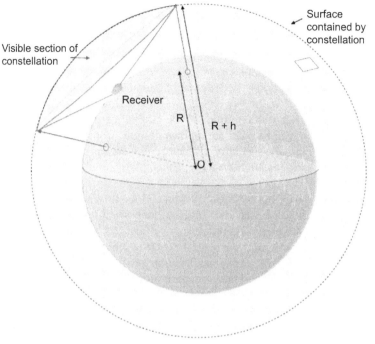

FIGURE 3.11

Estimation of satellites in space segment.

$$
S = \int_{0}^{2\pi} \int_{0}^{\cos^{-1}\left(\frac{R}{R+h}\right)} (R+h)d\theta (R+h)\sin\theta \, d\varphi
$$

$$
= \int_{0}^{2\pi} \int_{0}^{\cos^{-1}\left(\frac{R}{R+h}\right)} (R+h)^2 \sin\theta \, d\theta \, d\varphi \tag{3.39}
$$

$$
= 2\pi(R+h)^2 [\cos(\theta)]
$$

$$
= 2\pi(R+h)^2 [1 - R/(R+h)]
$$

$$
= 2\pi(R+h)\,h
$$

So, over the sky of area $S = 2\pi\,(R+h)\,h$, if we need nine satellites over the whole area covered by the constellation surrounding the earth, the total number of satellites that are required will be

$$N = 9/\{2\pi\,(R+h)\,h\}\cdot 4\pi(R+h)^2$$
$$= 18\,(R+h)/h$$
$$= 18\,(1+R/h)$$

This number increases with a decrease in the height of the constellation.

To verify this, let us use an example. For a GPS system, the height of the constellation is 26,500 from the center of the earth on average. Thus, here, $(R+h) = 26{,}500$, where $h = 20{,}000$, both in approximate figures. Using the previous numbers, we get

$$N = 18\,(26500/20000)$$
$$= 18 * 265/200$$
$$= 23.85$$
$$\sim 24$$

However, for a GEO height of about 42,500 km radius, this will become

$$N_g = 18\,(42500/36000)$$
$$= 18 * 425/360$$
$$= 21.25$$
$$\sim 22$$

Thus, with a 26,000-km radius, it is sufficient to have 24 satellites to provide about nine satellites visible between horizons. However, for a practical value of the cutoff of the elevation angle, generally eight satellites are seen with the same number of satellites in the constellation. The number will be 22 if they had to be in the GEO orbit. The reason behind the reduction in the requirement with increasing height of the satellites is that the part of the total constellation coming within the visibility span from a point on the earth's surface for the higher range of the satellites is more.

At orbits lower than the GEO, the velocity of the satellites will be more relative to that of the earth, and they will appear to be moving with respect to a point on the earth. However, because all of the satellites will be moving in the same fashion, when one satellite moves out of the visibility range of any point, another will appear by moving in from outside. Thus, statistically, the total number of visible satellites remains the same. Moreover, because it is nearer to earth, it will need less transmission power to produce the same power flux at the receiver compared with that of a satellite on the GEO. But, again, because they experience feeble but finite air drag, the life of satellites on this orbit is less than that in GEO.

Then, why do we not go for LEO? In addition to the larger numbers of satellite requirements, the intrinsic problem with the LEO system is that as it passes through a denser atmosphere, it experiences large drag, and hence its lifetime is very much limited. Moreover, the velocity of the LEO satellites is large; it rises and sets at smaller intervals, resulting in a rapid switchover requirement of satellites by the navigation receiver. Therefore, MEO satellites are chosen in general for the purpose.

Conceptual questions

1. Can a satellite move in a stable orbit contained on a plane that does not pass through the center of the earth?
2. Using Eqn (3.11a), derive the escape velocity, (i.e. the velocity that must be given to the satellite to escape from the influence of the earth). Is it necessary to provide this velocity to the rocket right from the takeoff condition?
3. Show that when the limit $L^2/(\mu a) = 2$ is crossed, the ε gets the generic requirement of hyperbola.
4. A satellite in a highly elliptical orbit moves partly through a dense atmosphere and loses part of its energy. What will happen to its semi-major axis, "a," and argument of perigee, ω?
5. A satellite in a circular orbit at a radius "a" is imparted with an excess energy of E. Find the position in the orbit after a time t. If the other orbital parameters are ω, i, and ω, find the position of the satellite in ECEF coordinates.
6. From the navigation standpoint, is it better to have a fast-moving satellite nearer to the earth or a slow-moving one farther from it?
7. Derive Eqn (3.24) from Eqn. (3.13a) using the definition of L and the relation between h and a.

References

Dorsey, A.J., Marquis, W.A., Fyfe, P.M., Kaplan, E.D., Weiderholt, L.F., 2006. GPS system segments. In: Kaplan, E.D., Hegarty, C.J. (Eds.), Understanding GPS Principles and Applications, second ed. Artech House, Boston, MA, USA.

Feynman, R.P., Leighton, R.B., Sands, M., 1992. Feynman Lectures on Physics, vol. I. Narosa Publishing House, India.

Maral, G., Bousquet, M., 2006. Satellite Communications Systems, fourth ed. John Wiley & Sons Ltd, U.K.

Pratt, T., Bostian, C.W., Allnutt, J.E., 1986. Satellite Communications, second ed. John Wiley & Sons Inc, USA.

Roychoudhury, D., 1971. Padarther Dharmo (In Bengali). Poschimbongo Rajyo Pushtok Porshod, Calcutta, India.

Rider, L., 1986. Analytical design for satellite constellations for zonal earth coverage using inclined circular orbits. The Journal of Astronautical Sciences vol. 34 (no. 1), 31−64.

Spilker Jr, J.J., 1996a. GPS navigation data. In: Parkinson, B.W., Spilker Jr, J.J. (Eds.), Global Positioning Systems, Theory and Applications, vol. I. AIAA, Washington DC, USA.

Spilker Jr, J.J., 1996b. Satellite constellation and geometric dilusion of precision. In: Parkinson, B.W., Spilker Jr, J.J. (Eds.), Global Positioning Systems, Theory and Applications, vol-I. AIAA, Washington DC, USA.

Strelkov, S.P., 1978. Mechanics. trans. Volosov, VM & Volosova, IG. Mir Publishers, Moscow. original work published 1975.

Tascione, T.F., 1994. Introduction to the Space Environment, second ed. Krieger Publishing Co, Malabar, Florida USA.

Walker, J.G., 1984. Satellite constellations. Journal of the British Interplanetary Society vol. 37 (12), 559−572.

Navigation Signals

Understanding Satellite Navigation. http://dx.doi.org/10.1016/B978-0-12-799949-4.00004-X

4.1 Navigation signal

One of the most important elements of the whole satellite navigation system is its signal. It is the means through which all information that the system wants to convey to the user is disseminated. This information includes the elements for satellite positioning, time and clock-related updates, corrections and notifications regarding the current state of the whole constellation. However, the utility of the signal is manifold and is not limited to just giving out this information. Here, we shall attempt to understand the multiple facets of the navigation signal. To appreciate its different aspects, we need to recall some fundamental concepts of communication. These will be discussed in brief for continuity and a better understanding.

4.1.1 Generic structure

The basic purpose of the navigation signal is to inform the user about some primary information and enable him to derive other necessary data required to fix his position. This information should reach the user in a convenient, reliable, and secure form. In recent times, there can be no method more convenient than the digital communication system to transmit information through a satellite channel fulfilling all of these criteria. Thus, the navigation signal is actually a sophisticated form of digital communication signal. Typically, it has a tiered form in which the tiers are constituted by the navigation data, followed by the ranging code and then the modulated carrier. These constituents are multiplied together to form the final signal structure. Each component of this product (i.e., the binary data, binary ranging code, and a sinusoidal carrier) has predefined characteristics designed to meet the service objectives.

    The navigation data provide the necessary information of the system, particularly of the space segment and timing, to the user in a binary format. These binary bits are multiplied with much faster binary ranging codes to enable receivers to carry out one-way range measurement. In addition, the orthogonal nature of these codes is exploited in Code Division Multiple Access (CDMA) systems. The relatively faster codes thus multiplied to the signal also spread the spectrum of the navigation data and help in better reception performance and security of the service. Furthermore, for data security and authentication purposes, encryptions in certain forms are also done on this part of the signal. This product of data and code then modulates an even faster sinusoidal carrier, and hence rides on it to propagate through the intermediate space from the satellite to the receiver. The most popular type of modulation for navigation is binary phase shift keying (BPSK). In certain cases, an intermediate binary subcarrier is also used. This type of modulation, called binary offset carrier (BOC) modulation, is used to avail certain advantages.

Navigation satellites transmit these modulated navigation signals in allotted bands with a predefined transmission power and defined polarizations, typically circular. Sometimes an additional pilot channel is also sent with the signal. This does not carry data but is the product of the code and the carrier only. The pilot signal helps the receiver in acquiring and tracking the signal mainly under poor signal conditions. The signals are transmitted by the satellites synchronously (i.e., when one satellite starts sending a data bit, so do the others). Furthermore, the code and the data, and also the code and the carrier phases are synchronized (Spilker, 1996).

The modulated carrier in this form experiences certain impairments while propagating through the medium from the satellite to the receiver, including power loss, excess delay, loss of coherency, multipath, and so forth. These impairments need to be properly handled by using appropriate designs or through proper correction of any error that gets added into the signal. A very low signal level is expected at the receiver over the targeted region with even lower power density owing to the spreading of the navigation signal spectrum. Navigation signals also experience interference from other navigation signals in these bands. This interference level needs to be minimized using certain techniques.

Now, we shall discuss in detail each the individual component of the signal with its related aspects.

4.2 Navigation data

The navigation data, D(t), is the component of the signal that carries all predefined information describing details about the current condition of the system that need to be broadcast to users. This information is transmitted as binary data arranged in a structured form. Here, we shall concentrate on the data for primary systems only. Besides that, our discussion will remain generalized and consolidated, and will describe only the key aspects of the navigation data (Global Positioning System Directorate, 2012a,b).

4.2.1 Data content

At this point, it is legitimate to seek the mandatory parameters that should be present in the data. The nominal set of navigation data is determined by the requirements of the receiver. We recall from Chapter 1 that in a primary system the positions are determined from knowledge of the reference positions (i.e., the positions of the satellites and the measured range of the satellites from the user).

Therefore, the user should be able to obtain satellite positions every time he wants to fix his own position. One simple option is to send the position of the satellites as a part of the navigation signal. However, that is not a pragmatic choice. The positions of the satellites vary significantly with time. The linear velocity of a satellite at a height of 20,000 km from the earth's surface is about 4 km/s, which is considerably high. Therefore, its position is expected to drift by a large amount within a specific update interval of the data, however small it is. In addition, estimations of the satellite's position are to be done onboard, since very frequent uploading of the satellite positions by the ground control system is not possible and it adds to the computational load. Moreover, within the finite update time of the satellite position in the signal, the satellite will move a distance large enough to add an appreciable error in the user's position estimate. Thus, a more practical option is to send the satellite ephemeris. Ephemerides are the Keplerian parameters of each satellite using which, along with the appropriate values of the current time, their positions can be estimated by the receiver employing correct physics. We have seen previously that there are six such basic parameters to derive the satellite position at any time. Thus, these six parameters need to be transmitted as a part of the data for partial fulfillment of the basic receiver needs.

Apart from satellite positions, satellite ranges are the next important parameter required for position and time fixing. The ranges are measured at the receiver and are derived by multiplying the signal velocity with its propagation time. The propagation time, which is the difference between the reception and transmission time of a particular phase of signal in turn is derived from the time stamp indicated on the signal. This time stamp, signifying the transmission time of a predefined phase of the signal is thus another parameter needed to be transmitted with the signal.

These parameters give the receiver all of the information needed to derive the necessary parameters for reference positioning, ranging, and time keeping. However, to increase the accuracy and reliability of the estimations done at the receiver, more data need to be transmitted through the message, which delivers the necessary precision.

These additional data are variables, which are included mainly for reasons such as the correction of satellite positions, prediction of the state of constellation, and correction of time and hence ranging errors. In subsequent chapters, we will read about details of the correction procedure. Here we shall only identify and mention their occurrences in the message structure.

The satellite position derived at the receiver is a sensitive parameter for the estimation of the user's position. Thus, the accuracy of reference positioning determines the accuracy of the user's position fix. The satellites, as mentioned previously, experience perturbations in their orbital motions because of the nonuniformity of many physical factors. Consequently, knowledge of these perturbation factors leads to better positioning. This is the reason why the perturbation parameters of the satellites are also accommodated in the signal.

The time stamps available in navigation messages are marked using satellite clocks. These clocks are derived from atomic standards and are stable in nature. Nevertheless, there is some minor drift in these clocks with respect to the system time, which may lead to affecting the positioning. Hence, to correct the effect of this drift, the total satellite clock deviation may be estimated by the ground system and transmitted through these messages for necessary corrections at the receiver. For the same reason that the ephemeris are transmitted instead of satellite positions, it is convenient to transmit the clock bias drift and drift rate at definite reference instant instead of the total shift. However, the total clock shift, Δt_{sat}, may be obtained from these parameters as

$$\Delta t_{sat} = a_0 + a_{f1}(t - t_0) + a_{f2}(t - t_0)^2 \tag{4.1}$$

where a_0, a_{f1}, and a_{f2} are the clock bias, drift, and drift rate, respectively. t_0 is any arbitrary reference instants for calculating these values, and t is the time of estimation.

There are also propagation impairments that alters the range measurements by modifying the time of transit of the signal. In addition to the clock and orbital correction terms, the correction factors corresponding to these propagation impairments may be disseminated to the user if the exact correction values are already known to the control system. Many of these errors depend on the location of the individual user, and hence cannot be explicitly known by the system. However, there are some parametric models of these impairments from which the receiver can estimate the required correction for himself. The parameters of these models are provided with the message to deal with this problem.

In a Frequency Division Multiple Access (FDMA) system, the composite signal received by the receiver may easily be analyzed for its constituent components using Fourier transform and segregated by passing through different filters. However, for a CDMA system, the individual signals need to be identified by recognizing the corresponding constituent codes. This is a time-consuming task. Therefore, it is always advantageous if it is possible at any location to approximately predict the possible set of satellites currently visible or those expected to rise or set over the horizon soon. Then, whenever the receiver is put on, the quick prior prediction makes it easier to specifically identify each individual satellite and hence the possible codes which can be present in the received composite signal.

To support this process, the system may include with the signal a reduced set of ephemeris parameters for all satellites. This list of selected ephemeris data is called an almanac. The approximate positions of the satellites in the constellation can be derived from the almanac. Hence it keeps the receiver prepared for the acquisition of new data or the switchover of satellites.

Apart from the information required at the user's receiver, certain telemetry data generated at the satellite are needed by the control segment for analysis. These data sets, which may include the integrity assurance, certain specific alerts, flags, and status data, may be combined in the form of telemetry data and be transmitted by the satellite.

Many other data may be included in the signal, depending on the design parameters and receiver requirements; these may be added to the navigation data structure. These data is said to update when the older values of parameters in the data are replaced by the newest ones. The updating rate of the parameters are different and it generally based on consideration of the rapidity with which the true physical parameter changes its values. Fast-varying parameters need to be updated frequently, whereas slow-varying ones are updated at larger intervals. In addition, because all data have an expiration term, the time of applicability may be associated with each type of data.

4.2.2 **Data structure**

Communication of navigation data occurs just as with any other digital communication system, with the basic objective of informing the user about the information required to fix his position. This information, as we have just learned, is a set of updated values of different parameters describing the current state of different elements of the system and constituting the navigation data. These values should reach the user in a convenient, reliable, and secure form without error; for this, digital binary communication is an excellent candidate.

The navigation data values need to be source coded and converted into binary bit sequences for the purpose. The number of bits used to represent a particular parameter is chosen so that the quantization error is less than the resolution of its values. For example, if the value of a parameter ranges from 0 to 1000, with a resolution of 0.01, the number of significant decimal digits required for the purpose becomes 6, with a scaling factor of 100. With this scaling factor, the values span from 0 to 100,000. Thus, it needs at least 17 binary bits for representation.

These binary bits can take either of two logical values, 0 or 1, typically represented by levels $+1$ and -1, respectively. After being mapped from parameter values to binary bits, these coded data need to be added with appropriate channel coding to defend against channel noise. Channel coding adds redundant bits to these data, making the latter robust against any error occurring during transmission of the signal through the channel.

These navigation data bits are organized in a structured manner in some suitable predefined arrangements for the efficient transmission of information and convenient extraction at the receiver. It is worth mentioning two alternative options for the transmission of data. In one, the different parameters are sent in a fixed frame format and all parameters have definite positions in the frame. Here, the whole framed structure with subframes of the data repeats in a fixed pattern. Each time the frame is repeated, the data is retransmitted with it. Thus, the repetition rate of the parameter is equal to the frame repetition rate. However, the same values of the parameters are retained until their expiry. The update interval of any parameter in the dataset can only be certain integral times of the frame repeat time. The other option is to transmit each parameter or a group of similar parameters in the form of an individual message. Proper indices may be used for identification of the message type. Each individual message type containing the associated parameters may be broadcast in a flexible order with a variable repeat cycle. Therefore, the message may be repeated at intervals, not necessarily equally, while it is updated, maintaining a fixed threshold time sufficient to achieve and maintain the required accuracy. The message sequence and update rate are mainly determined by the priority. The latter option, which has more flexibility and options to send more accurate and resolved data, leads to better performance (Global Positioning System Directorate, 2012a). In this option, there is ample scope to add integrity by sending statistical range errors and error checking facilities such as the Cyclic Redundancy Check (CRC). The flexible data message format leads to optimization of the transmission of the satellite specific data, resulting in minimized time for the position to be fixed for the first time, and thus increasing the speed of operation (Kovach et al., 2013). However, the flexible message structure requires higher storage capacity at the satellite compared with that required by the fixed frame format.

Irrespective of the data messaging approach, the individual units of the messages, whether subframes or indexed messages, need to carry the preamble bits to facilitate frame identification and synchronization.

4.2.3 **Data spectrum**

The navigation data is composed of some selected parameters, the values of which are encoded to binary. They thus form a random sequence of bipolar Non Return to Zero (NRZ) binary bits whose temporal variation may be represented by a series of positive and negative square pulses of amplitude "a" and constant width T_d. Thus, the navigation data s(t) at time t may be expressed as

$$
\begin{aligned}
s(t) &= a(\lfloor t/T_d \rfloor) \\
&= \pm a_k p(t - kT_d)
\end{aligned}
\tag{4.2}
$$

Here $a_k = \pm a$ is the amplitude of the kth bit of the sequence reckoned from an arbitrary instant taken as $t = 0$. k is equal to $\lfloor t/T_d \rfloor$, which represents the floor value of the argument t/T_d, i.e., the integral part of the ratio. T_d is the bit duration in the signal and depends on the rate, R_d, at which the bits are transmitted. p(t) defines the shape of the bit. Here, it is a unit positive pulse defined within the time $t = 0$ to $t = T_d$. The time variation is shown in Figure 4.1(a). Here, because we are considering the bits to be of finite width and constant amplitude, different types of pulse shaping are possible. However, they are not popular in navigation.

The frequency spectrum of this form of signal and the bandwidth occupied by it depend on the bit rate of the data. We know from communication theory that such a signal constituted by pulses of finite width T_d and amplitude "a" have as their spectrum, given by

$$
\begin{aligned}
S(f) &= aT_d \, \sin(\pi f T_d)/(\pi f T_d) \\
&= aT_d \, \mathrm{sinc}(\pi f T_d)
\end{aligned}
\tag{4.3a}
$$

where S(f) is the spectral amplitude of the signal at frequency f.

The spectrum of such a signal is shown in Figure 4.1(b) for a binary sequence in terms of its bit rate of $R_d = 1/T_d$. It is clear from the figure that the primary band of this spectrum, defined by the first nulls of the main lobe about the primary peak, is of width W, where

$$
\begin{aligned}
W &= 2/T_d \\
&= 2R_d
\end{aligned}
\tag{4.3b}
$$

Thus, this bandwidth is directly proportional to the bit rate. The faster the bit rate, the smaller the bit duration T_d, and hence wider the band width.

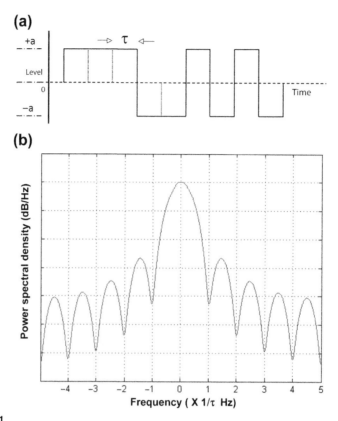

FIGURE 4.1

(a) Time variation and (b) relative amplitude and power spectrum for a random bipolar NRZ binary signal.

There is a mathematical theorem called Parseval's theorem, which states that the sum (or integral) of the square of a function is equal to the sum (or integral) of the square of its transform. We can apply this theorem for Fourier transform between the time and frequency domain to use the signal spectrum to obtain the power density. The theorem may be stated mathematically as (Lathi, 1984)

$$\int s^2(t)dt = \int S^2(f)df \qquad (4.4)$$

You can also understand this intuitively simply by considering the total signal as the sum of independent oscillations at all possible components of sinusoidal frequencies. Then, the total power of the signal is

$$p(t) = 1/T \int s^2(t)dt$$

$$= 1/T \int S^2(f)df \qquad (4.5a)$$

So, the power spectral density is given by

$$P(f) = d\,p(t)/df$$
$$= (1/T)\,S^2(f) \tag{4.5b}$$

The total power of the signal is the sum of the power carried by the individual orthogonal independent frequency component of this signal, and the power spectral density at any frequency f is nothing but the power carried by the spectral component at this frequency. It follows from this argument that the power spectral density function, representing the power P(f) carried by the individual frequency component f of the signal, is given by

$$P(f) = 1/T_d\left\{a^2 T_d^2 \sin^2(\pi f\,T_d)/(\pi f\,T_d)^2\right\}$$
$$= a^2 T_d \sin^2(\pi f\,T_d)/(\pi f\,T_d)^2 \tag{4.5c}$$
$$= a^2 T_d\,\mathrm{sinc}^2(\pi f\,T_d)$$

Because this is the square of the amplitude spectrum, it is always positive. Obviously, the positions of the nulls remain the same as those in the amplitude spectrum. Also, note how the power spectrum falls off rapidly about the center frequency. The power content at frequencies away from the DC is small. More than 90% of the power remains within the first null about the center of the spectrum (i.e., within the primary bandwidth). We will find in Box 4.1 how much of the total power of the signal remains within the first null of the spectrum. This will help us in explaining other features of the signal at a later stage of this chapter.

BOX 4.1　DATA SPECTRUM

The MATLAB program data_psd.m was run to generate the power spectral density plot of a random binary sequence with a data rate of 1 kbps. The plot obtained as a function of the frequency is illustrated in Figure M4.1, which shows that the normalized power has low absolute values to make the total power, obtained by integrating this density across the frequency unity. Observe how the side lobes vary across frequency with respect to the central lobe. It exhibits a $\mathrm{sinc}^2(x)$ variation, as seen in Figure 4.1. The power density values are logarithmic (in dB). The values go to zero at the nulls, as seen in Figure 4.1. The nulls are obtained at 1-kHz intervals and are in accordance with the fact that the bandwidth is proportional to the bit rate. Furthermore, the total power content in the side lobes is significantly lower than that in the main lobe.

Go through the program. Use the program to obtain the spectra for different data rates.

The spectral data and their corresponding frequencies are in the structure HPSD. Type HPSD on the MATLAB workspace after running the program to see its structure. Type "HPSD.data" and "HPSD.frequencies" to obtain the spectrum and corresponding frequency values. The values are in linear scale and can be plotted using the script [plot (HPSD.frequencies, HPSD.data)]. You can verify that the total power integrates to unity by using the command:

[sum (HPSD.data(2:end).*(HPSD.frequencies(2:end)-HPSD.frequencies(1:end-1)))]

The total power content within different frequency ranges is generated by the same program and is shown Figure M4.1(b). The power within the first lobe of 1000 Hz is more than 90%. Use the program to obtain the power content in different ranges of frequency.

Continued

BOX 4.1 DATA SPECTRUM—cont'd

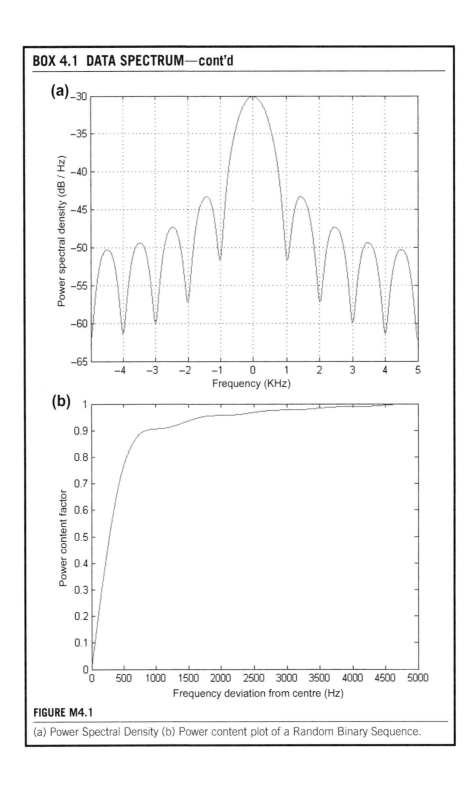

FIGURE M4.1

(a) Power Spectral Density (b) Power content plot of a Random Binary Sequence.

4.2.4 **Error detection and correction**

As the signal propagates through the channel, it picks up noise. Noise is composed of random electrical disturbances that add unwanted variations to the signal. This additive noise may corrupt the signal to the extent that the receiver fails to identify the true level of the binary signal. To recover the true content of the signal, considerably high signal power needs to be transmitted so that the received signal levels are adequately large for the noise to distort it beyond recognition. In other words, the signal to noise ratio needs to be large.

Alternatively, like any other digital communication system, channel coding is applied to the data. This limits effective errors in the data to a level that can be accepted by the system without sacrificing its minimal performance.

The foundation of channel coding rests on the Shannon–Hartley theorem (Mutagi, 2013; Proakis and Salehi, 2008), from which it can be proved that for an information rate R that is less than channel capacity C of a noisy channel, there exists a code that enables information to be transmitted with an arbitrary small probability of error despite the presence of noise in the channel. Capacity C is directly related to the channel bandwidth.

These codes, generally referred to as channel codes, are generated from data bits by adding some redundant bits to them. These extra bits are linear function of the data and are redundant in the sense that they carry no additional information in themselves. However, their addition makes it possible to detect and even correct a limited number of errors in the message regenerated at the receiver. The code bits use the difference between channel capacity C and information rate R, and can have a maximum rate of C-R. So, for a band-limited channel, lower bit rate R allows more redundant bits to be accommodated and makes the system more robust to errors owing to noise. For navigation, the data rate is small, so the addition of extra code bits raises no issues. The symbol rate, as the combined rate of data and code is generally referred to, consequently depends on the numbers of code bits added. Adding code bits and keeping the data bit rate unchanged, however, results in an increase in signal bandwidth.

The code bits may be used at the receiver end to identify any error occurring during transmission and to recover the true data. The necessary immunity against the noise depends on how correct one needs the data to be at the receiver for the required performance. This is again determined by the extent to which the errors affect the user in any specific application. The correctness of the received data is quantified by the ratio of erroneous bits to the total numbers of bits received by the receiver. This is called the bit error rate (BER).

It follows directly from this, that a tradeoff must happen between the transmitted power, the information rate or signal bandwidth, and the performance of the transmission system in terms of BER. For the same transmission power, the BER can be improved by lowering the information bit rate and using additional code bits for a given channel capacity. Performance may also be improved by maintaining the information rate but increasing the bandwidth owing to the addition of code bits with the same transmission power.

There are two different approaches to handling the errors. The data may be transmitted in a definite structure, as mentioned in Section 4.2.2. The errors in every unit of the transmitted messages may be detected at the receiver and discarded to get the same data over again. Because the navigation data are transmitted repeatedly in either fixed or varying intervals between updates, the receiver can readily receive the same message over the channel once again if it discards the one received erroneously.

However, rejecting the data and waiting for their scheduled retransmission every time a bit error is encountered will result in a large lead time. Therefore, forward error correction (FEC) may instead be done on the data that will correct most of the errors in the received data bits concurrently, and hence reduce the effective errors introduced as a result of channel noise. Consequently, only the data that are not rectified even with FEC will need to wait for retransmission, which reduced the overhead time for waiting.

Different types of error detection and correction methods may be used for navigation signals. Two important variants of channel coding for navigation are systematic cyclic codes and convolutional codes. Because the theory for these codes is involved, here we shall discuss only the basic underlying principle of error detection and correction. The basic components in such an error detection or correction technique consist of three main operations: code generation at the transmitter, code validation, and error identification at the receiver. We next explain these major activities (Lin and Costello, 2010).

4.2.4.1 Code generation

Generation of the code is done at the transmitter using the data bits as input. It results in the formation of a code, typically a linear function of the input bits, which is transmitted to the receiver. The main mathematical element of the generation process is the generator matrix.

4.2.4.1.1 Generator matrix

The operation of generating codes from data bits may be executed by multiplying the input data bits with some definite matrix of suitable dimension, called the *generator matrix*.

The k bit message may be regarded as a $[1 \times k]$ matrix M, or rather, a k bit row vector that, upon multiplying with a generator matrix G of dimension $[k \times n]$, produces an array of n code bits, i.e., a matrix C of dimension $[1 \times n]$. Thus,

$$C = M \cdot G \qquad (4.6)$$

Because the message word is m bits long, there are 2^m different combinations of bits of this length, whereas there can be 2^n $(>2^m)$ different possible code words out of an n bit code. The question is, which 2^m words out of these 2^n combinations will form the code? This is determined by the structure of G.

The generator G may be assumed to be formed by the basis set of k linearly independent vectors of dimension n defining the row space of G. Because there are

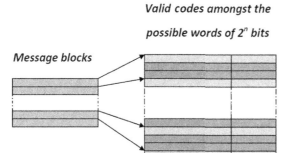

Valid codes amongst the

possible words of 2^n bits

Message blocks

FIGURE 4.2

Messages mapped to valid codes.

only k such vectors, they can span only a subspace of the complete n dimensional space. The exact code vector generated is determined by the explicit message bits. In simpler words, because there are as many rows in the generator matrix as there are bits in the message, a 1 at the jth position of the message selects the jth row vector in the code. Likewise, any combination of 1s in the message generates a code that is the combination of corresponding row vectors. Mapping from the possible messages to the legitimate codes is shown in Figure 4.2.

In navigation signal, the data containing the values of the required parameters need to be explicitly present in the navigation message, so the codes necessarily need to be systematic. For systematic codes, the first m elements are the message bits themselves, whereas the rest form the parity bits and the code appears as

$$C = m_1\ m_2\ m_3...m_k\ p_1\ p_2...p_{n-k} \qquad (4.7)$$

For a systematic code, the first $(k \times k)$ sub-matrix of G constitutes a unitary matrix. When any message bit is 1, the corresponding row of G is selected. Due to the leading unitary submatrix in G, it contributes an 1 in that respective position of the first m bits in the code. It thus constitutes the systematic part. The subsequent product of the data array from the $(k + 1)$th column of G and onward produces the $(n - k)$ parity bits. Thus, the generator matrix, G, for systematic codes may be represented as

$$G = [I\,|\,P] \qquad (4.8)$$

For example,

$$G = \begin{pmatrix} 1 & 0 & 0 & ....0 & 0 & 1 & 0 & 0 \\ 0 & 1 & 0 & ....0 & 0 & 0 & 0 & 1 \\ 0 & 0 & 1 & ....0 & 0 & 1 & 1 & 0 \\ \vdots & \vdots & \vdots & .... \vdots & \vdots & \vdots & \vdots & \vdots \\ \vdots & \vdots & \vdots & .... \vdots & \vdots & \vdots & \vdots & \vdots \\ 0 & 0 & 0 & ....0 & 1 & 0 & 1 & 1 \end{pmatrix}$$

For this example, the parity generator is arbitrarily taken as

$$P = \begin{pmatrix} 1 & 0 & 0 \\ 0 & 0 & 1 \\ 1 & 1 & 0 \\ \vdots & \vdots & \vdots \\ \vdots & \vdots & \vdots \\ 0 & 1 & 1 \end{pmatrix}$$

So,

$$C = [M \,|\, MP] \tag{4.9}$$

For each systematic part in the code, representing the k tuple message, there can be 2^{n-k} parity parts possible. Of these, only one will be valid for one definite message, as shown in Figure 4.2. Here, matrix P maps the message bits into the appropriate $n - k$ parity bits with the leading systematic part to get code C.

Effectively, the jth code bit is formed by the logical addition of the product of the message bits with the corresponding elements of the jth column. Therefore, it may also be assumed to be a selected sum of the message bits, where the elements of the jth column of G determines which message bits are to be selected. This may be thus implemented using registers in the following manner, as shown in Figure 4.3.

It is evident from the figure that the p_{th} code bit can be expressed as $C_p = \sum g_{jp} m_j$, which is nothing but the matrix multiplication of the message bit array with the p_{th} code of the generator matrix.

The code can also be written in polynomial of order $(n - 1)$ as

$$C = c_0 + c_1 X + c_2 X^2 + c_3 X^3 + \dots \; c_{n-1} X^{n-1} \tag{4.10a}$$

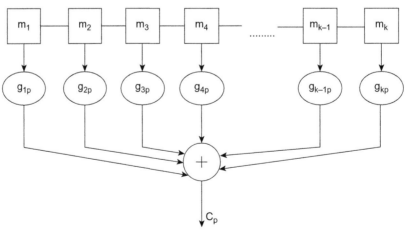

FIGURE 4.3

Implementation of code generator.

Similarly, the corresponding message may also be written as a polynomial of order k − 1 as

$$m = m_0 + m_1 X + m_2 X^2 + m_3 X^3 + \dots \ m_{k-1} X^{k-1} \qquad (4.10b)$$

Therefore, there can always be a polynomial g(X), called the generator polynomial, of order n − k, which, when multiplied with the message polynomial, will generate the code.

Once the n numbers of code bits are generated from a set of k data bits, it is also important to know how these k bits are updated or renewed. In block codes, after the generation of the code word from the data, all of the k data bits are removed and the next set of k data bits takes their place. The convolution codes, on the other hand, may be seen as those in which, once the k data bits generate a code, it gets shifted by m bits, where m < k. The shift is such that the oldest m bits with highest significance in the message get out and new m bits enter the sequence with least significance. Thus, k − m bits of the previous message remain in the array, acquiring higher significance owing to the shift.

4.2.4.2 Received code verification
This includes the verification process of the received code at the receiver. It checks whether the code received V is a valid code or whether it has been corrupted over the channel during transmission. The main mathematical element of this activity is another matrix called the parity check matrix.

4.2.4.2.1 Parity check matrix
For every [k × n] generator matrix G of the code, there exists a matrix H of dimension [n − k × n] such that the row space of G is orthogonal to the column space of H^T, i.e., $GH^T = 0$.

Therefore, for matrix G, matrix H is such that the inner product of a vector in row space of G and the corresponding rows of H is zero. Stated in mathematical notations, the matrix product $GH^T = 0$. So,

$$\begin{aligned} CH^T &= MGH^T \\ &= 0 \end{aligned} \qquad (4.11)$$

Thus we can affirm a received code V to be correct and it is a valid code C of generator G, i.e. V = C, if and only if $VH^T = 0$. The matrix H is called the parity check matrix of the code. Expressed element-wise, the ith element of this matrix product may be also written as

$$s_i = c_1 h_{i1} + c_2 h_{i2} + \dots + c_n h_{in} \qquad (4.12)$$

In a fashion, similar to the generation of the code bit, the parity checking may be implemented by storing the code bits in a shift register and adding them selectively in accordance with the elements of H.

For systematic codes, this orthogonality condition, $GH^T = 0$, is fulfilled when the H matrix is such that upon multiplication with the code, it selects the parity bit p_j from the parity part of the code and those systematic bits which constituted p_j, from the systematic part of the received code. It then XOR all the selected bits. This is equivalent to reconstructing the parity bit from its constituents and then XORing it with the received parity bit. If the code bits remain unaltered on transmission, the systematic part would recreate the p_j exactly and, on XORing with the corresponding parity bit, will result in a zero. So, the general form of the parity check matrix H for a systematic code will be

$$H = [P^T \mid I] \tag{4.13}$$

where P^T is a $(n - k \times k)$ matrix and I is a identity matrix of dimension $(n - k)$. This makes,

$$H^T = \begin{bmatrix} P \\ I \end{bmatrix} \tag{4.14}$$

If the received code V is a valid code, it may be represented by M'G. Multiplying C with the H^T matrix, the result becomes

$$\begin{aligned} CH^T &= M'GHT \\ &= M'[I \mid P]\begin{bmatrix} P \\ I \end{bmatrix} \\ &= M'[IP + IP] \end{aligned} \tag{4.15}$$

This should yield zero if the parity bit thus regenerated from the systematic part and that received are identical. This condition is only fulfilled if both parts are received correctly. Therefore, for the parity matrix shown in the example, the H matrix will be

$$H = \begin{pmatrix} 1 & 0 & 1 & \dots\dots 0 & 1 & 0 & 0 \\ 0 & 0 & 1 & \dots\dots 1 & 0 & 1 & 0 \\ 0 & 1 & 0 & \dots\dots 1 & 0 & 0 & 1 \end{pmatrix}$$

4.2.4.3 Error identification

The final step of the process is identification of the error. This is an extension of the code validation process using results obtained during the process. The key mathematical element of the error identification process is the syndrome.

4.2.4.3.1 Syndrome

Each bit of an error correction or detection code is a function of a set of data bits. Thus, when the systematic data bits involved in generating a definite code are combined at the receiver in the same way as to produce the parity bits of the code and the product is then compared with the corresponding parity bit actually received, both should ideally be identical.

However, if these contributing systematic data bits get corrupted during transmission, it cannot produce the same code bit again. Therefore, XORing it with the received parity bit at the receiver results in something not zero. The same argument holds if the received systematic bits are not correct.

This nonzero result is called the syndrome. The difference between the actually transmitted code bit and the one locally regenerated at the receiver from the received systematic bits is what generates the syndrome. A nonzero syndrome indicates an error in one or more constituent systematic bits received or the received parity bit. For a received code V, which is the true code combined with the error, the syndrome may be expressed as

$$
\begin{aligned}
S &= VH^T \\
&= (C + e)H^T \\
&= CH^T + eH^T \\
&= M[GH^T] + eH^T \\
&- 0 + eH^T
\end{aligned}
\tag{4.16}
$$

According to the definition of the parity check matrix, $GH^T = 0$. Thus, syndrome S becomes the product of the error sequence in the code with the parity check matrix, H. It can also be deduced from this that syndrome S will be 0 if $e = 0$, i.e., the received code has no error. Syndrome S helps the receiver to identify errors and correct them. It is the method used in detecting errors in systematic block codes and for hardware-based decoding of convolution codes. The process of generating the error correction parameters is illustrated in Focus 4.1.

Now that we have learned the key elements of error detection and correction codes, we understand how their features are used in navigation using a suitable one amongst the major classes of block codes and convolutional codes (Lin and Costello, 2010).

4.2.5 Addendum

The volume of data required to be transmitted carrying the total mandatory information for navigational purposes is usually small. Therefore, a lot more information can be transmitted through it, maintaining the requisite BER. Thus, it would be normal to put more information along with these navigation data that may serve for other applications. One can exploit the residual capacity by adding additional data for value added services apart from core navigation. Feasible options of such applications are search and rescue, and geotagged information dissemination including weather, natural calamity, and so forth. We will learn about some interesting applications in the final chapter.

Mere transmission of this information is not enough for navigational purposes because it requires extended reliability and security of data. The data need to reach the user not only errorless under all channel conditions, but also with the indemnity of signal security. The signal should be transmitted in such a manner that it cannot be figured out by an unauthorized user and nobody can hinder real users from getting the actual message. Moreover, propagation impairment such as multipath and interference should also have the least effect on the signal. These issues are handled using the other components of the signal and will be discussed in the next section.

FOCUS 4.1 ERROR CORRECTION CODING

We will demonstrate all of these processes using the generator of a 7-bit code word from a 4-bit message. A particular generator matrix was using with the following parity:

$$P = \begin{bmatrix} 1 & 0 & 0 \\ 0 & 0 & 1 \\ 1 & 1 & 0 \\ 0 & 1 & 1 \end{bmatrix}$$

The systematic code generator was

$$G = [I|P]$$

$$= \begin{bmatrix} 1 & 0 & 0 & 0 & 1 & 0 & 0 \\ 0 & 1 & 0 & 0 & 0 & 0 & 1 \\ 0 & 0 & 1 & 0 & 1 & 1 & 0 \\ 0 & 0 & 0 & 1 & 0 & 1 & 1 \end{bmatrix}$$

If two 4-bit message words are chosen as the following,

$$m = \begin{bmatrix} 0 & 1 & 1 & 0 \\ 1 & 1 & 1 & 0 \end{bmatrix}$$

The two 7-bit code words generated from the two selected messages are

$$C = m * G$$

$$= \begin{bmatrix} 0 & 1 & 1 & 0 & 1 & 1 & 1 \\ 1 & 1 & 1 & 0 & 0 & 1 & 1 \end{bmatrix}$$

We take a random sequence of seven error bits for the two 7-bit code words as the following:

$$e = \begin{bmatrix} 1 & 0 & 1 & 1 & 1 & 1 & 1 \\ 0 & 1 & 1 & 0 & 1 & 1 & 0 \end{bmatrix}$$

The resultant received words for the two codes after adding the errors are

$$R = c \oplus e$$

$$= \begin{bmatrix} 1 & 1 & 0 & 1 & 0 & 0 & 0 \\ 1 & 0 & 0 & 0 & 1 & 0 & 1 \end{bmatrix}$$

The resultant matrix is obtained from the generator as

$$H = [P^T | I]$$

$$= \begin{bmatrix} 1 & 0 & 1 & 0 & 1 & 0 & 0 \\ 0 & 0 & 1 & 1 & 0 & 1 & 0 \\ 0 & 1 & 0 & 1 & 0 & 0 & 1 \end{bmatrix}$$

Finally, the syndrome is obtained by multiplying the received code R with HT as

$$S = R\,H^T$$

$$= \begin{bmatrix} 1 & 1 & 0 \\ 0 & 0 & 1 \end{bmatrix}$$

4.3 Ranging codes

It is a requirement for navigation systems that the distance of the transmitting satellite from the receiver be estimated by the latter. From this measured distance, position fixing is done. Plus, in CDMA-based systems, signals from multiple satellites need to be received by the receiver simultaneously in the same carrier frequency without interference between them. These requirements must be fulfilled in addition to the basic criterion of secured transmission of the data.

All these may be easily achieved by using a simple technique: by multiplying the data bits with a pseudo random binary sequence. This is a term we are using for the first time. We will learn it in the next section in more details.

Broadly, the pseudo random binary sequence consists of a fixed and finite length of random binary bits called chips. One such complete length of chip constitutes a code. These chips are repeatedly generated after the fixed length of N chips so that the same bit sequence, i.e., the same code, repeats itself recurrently. These chips are generated at a very fast rate that is, in general, many integral multiple times of the data rate. Thus, within a data period, not only many chips, but many code sequences are produced. If the data rate is R, the code repetition rate will be p × R, allowing p complete codes to be accommodated within a data bit period. If N is the code length, the chipping rate will be N × p × R.

These chips are generated synchronously with the data and multiplied with them. This means the leading edge of a data bit exactly matches the leading edge of a chip. Again, with the definite rate mentioned previously, the trailing edge of the same data will match the trailing edge of the (N × p)th chip, counted from the former. The resultant product after multiplying the data with the code chips also becomes a binary sequence with a bit rate equal to the rate of this sequence of chips whenever the chips and the data bits are synchronized. If the binary sequences are represented by levels +1 and −1, the sequence remains same as the original pseudo random (PR) sequence over the length where the data are +1 and gets inverted where the data are −1 (Figure 4.1). The bandwidth of the signal increases accordingly.

4.3.1 **Pseudo random noise sequence**

When binary bits represented by logical 0 and 1s or their algebraic form, given by +1 or −1, appear in a random fashion to form a sequence, it is called a random binary sequence. The individual bits of this sequence are called chips, to distinguish them from the information-carrying bits of the data or the bits of the error-correcting codes. Figure 4.4 illustrates a section of such code with chip period Tc.

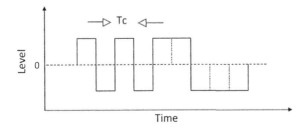

FIGURE 4.4

A random sequence.

Before we start describing a pseudo random sequence, it is important to understand the difference between logical and algebraic variables and operations. The notion of binary numbering is to logically represent contrasting states by two well-distinguished terms. Typically, they are represented by logical 1 and 0. These could be represented as well by any other form such as "good" and "bad" or "yes" and "no," or "true" and "false." The representation with numerical 0 and 1 is for convenience.

No algebraic operations such as multiplication, integration, or addition are valid in the logical states, considering that these logical variables are different from algebraic ones. However, our conventional processing needs algebraic operations to be done on the data bits. So, to carry out algebraic operations such as correlation on the sequence, it is necessary to represent logical data in some algebraic form so that the algebraic operations can be made with them. Simultaneously, we shall also be able to represent the logical operations by some equivalent algebraic operations.

The XOR is a logical operation upon the logical variables A and B in which the result of the operation is also a logical variable represented by A $\oplus$ B. It is the operation for measuring the similarity of the sequence of logical states that these two variables acquire. Notice in the truth table for the XOR operation in Table 4.1 that a similarity of two variables results in logical 0, but a dissimilarity, in logical 1. Thus, the total count of excess 0s over the count of 1s across the sequence length represents the effective similitude.

If the logical values are to be substituted by some algebraic representations, the logical operations on these values must also have matching results when the logical operation is substituted by an algebraic equivalence. This is the case when we replace a logical 1 by algebraic -1 and a logical 0 by algebraic $+1$. The operation of XOR is substituted by the algebraic multiplication. The logical truth table and the algebraic product table for these operations are given below in Table 4.1 to establish the fact.

Comparing the two operations, we see that the result of logical operation of XOR on logical representations of the binary states of two variables is the same as that obtained for the algebraic operation of multiplication on the algebraic representation of these states. So, we can conclude that when logical 0 and 1 are represented by algebraic $+1$ and -1, respectively, the XOR operation can be replaced by multiplication on these data.

The effective similarity between two sequences of logical variables is measured by the sum total of the numbers of 0s over 1s after XOR-ing two sequences. Similarly, in the algebraic domain it is done by summing of the counts of $+1$s over -1s after

Table 4.1 Logical and Algebraic Form of XOR Operation

A	B	A $(\oplus)$ B	A	B	A*B
LOGICAL VALUES			ALGEBRAIC VALUES		
0	0	0	+1	+1	+1
0	1	1	+1	−1	−1
1	0	1	−1	+1	−1
1	1	0	−1	−1	+1

multiplication, by summing up the product. When averaged, this total process effectively represents the correlation between the two sequences. One important aspect of a random sequence is its correlation property. Correlation A of any two time signals S_1 and S_2 for a finite delay τ between the signals is defined by the equation

$$R_{x1x2}(\tau) = 1/T \int_0^T S_1(t)S_2(t+\tau)dt \qquad (4.17a)$$

To understand this for a binary random sequence, let us take an infinitely long sequence of binary chips, that takes values $+1$ or -1. Each of these chips is of width Tc extended over time. The sequence is then multiplied by itself. Now, we define a variable that represents the product of the chips of a sequence with itself with a relative delay of n chip length, averaged over the sequence length N, and call it normalized autocorrelation of the sequences, with delay nTc. It is given by

$$R_{xx}(\tau) = 1/(NTc) \int_0^{NTc} S(t)S(t+nTc)dt \qquad (4.17b)$$

If we keep no relative delay between the two sequences, i.e., n = 0, then the same chips get placed on each other. Because of this exact superimposition of the two sequences, where there is $+1$ for one sequence there will be also $+1$ for the other, and where there is a -1 for one there will be a -1 for the other. Consequently, when we multiply the corresponding chips now, it always yields $+1$ at every position. When we sum up the product over the length of N chips of the sequence, the sum becomes NTc, where Tc has already been defined as the width of each chip on the time scale. The time interval of N chips is also NTc. This is true for any such binary sequence. So, from the definition, the correlation of such a sequence S_1 with itself with zero delay becomes

$$R_{xx}(0) = 1/T \int_0^T S_1(t)S_1(t)dt$$

$$= 1/(NTc) \int_0^{NTc} a_k a_k dt$$

$$= 1/(NTc) \sum_{k=1}^N \left[a_k^2 \times Tc \right] \qquad (4.18)$$

$$= 1/(NTc) \sum_{k=1}^N [1 \times Tc]$$

$$= 1/(NTc)[NTc]$$

$$= 1$$

Let us consider a situation in which the chips of such a sequence are placed on themselves with some finite delay τ. This situation is shown in Figure 4.5.

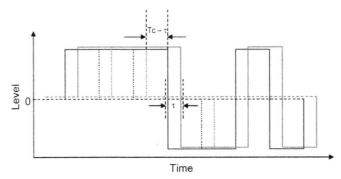

FIGURE 4.5

Signal condition for autocorrelation with finite delay.

Here, the delay $\tau = nTc$. If N is the number of chips over which the integration is made, for integral values of n, the Eqn (4.17a) for a discrete autocorrelation function becomes

$$R_{xx}(nTc) = \{1/(NTc)\}\ S(t)\ S(t + nTc)\ dt$$
$$= \{1/(NTc)\}\ Tc\sum a_j a_k \qquad (4.19)$$
$$= (1/N)\sum a_j a_k$$

where $k = j + n$, i.e., the chip represented by a_k comes n bits later than the a_j in the sequence and the chip width is Tc.

For relative delay of the sequence of 1 bit, i.e., n = 1, the bits in a sequence get multiplied with the bit just adjacent to it, which holds true whichever way we move one with respect to the other. Then, the autocorrelation value becomes

$$R_{xx}(1Tc) = 1/(N)\sum_{i=1}^{N} a_k a_{k+1} \qquad (4.20a)$$

Because the sequence of bits is random, the adjacent two chips may be same or different. If they are same, i.e., $[+1\ \&\ +1]$ or $[-1\ \&\ -1]$, the product becomes $+1$; when they are different, i.e., $[+1\ \&\ -1]$ or $[-1\ \&\ +1]$, the product becomes -1. For a random sequence, there is an equal chance these two cases will happen. Therefore, when they are added up, it makes the net sum zero in the summation. It makes the autocorrelation

$$R_{xx}(1Tc) = 1/(NTc)\sum_{i=1}^{N}[0.5 \times (+1) + 0.5(-1)] \times Tc = 0 \qquad (4.20b)$$

The same argument may be extended when the delay is more than that of 1 bit length.

When the delay is a fraction of a chip length, i.e., $0 < \tau < Tc$, for the relative displacement of τ in time, the argument n becomes a fraction (τ/Tc). Here, only a length $(Tc - \tau)$ of the chip gets positioned on the same chip and the rest of length

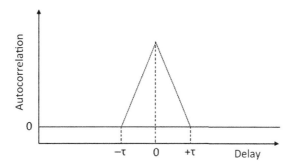

FIGURE 4.6

Autocorrelation function of a random sequence.

τ is placed over its adjacent chip. This is shown in Figure 4.5, and it is true whichever way you move the sequence. So, the matched portion of $(Tc - \tau)/Tc$ of each chip will contribute to the sum an amount equal to $(1 - \tau/Tc)$ in Eqn 4.19, whereas the bits laid on their adjacent ones will add up to zero due to the same reason that we described in relation to equation 4.20. The autocorrelation will thus become

$$R_{xx}(\tau) = 1/(NTc) \sum_{i=1}^{N} \{(1 - \tau/Tc) * 1 + \tau/Tc * 0\}Tc$$

$$= (1 - \tau/Tc)N/N$$

$$= (1 - \tau/Tc)$$

(4.21a)

Because this is true for both positive and negative delays, i.e., for both right and left excursion of one code with respect to the other, the expression for autocorrelation for a relative delay "τ" will be modified to

$$R_{xx}(\tau) = (1 - |\tau|/Tc)$$

(4.21b)

The value of the autocorrelation is reduced at the rate $1/Tc$ from one by an amount proportional to the delay between the two signals. Verify the expression in Eqn (4.21b) with our previous arguments by putting $\tau = 0$ and $\tau = Tc$, i.e., when they are exactly superimposed and when shifted by Tc, respectively. Figure 4.6 shows the autocorrelation function of a random sequence and box 4.2 illustrates it.

When similar averaged sums of the product of the bits are taken with two different sequences, it is called cross-correlation of the sequences. The normalized discrete cross-correlation for two sequences, a and b, for a delay τ, which is an integral multiple of the chip width Tc, is given by

$$R_{xy}(\tau) = 1/(NTc) \sum_{i=1}^{N} a(t)b(t + \tau)Tc$$

$$= 1/N \sum_{i=1}^{N} a(t)b(t + \tau)$$

(4.22)

BOX 4.2 MATLAB AUTOCORRELATION

The MATLAB program autocorr_rand was run to generate the autocorrelation function of a random binary signal. The following figure was obtained using the MATLAB internal function xcorr() and random binary sequence of 1000 bits is sampled at a rate of 10 samples per bit (Figure M4.2).

    The function is symmetric on both sides of 0. Run the program for different lengths of the sequence by changing the argument of S_1, but not too high to limit the computational load. Also see the effect of different sampling rates by changing the upper limit of t. The plot shown will be asymmetric in general owing to the default nature of the xcorr function.

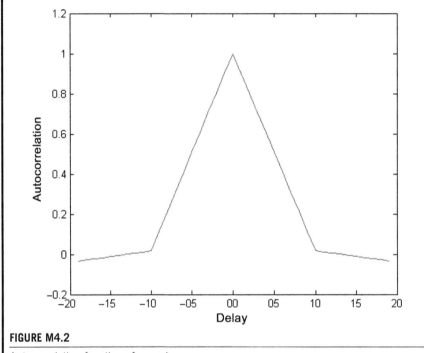

FIGURE M4.2

Autocorrelation function of a random sequence.

The value of the correlation depends on the bits of the individual sequences.

We have considered fully random sequences until now to define the concepts of the correlation. The integration carried out there or the number of bits N over which the summation has been taken for deriving these parameters were assumed to be very large. Because they are random, there were no restrictions in choosing this length and the parameter values were defined within this range only.

A pseudorandom sequence, also called a PRN code, is like a random sequence with a little difference. Instead of the random bits continuing up to infinity, these sequences run randomly for a finite length and then the same sequence is repeated again and the process continues recurrently. The random portion of the bits of finite length constitutes a definite

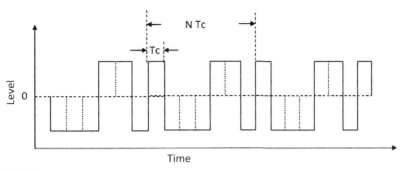

FIGURE 4.7

Pseudo random sequence.

pattern and called a *code*. Every bit of a code is called a *chip*, to distinguish it from the information carrying data bits. The length of the sequence expressed in the number of chips after which it repeats itself, i.e., the numbers of chips in a code, is called the *code length* and is denoted typically by N. The time interval of repetition of the entire code is called the *code period*. This is equal to NTc, in which Tc is the *chip width*. The chip width Tc depends on and is actually the inverse of the *chip rate*, R_c, at which the chips are generated. Given a definite chip rate, the code period depends on the numbers of chips, N in the code. The code repetition period can also be expressed as NTc = N/R_c (Cooper and Mc Gillem, 1987; Proakis and Salehi, 2008).

As a result of this definite repetition, once a single run of the code is identified in terms of its chip sequence and chip rate, the rest of the sequence becomes predictive. Figure 4.7 shows a typical pseudorandom sequence.

Besides the structure, the repeating character of the sequence and their predictive nature cause differences in their autocorrelation properties and their frequency spectrum, as well from the purely random sequences.

4.3.1.1 Maximal length PRN sequence

A useful form of orthogonal pseudorandom sequence code is the *maximal length PRN sequence*. It is also called the *shift register code*, because these codes may be generated with shift registers with proper feedback. It produces a code of length $2^m - 1$, with "m" shift registers; hence, it is also known as *m-sequence code*. The term "maximal length" (ML) indicates that the sequence repeats only after a length that is the maximum sequence length that can be produced using m shift registers. Because only $2^m - 1$ discrete states are possible for m bits, excluding all zeros, the sequence generates $2^m - 1$ chips before repetition.

Two important properties of an ML sequence are:

1. Balance nature of the sequence, i.e., of its $2^m - 1$ chips, which is essentially an odd count, the number of −1s in the sequence is one greater than the number of +1s. So, in the sequence, there are 2^{m-1} numbers of −1, whereas the count of +1s is $2^{m-1} - 1$.

2. The XOR of the sequence with any shifted version of the same sequence generates the same sequence with a different shift.

Using these two characteristics, let us resume our discussion on the autocorrelation process. The autocorrelation for a PRN sequence is defined as the integral of the products of the same sequence with some definite shift (τ), averaged over a code length. So, the autocorrelation may be expanded as

$$R_{xx}(\tau) = 1/(NTc) \int_0^{NTc} s(t)s(t + \tau)dt \tag{4.23}$$

It can be easily understood that even here, the autocorrelation with zero delay remains equal to 1 owing to exact superimposition of the chips. However, because the sequence repeats after a finite length of N chips, the sequence is indistinguishable from the sequence delayed by that length. Thus, the superimposition condition gets on repeating in the whole sequence after every N chip shift, i.e., after every code period of NTc. Therefore, the autocorrelation values have a repetition period equal to the code period.

Now, let us consider the case of the delay of integral chip width. When the chips are shifted by 1 bit delay, each bit of the code is positioned on the next bit in the code within the code length, with only the last chip lying on the first of the next recurrence of the code. For a relative shift of 2 bits, two trailing bits of one code get placed on two leading the next. It goes on in this manner. Because the sequence is not infinitely random, the total probability of like chips pairing is not equal in count to the number of unlike chips pairing. Hence, the sum of their product is not zero. It gives a nonzero value to the autocorrelation for such a delay of integral numbers of bits (but not equal to NTc). The nature of the sequence, i.e., the arrangement of +1 and −1s in the finite sequence length, determines what the exact value of this autocorrelation will be.

For any shift of integral numbers of chip length, it becomes the inner product, i.e., the bit by bit algebraic product, of the sequence. In logical terms, this is nothing but the XOR operation on the shifted version of the same sequence in logical terms. The XOR operation of such sequences results in the generation of the same sequence with a different shift, which is evident from the second property. So, the algebraic inner product of the sequence formed during autocorrelation thus produces the sequence itself with a delay. From the first property, the product sequence has one −1 excess over the numbers of +1s. Upon integrating this product over a time of the complete code duration of N bits, and averaging, the autocorrelation value it yields is

$$R_{xx}(NTc) = 1/(NTc) \sum_{i=1}^{N} \left(a_j a_k Tc\right)$$

$$= 1/(NTc) \sum_{i=1}^{N} a_m Tc \tag{4.24}$$

$$= -1/N.$$

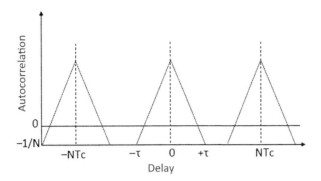

FIGURE 4.8

Autocorrelation function of a PN sequence.

The same nature of the correlation function will be repeated over the whole sequence with the exact superimposition condition returns back after N chips. With increasing N, i.e., number of chips in the code, as the sequence approaches a purely random nature, the autocorrelation for delays of integral multiples of chip period moves toward triviality. Obviously, the repeating period of the function also increases accordingly, which is evident from Eqn (4.23), too. From this, we understand that as the length of the code N increases, the autocorrelation values nears zero and the sequence behaves more like a purely random one. Figure 4.8 shows the autocorrelation of a pseudo random function.

Readers may find the concepts of autocorrelation illustrated in Box 4.3.

1. Spectrum of the codes:

This section discusses in detail the spectral characteristics of a PRN sequence. The spectral characteristics will be used in subsequent discussion, and hence this topic may be viewed as a precursor for the conceptualization of those ideas. First, we will describe the amplitude spectrum, and then we will mention the corresponding power spectrum of the sequence.

The amplitude spectrum of a PRN code is slightly different from that of a fully random one. To understand it, let us consider a code of code length N chips with chip width Tc. The code period is consequently NTc.

BOX 4.3 MATLAB AUTOCORRELATION

The MATLAB program autocorr_prn was run to generate the autocorrelation function of a PRN sequence for different positive delays between the two sequences. The following one-sided autocorrelation function is obtained using the definition of the autocorrelation. The samples per chip and the numbers of recurrences are made configurable. The same variation is obtained for the negative delays between the two (Figure M4.3).

Note the differences in Figure of M4.2 for a complete random sequence. Also notice how the circular autocorrelation has been achieved.

Run the program for other sequences and for other values of the configurable parameters.

Continued

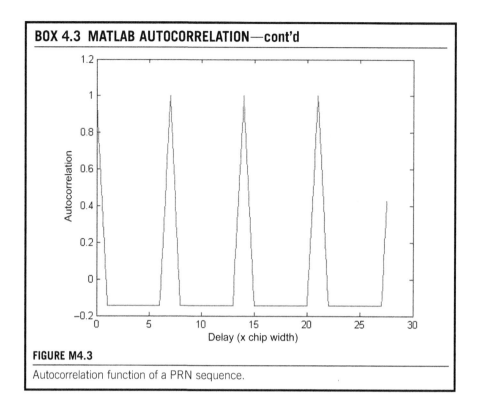

BOX 4.3 MATLAB AUTOCORRELATION—cont'd

FIGURE M4.3

Autocorrelation function of a PRN sequence.

Let us consider one set of complete block of code of N chips, each of width Tc, occurring over any interval of NTc across the time line. This sequence with the particular occurrence has an amplitude spectrum whose shape can be represented by a sync function of nature

$$S(f) = a[\{\sin(\pi f\, Tc)\}/(\pi f\, Tc)\}]$$
$$= aTc\mathrm{sinc}(\pi f\, Tc) \tag{4.25}$$

Thus, this spectrum is a continuous sinc function about $f = 0$ with nulls at $f = 1/Tc$ and its multiples, the plot for which is shown in Figure 4.9(a). The corresponding power spectrum is also shown.

To explain the nature of the spectrum, we need first to appreciate that the repeating form of this signal in the time domain may be obtained by convolving a single code of N chips with an infinite train of unit impulses, the impulses being separated by the distance of the code width, i.e., NTc. The expression for this train of impulses may be

$$\tau(t) = \sum_{k=-\infty}^{\infty} \delta(t - kNTc)$$

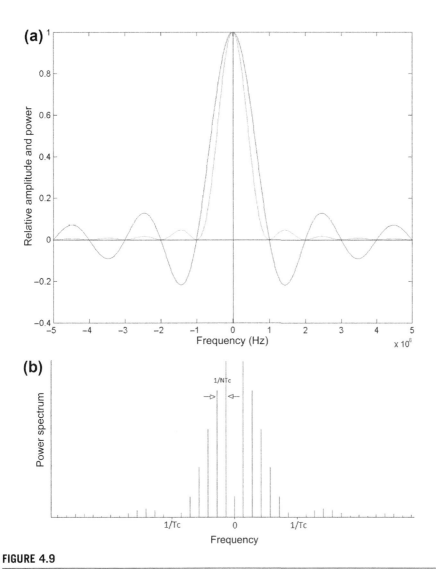

FIGURE 4.9

Spectrum of a pseudo random code sequence.

where $\delta(.)$ is the Kronecker delta such that $\delta(x = 0) = 1$ and $\delta(x \neq 0) = 0$. Convolution of two signals s_1 and s_2 is defined by the expression

$$h(t) = s_{1(t)} \otimes s_{2(t)}$$
$$= \int s_1(\tau) s_2(t - \tau) d\tau \tag{4.26}$$

This spectrum of a train of unit impulse in time domain transforms to a similar unit impulse train in frequency domain separated by 1/NTc. The corresponding nature of this impulse train may be represented by

$$T(f) \sim \sum_{k=-\infty}^{\infty} \delta(f - k/NTc) \qquad (4.27)$$

Because the actual signal is formed by the convolution of the two component signals in the time domain, it forms a product in the frequency domain. Thus, the resultant spectrum is the product of a unit pulse train with the continuous sinc function. It turns the product spectrum into discrete lines at an interval of 1/NTc and with an envelope of the sinc function of the first null width of 1/Tc. The resultant spectrum is shown in Figure 4.9(b). The characteristic expression is

$$S(f) \sim \{\sin(\pi f\, Tc)/(\pi f\, Tc)\}\delta(f - k/NTc) \qquad (4.28)$$

Then, the corresponding power spectral density becomes

$$P(f) \sim \mathrm{sinc}^2(\pi f\, Tc)\delta(f - k/NTc) \qquad (4.29)$$

However, all previous discussions used relative representation to show the discrete impulsive nature of the spectrum. The exact expression is required to be adjusted so that it represents the correct power density at zero frequency, as well as at others. This exact expression can be shown to be (Cooper and Mc Gillem, 1987) (Box 4.4).

$$P(f) = (N+1)/N^2[\{\sin(\pi f\, Tc)\}/(\pi f\, Tc)]^2\delta(f - k/NTc) - 1/N\delta(f) \qquad (4.30)$$

2. Generation of PR codes

The theory behind the generation of the PRN sequences is based on the background theory of the finite Galois fields and that of the cyclic codes. We start our discussion by briefly introducing the elements of a Galois field.

For any prime number p, there exist a finite field denoted by GF(p) containing p elements. For example, the binary field is a Galois field GF(2). Now, the elements of a binary field with only two elements {0, 1} cannot satisfy the equations such as $X^3 + X + 1 = 0$, because these equations have three roots and the field has only two distinct elements.

BOX 4.4 MATLAB PSD OF PRN CODE

The MATLAB program prn_psd was run to generate the power spectral density (PSD) function of a PRN sequence. The sequence is repeated for a large number of times in the code to replicate infinite recurrence. The following one-sided PSD function is obtained using the definition of the PSD employing the square of the distribution of spectral amplitudes obtained through the FFT function. The samples per chip and the numbers of recurrences are made configurable.

Note the differences with in Figure of M4.4(a) and (b). The first is the PSD for random pulses, whereas the second is for the PR sequence. The repeating nature makes the spectrum discrete. Run the program for other sequences and for other values of the configurable parameters.

BOX 4.4 MATLAB PSD OF PRN CODE—cont'd

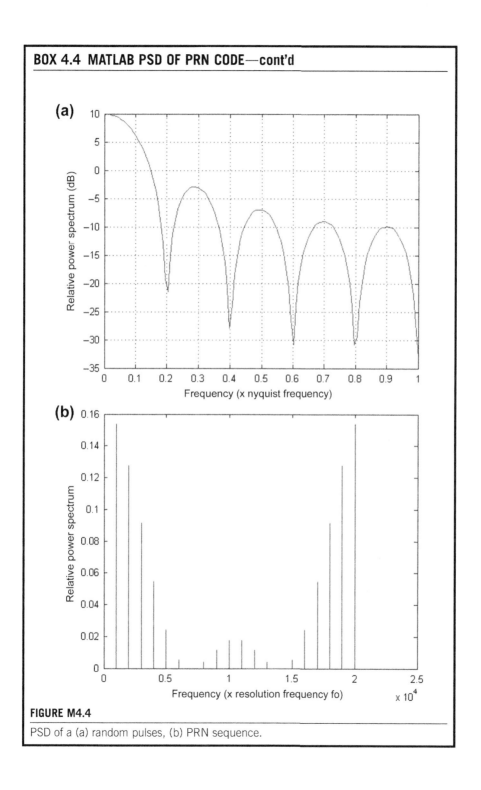

FIGURE M4.4

PSD of a (a) random pulses, (b) PRN sequence.

The GF(2) may be extended to a higher dimension of m to form an extension field $GF(2^m)$. This is the same way in which a one-dimensional number line of real numbers $\{R^1\}$ extends into a three-dimensional vector space of $\{R^3\}$ representing vectors. However, in $GF(2^m)$ space, each dimension can assume either of the two values, 0 or 1.

It follows from this that a GF (2^3) field has only $2^3 = 8$ distinct elements. These elements can be represented by s_k, where k runs from 0 to 8. It forms a finite field in which the eight elements can be represented with basis $\{1, X, X^2\}$; for example, $s_j = \{1X^2 + 0X + 1\} \equiv 1\ 0\ 1$, as shown in Figure 4.10.

This finite field is said to be closed under the operation of multiplication of any two elements of this field, say s_k and s_j, if the product $(s_k s_j)$ fold back to one of the elements in this field. There must be some mapping function $f(X)$ that defines how this fold is to happen.

It is apparent from the previous discussion that the elements of the $GF(2^m)$, save $\{0, 0, 0\}$, may be used as representations of the seven possible states during the generation of a 7-bit PRN. We can arrange these seven states as some ordered set as $F = \{\alpha^0, \alpha^1, \alpha^2, \alpha^3, \alpha^4, \alpha^5 \ldots \alpha^{n-1}\}$, where $n = 2^m - 1$, and each state is represented exclusively by an element of the $GF(2^m)$, i.e., $\alpha^k \equiv s_k$. Therefore, the PRN generator should

1. Assume value of one definite element, s_n of the field for one exclusive state, α_n.
2. Arrive once at every possible state in ordered succession upon *actuation*.
3. Ensure that the same order of states is reiterated in a cyclic fashion.

Consequently, it is necessary to address the following for the purpose:

1. What is the actuation process?
2. How may these criteria be fulfilled?
3. How may the whole thing be implemented?

We attempt to answer them in sequence. Suppose one element of the $GF(2^m)$ is taken as a state in the process of generation of the PRN. There must be some

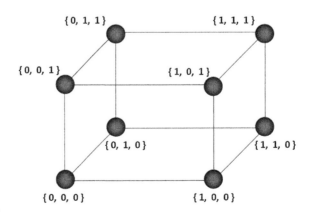

FIGURE 4.10

Elements of a finite field, $GF(2^3)$.

mathematical operation done on this that will ensure the change of the current state to the next. That mathematical operation is the *actuation* process and is much more obvious when implemented in real hardware.

It is evident that, if during the process of generation, it is always the next higher ordered state that is assumed upon actuation and in the process and the first state is cyclically taken up after reaching the highest available, then the states always trace the same path in every recurrence. This can be represented in mathematical form as

1. $\alpha^j \xrightarrow{\text{Actuation}} \alpha^{j+1}$
2. $\alpha^n = \alpha^0$ where $n = 2^m - 1$

The ordered set of the states assumed in a cyclic manner may be warranted if we ensure that the successive elements in the ordered set of states are someway mapped to the consecutive codes of a $2^m - 1$ bit cyclic code. Then the actuation reduces to the process that leads to the successive change in the states and results in the cyclic rotation of the cyclic codes in the mapped field. There must be some constraint equation that determines this action. To obtain the exact equation, we now use our knowledge of cyclic codes, in which a cyclic shift of a valid code generates another valid code.

For convenience of understanding, if we take $m = 3$ for the generation of a $2^3 - 1 = 7$ bit PRN code, the extension field will have seven elements. The corresponding cyclic code onto which these elements will be mapped should also be a 7-bit code. So, for a 7-bit code, the same initial code reappears after seven cyclic shifts, and hence the same state with which the code is mapped is arrived at. If $g(x)$ is a valid generator polynomial that generates a (7, 3) cyclic code, and $c_i(x)$ is the corresponding code generated from the state α_i, representing α_i in polynomial form, in terms of its coordinates, $\{1, X, X^2\}$ we get

$$\begin{aligned} \alpha_i &= \alpha(X) \\ &= a_2 X^2 + a_1 X^1 + a_0 \end{aligned} \tag{4.31}$$

Again, the generator may also be expressed as a polynomial as

$$\begin{aligned} g &= g(X) \\ &= g_4 X^4 + g_3 X^3 + g_2 X^2 + g_1 X^1 + g_0 \end{aligned} \tag{4.32}$$

So, the code word is given by

$$c_i = g(X)\alpha_i(X) \tag{4.33}$$

From the characteristics of the cyclic codes, a cyclic rotation will generate a new valid code. Thus, seven cyclic shifts will make the same code reappear. One such shift is executed by multiplying with x and modulo dividing by $x^7 + 1$. If $Rm[a/b]$ denotes the remainder of the division a/b, then

$$\begin{aligned} c_{i+1}(x) &= \{x c_i(x)\} \text{Mod}(x^7 + 1) \\ &= Rm\big[\{x g(x)\alpha_i(x)\}/(x^7 + 1)\big] \\ &= Rm\big[\{g(x)x\,\alpha_i(x)\}/(x^7 + 1)\big] \end{aligned} \tag{4.34}$$

Again, as from the theory of cyclic codes, the generator polynomial is a factor of $x^7 + 1$, we can write

$$x^7 + 1 = g(x)h(x) \qquad (4.35)$$

Because $\alpha_i(x)$ is a polynomial of order 2 and the code is of order 6, it makes $g(x)$ a polynomial of order 4. Hence, $h(x)$ has an order of 3. Generalizing this observation, we find that $h(x)$ is one order higher than of that of $\alpha(x)$. It can be shown that $h(x)$ is also a generator polynomial for a cyclic 7-bit code (Proakis and Salehi, 2008). Eqn 4.34 can be written as

$$\begin{aligned} c_{i+1}(x) &= \mathrm{Re}[\{g(x)x\alpha_i(x)\}/\{g(x)h(x)\}] \\ &= g(x)\mathrm{Re}[\{x\alpha_i(x)\}/h(x)] \end{aligned} \qquad (4.36\text{a})$$

For the orderly sequence of the states to be followed, this code c_{i+1} must be generated by α_{i+1}. It requires

$$c_{i+1} = g(x)\alpha_{i+1}(x) \qquad (4.36\text{b})$$

Thus, comparing Eqns (4.45a and b), we get,

$$\begin{aligned} g(x)\alpha_{i+1}(x) &= g(x)\mathrm{Re}[\{x\alpha_i(x)\}/h(x)] \\ \text{or,}\quad \alpha_{i+1}(x) &= \mathrm{Re}[\{x\alpha_i(x)\}/h(x)] \end{aligned} \qquad (4.37)$$

This equation answers two of our questions. First, it states that during generation of the sequence of the code, the next state is achieved by multiplying it with x and modulo dividing it by $h(x)$. Shifting the elements of α_1 of the $GF(2^m)$ to one order higher place and modulo dividing by $h(x)$ generates α_2, and so on. Thus, this operation comprises the actuation process.

Second, because the cyclic codes resulted from the states, sequentially assume the seven possible values and return to the first, and the states are mapped exclusively to the codes by Eqn 4.33, the states also follow cyclic rotations. To makes the codes assume the states in a cyclic fashion, we need the feedback of the bit at position x^7 to the position $x^0 = 1$. This can be represented as $x^7 + 1 = 0$. Now, let us see how does this determine the generator for implementing the codes.

We have mentioned that $h(x)$ is a generator polynomial of a $2^m - 1$ bit cyclic code and of the same order of the extension field $GF(2^m)$. Thus, this function $h(x)$ should be chosen as an irreducible, primitive factor of $x^n + 1$. $h(x)$ is said to be primitive when there is no other value $n' < n$ such that $h(x)$ can exactly divide $x^{n'} + 1$. For a 7-bit PRN code, we need states of $GF(2^3)$, and hence a generator of order 3. For $m = 3$, $h(x)$ must be a function of $x^7 + 1$, which may be factored as

$$\begin{aligned} x^7 + 1 &= (x + 1)(x^3 + x^2 + 1)(x^3 + x + 1) \\ &= g(x)h(x) \end{aligned} \qquad (4.38)$$

Factors $(x^3 + x^2 + 1)$ and $(x^3 + x + 1)$ are irreducible as well as primitive and can be used here. Taking one of these two primitive polynomials, say

$h(x) = x^3 + x^2 + 1$, as our constraint equation, we appreciate that equating this to zero implies $x^7 = 1$, which is equivalent in binary system to $x^7 + 1 = 0$. This ensures the cyclic rotation of the codes and hence we get the sequence of the states. Therefore, for implementing $x^3 + x^2 + 1 = 0$, which is equivalent in binary to $x^3 = x^2 + 1$, the bit that attains the position of x^3 on the code by multiplying with x, as per this equation, folds back to the position of x^2 and x^3 and gets XOR-ed to the bit that is supposed to attain the position owing to the up shift.

The codes will start generating from any initial state, called "seed," selected from the elements of $GF(2^3)$, and they will go on assuming the values cyclically. If we arrange the corresponding states in the ordered manner expressed in accordance to the bases $\{1 \quad X \quad X^2\}$, this will follow as

$$S = [\{1 \quad 1 \quad 1\}, \{1 \quad 1 \quad 0\}, \{0 \quad 1 \quad 1\},$$

$$\{1 \quad 0 \quad 0\}, \{0 \quad 1 \quad 0\}, \{0 \quad 0 \quad 1\}, \{1 \quad 0 \quad 1\}]$$

These codes being systematic, the states are only definite portions of the code. Moreover, the codes being cyclic, taking any one particular bit of the state over the sequence of codes will lead to the generation of the PRN. For example, using the MSB will lead to the PRN sequence of $\{1 \quad 0 \quad 1 \quad 0 \quad 0 \quad 1 \quad 1\}$. Because this is cyclic, the same sequence will be obtained if we take any other bit, but with a delay.

To implement this, we first note that the state, α_i, an element of $GF(2^3)$, is 3-bit words. It can be contained in a 3-bit shift register. To attain the next state, the sequence first must be multiplied by x. This is achieved by up shifting the contents 1 bit toward the MSB. This needs to be divided by $x^3 + x^2 + 1$, and the remainder forms the next state. This modulo division is realized by providing feedback from the output of the MSB register and adding it to the content of x^2 and x^0, modulo 2. The implementation is shown in Figure 4.11(a).

Similarly, the other primitive polynomial, $x^3 + x + 1$, which is also a factor of $x^7 + 1$, may be used to generate another sequence. Here, the modulo divisor is $x^3 + x + 1$, the value of the MSB output on up shifting a sequence, representing the quotient, and must be fed back to the contents of x^1 and x^0 for modulo two addition. This implementation is shown in Figure 4.10(b) (Box 4.5).

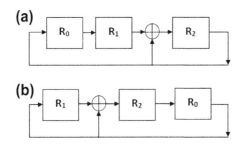

FIGURE 4.11

Implementation of a pseudo random sequence generation.

BOX 4.5 MATLAB: GENERATION OF ML PRN CODES

The MATLAB program m_sequence.m is run to generate an m-sequence of 7 bits with the primitive polynomial $h = x^3 + x + 1$. It was started with the state $S = [1 \quad 1 \quad 1]$ and the final repeating sequence generated was

$$P1 = [1 \quad 1 \quad 0 \quad 0 \quad 1 \quad 0 \quad 1]$$

The variable S represents the state variable that is a three tuple word and carried by the three elements of S. The arrangement and the variation of the elements S(1), S(2) and S(3) of the state variable S following the given relation are shown in Figure M4.5(a).

We need to see whether, if we start with another definite state for this arrangement, the same set of states will be repeated every seven steps. Run the program by changing the initial value of S from $[1 \quad 1 \quad 1]$ to $[1 \quad 0 \quad 0]$ and check the final sequence. It is apparent from Figure M4.5(b) that for this case, the criterion is fulfilled starting from the initial state of $[0 \quad 0 \quad 1]$. In this figure, the feedback to the position of X^1 is modulo-2 added with the other input coming through the up shift from X^0. The other feedback is directly placed in X^0 because there is no other input.

Thus, we see that both the arrangements lead to the same sequence that recurs every seven steps. The phase of the sequence may differ depend on which point is considered for getting the output.

Also, change the polynomial used by changing the connections in the code and check the sequence it generates.

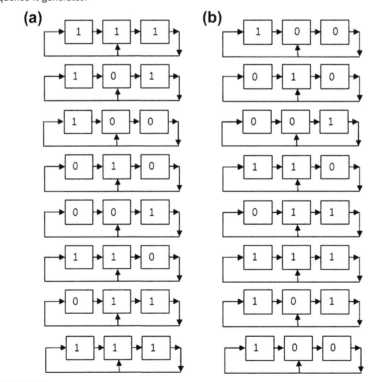

(a)   **(b)**

FIGURE M4.5

States of 7 bit m-sequence for primitive polynomial $h = x^3 + x + 1$.

Table 4.2 Valid States

Order		x^2	x	1	$x(\oplus)x^2$
α	$=$	0	0	1	0
α^2	$=$	0	1	0	1
α^3	$=$	1	0	0	1
α^4	$=$	1	0	1	1
α^4	$=$	1	1	1	0
α^6	$=$	0	1	1	1
α^7	$=$	1	1	0	0
α^8	$=$	0	0	1	0

If we represent a definite valid state of the generator by αj, αj is defined by the contents of the three placeholders. For primitive polynomial, equation, $h = x^3 + x^2 + 1$, if we consider the state $\begin{bmatrix} 0 & 0 & 1 \end{bmatrix}$ to be α_j, the higher orders of α go as shown in the following table (Table 4.2).

In these three place holders, x^2, x and 1, all seven possible different combinations of bits have occurred. If we consider the bit at any one place as the code, for each such combination there is a code bit generated, justifying the name maximal sequence, or m-sequence.

Furthermore, we can see that the last column that represents the XOR value of the first two columns is only a shifted version of the others. This proves that the modulo-2 sum, which is also the XOR operation of two time offset versions of the same m-sequence, is another offset version of the same sequence. This is an important property that we shall be using in our explanations later in this chapter and the next.

One question we still need to answer is, because one code set will be produced from one irreducible primitive factor, how many of such irreducible polynomials are possible for a code length of $2^n - 1$? It can be easily calculated from the value n. For the purpose, we use Euler's totient function φ. The function, $\varphi(x)$, represents the count of numbers that are less than x and relatively prime to x. It can be shown that the number of codes that are possible with m is

$$N = (1/n)\ \varphi(2^n - 1) \tag{4.39}$$

For $n = 3$, $2^n - 1 = 7$, and so $\varphi(7) = 6$. Thus, $N = 6/3 = 2$, which we have already seen in our discussion. Similarly, for $n = 4$, $\varphi(15) = 8$ and hence $N = 8/4 = 2$, whereas for $n = 5$, $N = 6$. Comparing the numbers, you can derive the fact that the numbers are more when the argument of Euler's function is itself prime. This is obvious because it gives large numbers of mutually prime numbers less than itself.

4.3.1.2 Gold codes
Of all orthogonal binary random sequences, the Gold code, owing to its important and exclusive features of bounded small cross-correlations and ease of generation, is extensively used in telecommunications and navigation systems. In the next

subsection, we will find how this helps in CDMA systems to keep interference noise to a minimum.

It is known from Euler's function that the number of m sequences increases rapidly with the increase in the code length. Although these m-sequences have suitable autocorrelation properties, most sequences do not have the desired cross-correlation appropriate for navigational purposes, i.e., low cross-correlation between different codes.

Gold (1967) showed that there are certain m-sequences with better periodic cross-correlation properties than other m-sequences of the same length. Some pairs of m-sequences yield three valued cross-correlation with corresponding integrals over the code period given by

$$
\begin{aligned}
R_{xy} &= 1/N\big[-1, -2^{(n+1)/2}-1, 2^{(n+1)/2}-1\big] \quad \text{when n is odd.} \\
&= 1/N\big[-1, -2^{(n+2)/2}-1, 2^{(n+2)/2}-1\big] \quad \text{when n is even}
\end{aligned}
\tag{4.40}
$$

Such m-sequences are designated *"preferred sequences."* When these preferred m sequences of the same code length are taken in a pair and are XOR-ed, a definite new sequence is obtained, known as a Gold and Kasami sequence, or more popularly, "Gold codes".

Changing the phase differences between the same two participating m-sequences give a different Gold code of the same family. Stated differently, Gold codes generated from the same parent set of m-sequence are said to constitute a family of codes. They have the same three-value cross-correlation between them. Thus, from a single pair of m-sequence of length n with preferred cross-correlation, we can generate n numbers of Gold codes with similar cross-correlation properties between them.

To understand the features of the Gold codes, we recall that XORing the same m-sequence with two different phase offsets generates the same m-sequence again, but with a new phase offset. Because two Gold codes of the same family are produced by XORing the same two definite m-sequences of a preferred pair, the same components reside in these two Gold codes, but with different relative phase offsets. Thus, the two Gold codes may be represented as

$$
\begin{aligned}
G_{1k} &= M_1(t) \oplus M_2(t+k_1) \\
G_{2k} &= M_1(t) \oplus M_2(t+k_2)
\end{aligned}
\tag{4.41}
$$

The constituent m-sequences are one of those selected pairs that generate a three-valued cross-correlation.

Here, M_1 and M_2 are the logical bit sequence of the constituent m-sequences. Because the logical values can be replaced by the algebraic surrogates m_1 and m_2, respectively, and the operation of XOR by multiplication, we can say

$$
\begin{aligned}
g_1(t) &= m_1(t)m_2(t+\delta_1) \\
g_2(t) &= m_1(t)m_2(t+\delta_2)
\end{aligned}
\tag{4.42}
$$

where $\delta j = k_j T_c$ and k_j is an integer. So, when the cross-correlation is obtained between these two Gold codes, the same pair of constituents present in the two Gold codes gets multiplied with different phase offsets. When the above two Gold codes are cross-correlated, we get

$$R_{xy}(g1, g2, d) = 1/T \int (g_1(t)g_2(t+d)dt$$
$$= 1/T \int \{(m_1(t)m_2(t+\delta_1)m_1(t+d)m_2(t+d+\delta_2)\} \quad (4.43)$$

The product may also be seen as the multiplication of two self-products of the constituent m-sequences. So,

$$R_{xy}(g_1, g_2, d) = 1/T \int [m_1(t)m_1(t+d)][m_2(t+\delta_1)m_2(t+d+\delta_2)]dt$$
$$= 1/T \int \{[m_1(t+\Delta_1)m_2(t \mid \Delta_2)\}dt \quad (4.44)$$

This is nothing but the crosscorrelation of the same constituent m-sequences with some new phase difference $\Delta_2 - \Delta_1$. Hence, this cross-correlation is also a three-valued function. However, for different values of d, phase differences are produced.

Furthermore, because the phase offset product of m_1 and m_2 is a new Gold code of the same family, i.e.,

$$R_{xy}(g_1, g_2, d) = 1/T \int \{m_1(t+\Delta_1)m_2(t+\Delta_2)\}dt$$
$$= 1/T \int g(\tau)dt \quad (4.45)$$

It is thus apparent that through the crosscorrelation of two Gold codes from the same family, we actually integrate a new Gold code over its code bits and divide by

---

BOX 4.6 MATLAB CORRELATION OF PRN CODE

The MATLAB program "autocorr_gold" was run to generate the Gold codes from two 7-bit m-sequences and then estimate their autocorrelation.

The two m-sequences selected are $[1 \ 1 \ 1 \ 0 \ 0 \ 1 \ 0]$ and $[1 \ 1 \ 1 \ 0 \ 1 \ 0 \ 0]$. These codes are converted to their equivalent algebraic forms and produced a three-valued cross-correlation, as given in Figure M4.6(a).

Check that the levels of the cross-correlation are -0.7143, 0.4286, and -0.1429. These values exactly match our expression of $(2^{(n+1)/2} - 1)/N$ and $-1/N$ for $n = 3$, i.e., $N = 7$.

All possible XOR results of the selected sequences are

$$G = \begin{bmatrix} 1 & 1 & 1 & 1 & -1 & -1 & 1 \\ -1 & 1 & 1 & -1 & 1 & 1 & 1 \\ -1 & -1 & 1 & -1 & -1 & -1 & -1 \\ 1 & -1 & -1 & -1 & -1 & 1 & 1 \\ -1 & 1 & -1 & 1 & -1 & 1 & -1 \\ 1 & -1 & 1 & 1 & 1 & 1 & -1 \\ 1 & 1 & -1 & -1 & 1 & -1 & -1 \end{bmatrix}$$

These seven codes form the Gold codes of the same family. Of all the codes generated, two were arbitrarily selected and cross-correlations were found for different delays. The

Continued

BOX 4.6 MATLAB CORRELATION OF PRN CODE—cont'd

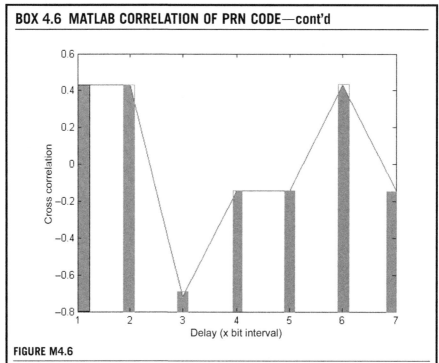

FIGURE M4.6

Cross-correlation of two Gold sequences.

cross-correlation obtained for different delays is shown in Figure M4.6. This cross-correlation is also three-valued following the given expression for $n = 3$.

The code arrays produced on XORing two selected Gold codes for different relative delays are shown below. The last two codes in the previous was selected for the purpose. It follows that the codes thus produced are only the shifted versions of the codes of the family or the constituent m-sequences.

$$GX = \begin{bmatrix} 1 & -1 & -1 & -1 & 1 & -1 & 1 \\ 1 & 1 & -1 & -1 & 1 & 1 & 1 \\ 1 & 1 & 1 & -1 & 1 & 1 & -1 \\ -1 & 1 & 1 & 1 & 1 & 1 & -1 \\ -1 & -1 & 1 & 1 & -1 & 1 & -1 \\ -1 & -1 & -1 & 1 & -1 & -1 & -1 \\ -1 & -1 & -1 & -1 & -1 & -1 & 1 \end{bmatrix}$$

Run the program by selecting other Gold sequences.

---

the code period NTc. Thus, the crosscorrelation is nothing but the averaged time integral value of this resultant generated Gold code over the code period.

When the new Gold code thus generated is balanced, the crosscorrelation is $-1/N$. However, among all the resultant Gold codes generated, not every one is balanced. When unbalanced codes are generated, the crosscorrelation assumes one of the other two possible values among $[\pm 2^{(n+1)/2} - 1]/N$ for odd n and $[\pm 2^{(n+2)/2} - 1]/N$ for even values of n (Box 4.6).

4.3.2 Effect of multiplying ranging code on signal

In the navigation signal, the data that repeat at the rate $R_d = 1/T_d$ are multiplied synchronously to the individual chips of the code that repeats at a rate of 1/Tc, which is much faster compared with that of the data. Consequently, the product sequence has the bit repetition rate equal to the chip rate of the code. We shall henceforth call these products of the information carrying data with the code chips "encoded data chips" (EDC), to distinguish them from the normal data bits as well as from the code chips.

The benefit of the PRN resides not only in the sequence of the chips in the code, but also in the relative rate at which the data and the code bits vary in the signal. Typically, the code chips are used at a rate that is a million times greater than the data rates. Multiplication of the codes with the data has two effects: the effect on the frequency domain is the *spreading* of the signal, whereas the effect on time domain is that, it provides the signal with a suitable *correlation* property. These two notions form the basis of the important activities carried out with the signal, including ranging and multiple access, in addition to its effect on data security. We will discuss each of these effects in details.

4.3.2.1 Spreading

We have learned about the spectrum of the PRN sequence in Section 4.3.1. We know that when the code is multiplied with the data sequence, in time domain, their spectrum gets convolved in the frequency domain. The code sequence, which is repetitive in time after a period of NTc, has a spectrum that is a sequence of discrete lines at interval 1/NTc, the full spectrum given by Eqn. 4.30. The data spectrum is a continuous sinc function of the form

$$Sd(f) = AT_d[\{\sin(\pi f\, T_d)\}/(\pi f\, T_d)\}] \tag{4.46}$$

Because of the data bit width of T_d, this spectrum has nulls at $1/T_d$. So, in the frequency domain, on convolving these two spectra, each discrete frequency line of the code spreads out into a replica of the message spectrum.

In general, $T_d > NTc$, and so, $1/NTc > 1/T_d$. Therefore, this null and hence the width of the main lobe of this data spectrum are much smaller than the separation of the discrete lines of the code spectrum and are fairly accommodated in this interval without aliasing.

In some cases, where the code length is much larger than the code bit width, 1/NTc is not wide enough to accommodate two halves of the data spectrum, making the total width of $2/T_d$. That results in aliasing and thus alters the shape of the total spectrum. Therefore, it is also clear that the resulting shape of the modulated signal will depend on the relative size of $1/T_d$ and 1/NTc.

If the repetition interval of the code is large owing to large code length, its spectral lines are placed so close to each other that they appear like a continuum and the total spectrum appears like a continuous sinc function with first nulls at 1/Tc. This is expected because high repetition interval tends toward random sequence whose spectrum is a continuous sinc function. For such a case, the effective spectrum is the convolution of these two sinc functions, one from the code sequence with width 1/Tc and the other from the data sequence with width 1/T_d. But as Tc $<<$ T_d, it follows that the latter spectrum appears to be almost a line compared with the former and hence the effective spectrum is almost identical to that of the code.

Finally, the data, which are themselves of spectral width 1/T_d, after code multiplication effectually spread out to 1/Tc. The ratio of the width of the spread spectrum to that of the original spectrum of the data is called the bandwidth expansion factor, Be (Proakis and Salehi, 2008). It is equal to the ratio of the chip rate Rc to the data bit rate R_d, as

$$\begin{aligned} Be &= (1/Tc)/(1/T_d) \\ &= T_d/Tc \\ &= Rc/R_d \end{aligned}$$

(4.47)

Multiplication of the data with the code effectively spreads the signal in the frequency domain as we have just seen, and hence is also known as the *spread spectrum system*. The spectrum thus produced is also called a *direct sequence spread spectrum*. Such systems are generally designed to permit communication of messages under the condition of very low signal to noise ratio or to combat interfering transmissions or multiple access communication.

4.3.2.2 Orthogonality and autocorrelation

The code has an important role in this spreading and dispreading process. The choice of proper code is important in any such system and is primarily determined by the code characteristics. We already learned in a previous section that the correlation of a signal is related to the inner product, i.e., chip-to-chip multiplication of two signals and averaging it over the entire code length. This signifies how one code is structurally similar to another. When the segment of one code is the same as that of the other, maximum correlation is achieved. For almost similar codes, this value is large. For middling similarity this is moderate, and for almost no similarity the value approaches zero. So, when the inner product of two codes is zero or near zero, the codes are said to be orthogonal to each other.

In statistical terms, orthogonal means completely uncorrelated. It is also evident from the previous definition that correlation is nothing but the inner product averaged over the code length. Therefore, drawing the analogy of vector inner (dot) products, if two codes have zero or very low correlation values, they are said to be orthogonal. The orthogonal properties of this sort are among the most desirable attributes of signal in a wide variety of situations. Like the way in which two signals modulated by sine and a cosine carriers of the same frequency do not interfere, here, two signals with two different data when multiplied with orthogonal codes cause no interference. So, orthogonal codes are vital in satellite navigation and techniques and permit a numbers of signals to be transmitted on the same nominal carrier frequency and occupy the same radiofrequency bandwidth, resulting in what is called as code division multiplexing. We will learn about this later in this chapter. The cross-correlation values of the codes to be used for navigation signal should be low, whatever the phase differences. Hence, orthogonal codes are very carefully selected for the purpose.

4.3.3 Navigational use of the effects

4.3.3.1 Ranging

Multiplication of the pseudo random ranging code with the data enables us to measure accurate distances from the satellite to the receiver. This is used not only in navigation, but also in other applications of one-way satellite ranging and in radar systems.

The geometric distance between a transmitter and a receiver may be obtained by measuring the propagation delay of a radio signal and then multiplying it by the velocity of the light. We will learn the details of ranging in Chapter 5. Here, we will only have an overview of it.

The codes are used in ranging using their autocorrelation properties. We know that the correlation of the code depends on the delay. If the receiver can synchronously generate the same code with the transmitter, both the satellite and the receiver can generate the same code phase at a definite instant simultaneously. The phase of a repeating ranging code may be defined as the current state of the code, i.e., which chip of the code and what fraction of it is currently being generated.

The code phase generated at the satellite is transmitted through the signal and takes some definite time to traverse through the path in between and reach the receiver. This delay occurring between the transmission and the reception, is the traverse time of the code phase. When the signal is received by the receiver, which is synchronously generating the same code with the transmitting satellite, it then generates a later phase of the code compared with the received one. The difference between these code phases is the local code phase progression during the traverse time and hence proportional to it. The relative code phase of the incoming signal is estimated through autocorrelation of the signal with time-shifted versions of the locally generated code. The time shift that results in the highest autocorrelation affirming the exact match of the two codes, is the relative delay between the local prompt and incoming code. This is equal to the time of propagation. The faster the chip rate of the code is, the steeper is the variation of the autocorrelation function profile with delay. This makes the delay estimates more precise. The estimated delay, when multiplied with the propagation velocity, which is the velocity with which electromagnetic waves traverse through a medium, gives the range of the transmitter from the receiver. However, to know more about this, we will have to wait until we reach the appropriate section in the next chapter.

4.3.3.2 Multiple access

In certain satellite navigation systems, all satellites transmit their signals on the same carrier frequency. This enables the user receiver to operate with a smaller bandwidth at the front end. The signals being multiplied by the different orthogonal codes, remain mutually separated and avoids interference. When the composite signal is again multiplied by the required code at the receiver, all other components of the total signal are rejected and the one associated with that code remains in the receiver with the code removed. This is possible by virtue of the good correlation properties of the codes selected for the purpose.

Because multiple access forms a separate aspect of the signal transmission, which includes other types of multiple access such as FDMA, we will consider this topic separately after the current section.

4.3.3.3 Processing gain

The spread of the signal bandwidth at the transmitter and the subsequent despreading at the receiver of the desired signal lead to a system processing gain. In navigation and any other spread spectrum system, as well, the term "processing gain" (PG) is important and is defined as the signal to noise ratio (SNR) improvement at the output with respect to the input of the receiver. It is expressed through the equation

$$
\begin{aligned}
PG &= 10 \log(SNR)|_{output}/(SNR)|_{input} dB \\
&= 10 \log(Bandwidth)|_{input}/(Bandwidth)|_{output} dB
\end{aligned}
\tag{4.48}
$$

The gain thus achieved through this process may be explained by the fact that in the encoded form the signal is spread over a large bandwidth. Thus, the associated signal and noise power to be considered is also over the same bandwidth. If P is the total signal power and N_0 is the noise power density, the SNR of the signal before despreading is

$$
\begin{aligned}
SNR^- &= P/(WN_0) \\
&= P/(RcN_0)
\end{aligned}
\tag{4.49}
$$

Here, we used our previous observation that the bandwidth after spreading is equal to Rc, the bandwidth of the code chips.

This signal is multiplied by the identical code at the receiver to recover the data. This removes the code variation of the signal, leaving only the data components. The associated power of the signal is brought back unabated by collating it into the reduced bandwidth of the original data. So, the signal power after despreading remains P. The random white noise component in the signal, which also is multiplied by the code at the receiver, retains its random nature. Its amplitude of variation remains the same while its phase may only get inverted when the local code is −1. Therefore, it remains unaltered in terms of its spectral nature and its power spectral density remains unaltered. Because the effective bandwidth of the signal, over which the power is collated, has now been reduced to that of the data, the total contribution to the noise attached to the signal is reduced by the same factor as the bandwidth. The total noise power after dispreading is $N = N_0 R_d$.

Whereas the total signal power remains the same at the input and output, the noise power is reduced by a factor equal to the factor of reduction in the effective bandwidth. The SNR now becomes

$$
\begin{aligned}
SNR^+ &= P/R_d N_0 \\
&= P/(R_c N_0) \times (R_c/R_d) \\
&= SNR^- \times (R_c/R_d)
\end{aligned}
\tag{4.50a}
$$

So,

$$
\begin{aligned}
SNR^+/SNR^- &= R_c/R_d \\
&= Bc
\end{aligned}
\tag{4.50b}
$$

where Be is the bandwidth expansion factor. Thus, the PG is achieved, which by definition can be expressed as

$$
\begin{aligned}
PG &= 10 \log\left(SNR^+/SNR^-\right) \\
&= 10 \log(R_c/R_d) \\
&= 10 \log(Be)
\end{aligned}
\tag{4.51}
$$

Processing gain is nothing but the bandwidth expansion factor expressed in dB. This can also be written as

$$
\begin{aligned}
PG &= 10 \log(Be) \\
&= 10 \log(R_c/R_d) \\
&= 10 \log(T_d/T_c)
\end{aligned}
\tag{4.52}
$$

Because typical R_c is million (10^6) times R_d, we get for such a case

$$
\begin{aligned}
PG &= 10 \log\left(10^6\right) \\
&= 60 \text{ dB}
\end{aligned}
$$

This enables the system to keep a low signal to power ratio of the signal during transmission, yet with an improved value of the same when received at the receiver. This is the basis of the data security availed in the navigation process.

4.3.3.4 *Data security and signal robustness*

From the discussion on the spectrum of the signals, we learned that the navigation data on multiplication with the codes is spread over a relatively larger bandwidth. Thus, the total power, which was initially contained within the data bandwidth of R_b, is now spread across the width of R_c. Consequently, the signal power density reduces for an obvious reason by the same factor.

When the data are multiplied by code with an adequate chip rate, the power density level may be reduced so much that for any frequency, it may even fall below the level of conventional noise added by the channel, the atmosphere, and the receiver, leading to low detectability of the signal.

On multiplying the signal with the same code, the signal is dispread and the spread signal power gets collated into the original signal bandwidth. For an authentic receiver knowing the exact code, the signal after the correlation is

$$\begin{aligned} S(k) &= a(k)1/T \int Ci(k)Ci(k)dt \\ &= a(k)R_{xx}(0) \\ &= a(k) \end{aligned} \tag{4.53}$$

Because the autocorrelation for the exactly synchronized codes is unity, it regenerates the original data bits for the authentic receivers. As a result, the power of the received signal that was spread over the code bandwidth on multiplication by the exact code can be brought back within the original data bandwidth to avail the PG. Thus, it results in an increment in SNR by the same factor. We have already seen how a processing gain is achieved for such a signal.

Now, this is only possible when the receiver knows the code. Otherwise, any other receiver, without the knowledge of the code, cannot dispread it. For such cases, multiplication of any code different from the one in the signal leads to the product given by

$$\begin{aligned} p(k) &= a(k)1/T \int Ci(k)Cj(k)dt \\ &= a(k)R_{xy}(\delta) \\ &\sim 0 \end{aligned} \tag{4.54}$$

Because well-selected codes are used for the purpose of navigation, which have low values of cross-correlation, this produces good rejection of the signal as a result. It remains a noise and the data cannot be recovered by any means.

Without knowing the actual code, the data cannot be retrieved, and thus the code used in the navigation signal also helps secure the data from unauthenticated users with no knowledge of the actual code.

Furthermore, the signal in such a form remains robust to interference and jamming conditions, provided they are narrow band. This can be achieved for the following reason. The original data in the signal, that remain in the spread condition, upon multiplying by the exact code, are dispread and turn into the original data sequence. But imagine what happens to the interference signal. These components actually added up with the spread signal. Therefore, during the process of dispreading, when the replica of the code is again multiplied by the composite signal, the interfering signal is multiplied with the code for the first time and hence gets spread. So, if I is the total interfering power, on receiving and subsequent correlation, it becomes I/PG.

Thus, the power of the original signal that was originally spread during the communication gets collated within the data bandwidth and hence its density rises while the interference and noise powers density gets abated, owing to the spread resulting from the multiplication of the code. Only the part of the interfering signal power density that lies within the data bandwidth is effectively incorporated and contributes to the SNR ratio. So, the SNR becomes

$$SNR = P_S/[N_0R_d + I/PG] \qquad (4.55)$$

where Ps is the total signal power, N_0 is the one-sided noise density, and Rd is the data bandwidth. I is the total interference power actually present within the relevant bandwidth, whereas PG is the processing gain. The larger the PG is for a given bandwidth, the more robust the system is to interference. With faster codes, with code rates many times more than the data rate, we get higher PG and hence better SNR, providing better protection of the data from interference.

Besides, we have seen in all of these applications that better performance is obtained when the codes are chosen with good correlation properties. As we have seen in the previous section, this can be achieved when the codes are orthogonal and long. Longer codes at a very high rate warrant better performance of the system. However, there are certain problems associated with its implementation, which we shall learn about under relevant topics.

4.4 Encryption

In the last section, we learned that multiplying the data by the code protects the signal from unauthentic users. But that is not enough. Some navigation services require the corresponding signals to be foolproof from any possible unauthenticated use. Hence, encryption is required.

Although encryption is typically applied to the baseband data bits in communications, we will learn about it at this point because in satellite navigation, encryption is generally done at this level on the encoded data bits and not on the bare navigation message data.

However, before we understand the encryption process, we need to know at which level the encryption is really required. In fact, what matters most is what you want to protect. Is it the data or the code or both? In most of the cases, the data are not only the concern; the code is, too. The code determines and delivers accuracy. Higher accuracy is obtained with faster code rates even in navigation the data remains the same. If we want accuracy to be delivered selectively, we must have two codes: one delivering normal accuracy and the other a precise one. The code that gives precise accuracy should be encrypted. This allows the same data to be used for both services simultaneously, whereas only the selective users can use the better encrypted code and have access to the superior accuracy. Unless the user knows how the code bits are encrypted, he cannot regenerate the exact codes, obtain the data, or use the code for ranging.

Nonetheless, the data may also be kept differently. Data for a particular service may carry some additional information compared with another. One may have more precise values of the satellite ephemeris, time stamps, clock correction, and so forth than the other. It ensures that when the satellite positions and clock corrections are derived from it, the results are precise. The other set may have these values deliberately perturbed to some extent and deriving parameters from these yields in comparatively inferior results. The former data may be encrypted to ensure only valid users have the associated precision, whereas the others use the less precise version of the data. The encrypted code may also have some more information that will provide value-added service to these selected users. Therefore, encryption may be put into the data, the code, or both to give selectively precise services to different users. Moreover, encryption is not only for secrecy, but also for authenticity. Navigation data may also be encrypted to protect their authenticity.

To understand encryption within the limited scope of the book, let us start with a definition of cryptography (Katz, 2007). Cryptography is the science and study of secret writing. Modern cryptography is generally used to protect data transmitted over any medium against unauthentic use or deliberate ratification. The original intelligible binary message is converted into an apparently random arrangement of constituent bits and is called a cipher. The process of transforming a plain message into a cipher message is called encryption; the reverse process is called decryption. The key element that controls both encryption and decryption is called a cryptographic key, and there can be one or more for the process of this conversion.

The encryption process consists of an algorithm and the key. The key is a sequence of binary bits that is independent of the message. It controls the flow of data and parameter selection inside an algorithm to produce the algorithm. Therefore, the algorithm will produce a different output depending on the specific key being used at the time and changing the key changes the cipher of the same message from the same algorithm. The cipher message is then transmitted, and upon receiving it, can be transformed back to the original message using a decryption algorithm and the same key that was used for encryption. There can be different keys for encryption and decryption, though.

Various encryption schemes, standards, and algorithms for encryption are readily available and have been extensively reported in the literature. The strength of the standard encryption scheme, its computation requirement, and the probability of breaking the ciphering are the factors which rate the encryption process. Typically, a cryptographic system has five components:

1. Plain message, $M = m$
2. Cipher message, $C = c$
3. Key, $K = k$
4. Encryption algorithm, E_k: $m \rightarrow c$ where $K = k$
5. Decryption algorithm, D_k: $c \rightarrow m$ where $K = k$

For a given k, D_k is the inverse of E_k, i.e., $D_k(E_k[m]) = m$ for every plain text message m. Figure 4.12 illustrates the encryption and decryption of data (Figure 4.12).

For efficient encryption to take place, it must satisfy three general requirements:

1. The key must result in efficient encryption and decryption transforms.
2. The system must be easy to use.
3. The security of the system should depend only on the secrecy of the keys and not on the secrecy of algorithms E or D.

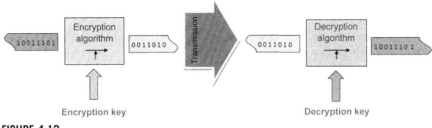

FIGURE 4.12

Encryption and decryption of data.

The last requirement implies that the encryption and decryption algorithms should be inherently strong. That is, it should not be possible to break a cipher simply by knowing the method of encryption. This is necessary because the algorithms may be in the public domain and known to all. However, there are specific requirements for secrecy and authenticity.

4.4.1 Secrecy requirements

1. It should be computationally infeasible for anyone to systematically determine the decryption transformation D_k from cipher message C, even if the corresponding plain message is known.
   This is important because for certain navigation systems in which the code is only encrypted and the same navigation message is transmitted to both general and selective users. Therefore, one is at liberty to compare the plain message and the encrypted message. This secrecy condition should be satisfied so that the decryption algorithm is not revealed through this comparison.
2. It should be computationally infeasible for anyone to systematically determine the plain message M from the cipher message C.

Secrecy requires only that decryption key to be protected. The encryption key can be revealed if it does not give away the decryption key. However, there is a threat of spoofing. Spoofs are message streams that looks alike a true satellite navigation message but carries the wrong data intentionally for distraction. Thus, some precautions are required to ensure the authenticity of the data.

4.4.2 Authenticity requirements

1. It should be computationally infeasible for anyone to systematically determine the encryption transformation E_k given C, even if the corresponding plain message m is known. Also, as in the previous case, the user should be barred from knowing the encryption transformation by comparing the open navigation data and the ciphered message.

2. It should be computationally infeasible for anyone to systematically find cipher message C' such that $D_k(C')$ is the plain message m.

Authenticity requires only that the encryption key be protected. Therefore, to ensure both authenticity and secrecy, both keys are required to be protected.

Cryptosystems can be classified as symmetric cryptosystems and asymmetric cryptosystems.

Symmetric or one-key cryptosystems: In symmetric ciphers, the encryption and decryption keys are the same or can be determined easily from each other. Hence, in symmetric ciphers, secrecy and authenticity cannot be separated.

Asymmetric or two-key cryptosystems: The encryption and decryption keys differ and it is computationally infeasible to determine one from the other. Here, the secrecy and authenticity can be provided separately by protecting D_k with the decryption key k_d for secrecy and E_k with the encryption key k_e for authenticity.

In a satellite navigation system, the data are encrypted by the system while the decryption is done at the users. Therefore, the decryption key, which is consistent with the encryption key, whether symmetric or asymmetric, needs to be disseminated to the authentic user. This must by some protective means that unauthentic users are not able to access. Key distribution generally remains an operation management issue.

4.5 Multiple access

Navigation satellites transmit signals that are received by receivers on the earth. Many satellites transmit the signal simultaneously which are spatially separated in their orbit. Multiple access (MA) is the ability of receivers to connect many satellites simultaneously and receive data from each without any affecting interference.

The receiver front end bandwidth cannot be increased in an unlimited fashion. An efficient MA scheme needs to be chosen so that the maximum numbers of satellites can be accessed by available resources. Mainly two types of MA are used for satellite navigation purposes: CDMA and FDMA. These are described below.

4.5.1 Code division multiple access

In previous sections, we already learned that by multiplying different signals by different orthogonal PRN codes during transmission, it is possible to select and separate them at the receiver and select the desired component out of the composite signal by rejecting the unwanted ones present in it.

This is achieved at the receiver by multiplying the composite signal by the corresponding orthogonal code. The required signal is selected by virtue of the large autocorrelation values with the required code. Other signals, forming cross-correlation with the multiplied code, result in low values and are rejected. This is the basic principle behind CDMA.

Considering the scenario for a navigation system, each satellite transmits its unique navigation data signal multiplying it by its particular unique code. All such signals transmitted by the different satellites multiplied by different orthogonal codes are received as a composite signal by the receiver. The composite signal consisting of all spread signals can be represented by the form

$$S(t) = D_1(t)g_1(t) + D_2(t)g_2(t) + D_3(t)g_3(t) + \ldots\ldots$$
$$= \sum D_i(t)g_i(t) \tag{4.56}$$

where S(t) is the composite signal, Di(t) and gi(t) are the data and code components of the signal from the ith satellite, respectively, at time t. When this composite signal is multiplied by a valid code $g_k(t)$ that corresponds to the kth signal in a synchronous manner, the product becomes

$$P(t) = S(t)g_k(t)$$
$$= \sum D_i(t)g_i(t)g_k(t)$$
$$= D_1(t)g_1(t)g_k(t) + D_2(t)g_2(t)g_k(t) + \ldots\ldots + D_k(t)g_k(t)g_k(t) + \ldots\ldots \tag{4.57}$$

Upon integration of the product over a data bit period and dividing by the integral value integration period, we get

$$p(k) = (1/Td) \int P(t)dt$$
$$= (1/Td) \left[\int D_1(t)g_1(t)g_k(t)dt + \int D_2(t)g_2(t)g_k(t)dt + \ldots \right. \tag{4.58}$$
$$\left. + \int D_k(t)g_k(t)g_k(t)dt + \ldots \right]$$

But because the integration is done over a data period, the data bit remains constant. So, the integration turns into

$$p(k) = \sum (1/T_d) \left[D_i(t) \int g_i(t)g_k(t)dt \right]$$
$$= \sum D_i(t) \left[(1/T_d) \int g_i(t)g_k(t)dt \right] \tag{4.59}$$

Each of the components in the right hand side of the equation becomes the correlation of the individual component codes of the signal and the code used in the receiver.

For all components of this product except the kth one, the codes are different. Because the codes are orthogonal, they produce low correlation values and vanish owing to their triviality. These components of the signal are rejected and are only treated as noise, n. However, only for the kth component, the two codes are identical and they produce the unit autocorrelation value. Hence, the data bit is retrieved. This may be mathematically expressed as

$$
\begin{aligned}
p(k) &= \sum (1/T_d) \left[D_i(t) \int g_i(t)g_k(t)dt \right] \\
&= \sum D_i(t) \left[(1/T_d) \int g_i(t)g_k(t)dt \right] \\
&= D_1(t)R_{xy}(g_1, g_k) + D_2(t)R_{xy}(g_2, g_k) + \ldots + D_k(t)R_{xy}(0) + \ldots \\
&\quad + D_2(t)R_{xy}(g_2, g_k) \\
&= D_k(t)R_{xx}(0) + n \\
&= D_k(t) + n
\end{aligned}
\tag{4.60}
$$

Therefore, after this process, the SNR ratio, including the cross-correlation noise, is given by

$$
SNR_k = P_k \bigg/ \left[N_0 R_b + \sum_{i=1, i \neq k}^{N} R_{xy}(\tau) \right]
\tag{4.61}
$$

Considering equal power P of the transmitted signals with processing gain PG received at the receiver from n satellites, SNR_k can be more conveniently written as

$$
SNR_K = P/[N_0 R_b + (n-1)P/PG]
\tag{4.62}
$$

4.5.2 **Frequency division multiple access**

The medium between the transmitting satellites and the receiver can efficiently carry radio waves only within a certain limited bandwidth. Within this bandwidth limited by the physical characteristics of the medium, a part is assigned for the purpose of navigation. Again, each individual system is allotted a definite portion of this band. In an FDMA system, each satellite in the navigation system transmits signal in a discrete channel of frequency. The range of the frequency of each channel is different but contained within the total band allocated to its system and transmits a signal with definite carrier frequency within that range. The sinusoidal waves at carrier frequencies are made to carry the encoded data chips through a process called modulation. We shall learn about it in section 4.6. The concept is shown in Figure 4.13.

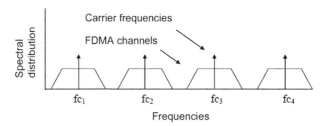

FIGURE 4.13

Signal spectral distribution for FDMA.

Satellites share the total bandwidth allocated to the system. However, there should be some separation between the bands of adjacent channels used by two different satellites, to avoid possible interference; this is called the guard band.

The receiver receives these discrete bands of signals with adequate bandwidth at the common front end to accommodate all available channels. The signal is then further distributed to individual receiving chains. Each of these chains have centre frequency aligned with the carrier frequency of transmission of the individual satellites and bandwidth adequate for an individual signal. It thus processes each of the single channels out from the combined received signal obtained from all the visible satellites.

Because each satellite needs to be given a separate width of the band, for large numbers of satellites in the fleet, the total band required for the system and hence the front end bandwidth of the receiver need to be very wide. This constant is somewhat alleviated by opting for frequency reuse among the satellites. Satellites that are located in antipodal locations in the constellation are never seen from the same point on the earth simultaneously. Therefore, they can be reallocated the same frequency. This reduces the band requirement by almost half. Nevertheless, the number of receiving chains required in an FDMA receiver still depends on the number of satellites to which it wants to have access simultaneously.

4.6 **Digital modulation**

The signal consisting of the product of the data and the code is to be transmitted from the satellite to the user. However, it cannot be sent in this binary baseband form, for two key reasons. First, the transmitting antenna needs to be of the order of the wavelength of the signal being transmitted for efficient transmission. The encoded data, which are mostly concentrated within the low frequency ranges and hence have very high wavelength, would require a very large transmitting antenna for the purpose. Second, the channel through which the signal passes cannot efficiently propagate the DC components and acts as a bandpass filter. This signal is ridden on a carrier wave of suitable high frequency. The technique of the baseband data being put on a high-frequency radio wave to be carried with it, so that it efficiently traverses from the transmitter to the receiver through the path in between, is called '*modulation*'. The high frequency of the carrier reduces the antenna size accordingly and carries the signal through the channel unattenuated, as well. Furthermore, modulation over a higher frequency makes it possible to use multiple access techniques such as FDMA and CDMA besides giving the signal better noise immunity from low-frequency noise sources.

For navigation, encoded navigation data form the baseband that is modulated on a carrier with a frequency much higher than the chip rate.

4.6.1 **Carrier wave**

The carrier is the sinusoidal radio wave of high frequency that carries the encoded data chips of the signal through the medium. The choice of carrier signal is basically choosing its frequency. For navigation, it depend on the overall system design. However, the broad frequency band from which the carrier frequency is to be chosen for navigation is specified by the International Telecommunication Union (ITU) through its recommendation.

For navigation systems using CDMA for multiple access, the carrier frequencies used by different satellites remain the same, while for those using the FDMA, the carrier frequencies are different but close to each other; the difference is determined by the bandwidth of the signal it carries. In CDMA, because different data are already separated by orthogonal codes, the same carrier can carry different signals at the same time without mutual interference. The sine and cosine signals of the same carrier frequency are also orthogonal, and hence can carry different signals, too.

For satellite navigation, relativistic corrections are done on the transmitting carrier signal. This is because the transmitting satellite and the receiver move relative with respect to the reference frame, and the satellite experiences variation in gravitational potential. To mitigate the relativistic effects of all of these factors that affect the satellite clock, the latter is adjusted to run at a frequency relatively lower than that of the design frequency on the earth surface.

4.6.2 Modulation techniques

The common question to be addressed at this point is, "In what form should the carrier carry the information?" The main characteristics of the carrier waveform by which it is defined are its amplitude, frequency, and phase. As a result of modulation of the sinusoidal carrier, one of these features is varied in accordance with variation in the information carrying encoded data chips, acting as the baseband. Thus, through the variation of this particular feature of the carrier the information is carried through space.

For a digital signal, the information is represented by discrete levels. Not only the levels vary between discrete values but the variations take place at discrete intervals, as well. For binary data, only two possible levels can be assumed by the baseband signal. So, upon modulation, any one amongst the phase, frequency or the amplitude of the carrier, changes between two fixed values. These two values represent the two binary levels of the baseband signal. Further, these changes occur at discrete but defined intervals. Depending upon the parameter of the carrier which changes, the modulation may be termed as 'Binary Phase Shift Keying', 'Frequency Shift Keying' or 'Amplitude Shift Keying' respectively. There can be multilevel modulations, too, but we shall restrict our discussion to only the binary levels that are used for the communication of navigation data. Binary keying gives a better error performance but it has a lower data transfer rate in general. It matches the purpose because the navigation data are small but need correct transferring to the receiver.

At the receiver's end the original binary information is recovered by the reverse process known as demodulation. However, depending on the modulation type, demodulation can be complex because it has to recover signals accurately from the weak received signal corrupted by random noise.

The process of modulation splits and shifts the original frequency spectrum of the signal by an amount equal to the carrier frequency in both the positive and negative direction on the frequency axis, keeping the shape of the original signal spectrum unaltered in general. However, under some particular modulations, these shifted signals overlap each other partially so that the ensuing shape is different from that of the original signal.

Although the most popular type of modulation for navigation is the BPSK, other types of modulation including BOC, which gives some additional benefits for the purpose are also used.

4.6.2.1 Binary phase shift keying

One of the most popular modulation schemes used in the navigation system is the BPSK. This is because this bears constant amplitude, and maintains fixed frequency over each symbol interval. However, its phase changes discretely at every change in data symbol. So, the current phase of the modulated signal is determined by the value of the data current symbol. This technique is adequately easy to implement.

1. Time domain representation of BPSK:

The time domain representation of the BPSK modulated signal characterizes how the carrier signal varies with time; its phase is altered with the variation of the modulating signal. The modulating signal, in our case, is the encoded navigation data.

Because the phase of the signal can vary from 0 to 2π, the maximum separation in phase of two signals in the same frequency can be π. Thus, in BPSK modulation the carrier phase changes between two different values separated by a phase π to represent the two possible states of the binary data, while its amplitude and frequency are not affected by the data. For a carrier represented by the signal $S = \cos \omega t$, after modulation, the time domain representation of the BPSK signal becomes

$$s(t) = \cos\{\omega t + \varphi(t)\} \tag{4.63}$$

where ω is the angular frequency of the carrier and $\varphi(t)$ is the phase at time t. This $\varphi(t)$ is the variable that actually carries the information and varies according to the baseband data. The baseband signal is binary and varies between representative levels $a_k = -1$ and $a_k = +1$. Accordingly, φ can take discrete values of $\varphi = \varphi_0 + 0$ and $\varphi = \varphi_0 + \pi$, to represent the two binary states of the data, a_k, where φ_0 is the arbitrary initial phase. So, the modulated signal becomes

$$\begin{aligned} s(t) &= +\cos(\omega t + \varphi_0 + 0) \quad \text{when } a_k = +1 \\ &= +1 \cos(\omega t + \varphi_0) \\ &= a_k \cos(\omega t + \varphi_0) \end{aligned} \tag{4.64}$$

and

$$\begin{aligned} s(t) &= +\cos(\omega t + \varphi_0 + \pi) \quad \text{when } a_k = -1 \\ &= -1 \cos(\omega t + \varphi_0) \\ &= a_k \cos(\omega t + \varphi_0) \end{aligned} \tag{4.65}$$

The BPSK modulated signal may thus be represented as

$$s(t) = a_k \cos(\omega t + \varphi_0) \qquad (4.66a)$$

a_k is the kth encoded data chip prevailing at the time t $(k = \lfloor t/T \rfloor)$.

From these equations, we can see that the BPSK modulation may also be represented as the product of the data bit with the carrier. This is because changing the phase by $\pi/2$ is just reversing the polarity of the signal, which is equivalent to multiplying the original signal by the complementary unitary data bit. In any case, the amplitude of the modulated signal will remain constant (Figure 4.14).

Similarly, if the phase values are shifted between $\varphi = \varphi_0 + \pi/2$ and $\varphi = \varphi_0 - \pi/2$, for $a_k = -1$ and $a_k = +1$ respectively, it is phase orthogonal with the first case. Therefore, it may be considered as a cosine carrier with initial phase $(\varphi_0 - \pi/2)$ instead of φ_0. So, the effective carrier may be assumed to be a sine signal instead with initial phase φ_0. The signal can be represented as

$$s(t) = a_k \sin(\omega t + \varphi_0) \qquad (4.66b)$$

2. Spectrum and bandwidth:

Expanding Eqn (4.66a), we get (Chakrabarty and Datta, 2007)

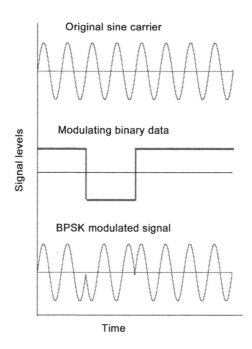

FIGURE 4.14

BPSK modulation.

$$s(t) = a_k[\cos(\omega t)\cos(\varphi_0) - \sin(\omega t)\sin(\varphi_0)]$$
$$= [a_k \cos(\varphi_0)]\cos(\omega t) + [-a_k \sin(\varphi_0)]\sin(\omega t) \quad\quad (4.67a)$$

This means that the signal is equivalent to the combination of a sine and a cosine wave at frequency ω with the effective amplitude a_k resolved into components $a_k \cos(\varphi_0)$ and $a_k \sin(\varphi_0)$, respectively, on I and Q planes, defined by the orthogonal functional axes of $\cos \omega t$ and $\sin \omega t$.

It is known that the encoded data chips, representing the binary baseband for the final carrier modulation, have a definite frequency spectrum. When this modulates a carrier, the resultant spectrum is the same as the original one, but only gets shifted by an amount equal to the carrier frequency and is given by

$$S(f) = aTc \text{ sinc } \{\pi(f - fc)/Rc\} \quad\quad (4.67b)$$

As the signal is an encoded data chips have a chip rate of $Rc = 1/T_c$, the spectrum has the null of its envelope at $f = fc \pm Rc$ and then successively at intervals of Rc on both sides of fc. So, the first lobe has a total width of $2Rc$ followed by side lobes of width Rc each. We have already seen that most of the energy remains contained within the first lobe of the envelope, and it is sufficient to take the first null to null width as the bandwidth of the signal. This makes the total bandwidth of the modulated signal $2Rc$. Box 4.7 explains the BPSK modulation in both time and frequency domain.

4.6.2.2 Binary offset carrier modulation
The BOC signal is actually an extension of the BPSK modulation that is produced when a BPSK signal is multiplied by a square wave subcarrier (Betz, 2001). The square wave may be sine-phased or cosine-phased. The more conventional sine-phased BOC signal can be mathematically represented as (Binary Offset Carrier (BOC), 2011)

$$s(t) = c(t)\text{sign}[\sin(2\pi fst)]\cos(2\pi fct) \quad\quad (4.68)$$

BOX 4.7 MATLAB BPSK MODULATION

The MATLAB program bpsk_mod was run to generate the time variation of the BPSK modulated signal. In this program, the sine wave is modulated by a square wave representing the carrier and the modulating encoded data chips, respectively. Figure M4.7 shows the temporal variation of the modulating and the modulated signal.

In the program, the carrier frequency is taken as four times the chip rate. Run the program with a higher frequency compared with the chip rate. Also, use different modulating chip values. Replace the sine carrier with a cosine carrier. The initial phase fi_0 is taken as zero. Change this value and run the program.

Figure M4.7(b) shows the power spectrum of this modulated signal. Notice the shift of the spectrum from zero to a higher frequency. This is the effect of modulation. Change the relative frequency of the carrier with respect to the chips and observe the change in this shift.

Continued

BOX 4.7 MATLAB BPSK MODULATION—cont'd

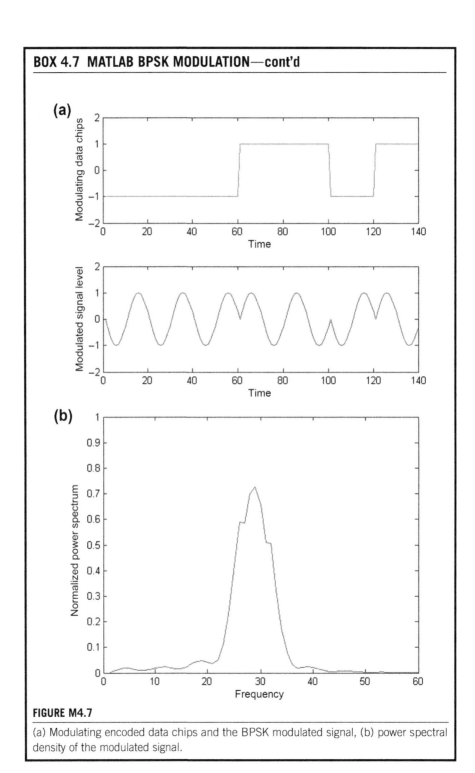

FIGURE M4.7

(a) Modulating encoded data chips and the BPSK modulated signal, (b) power spectral density of the modulated signal.

where s(t) is the signal at time t, c(t) is the encoded data, and fs is the frequency of the subcarrier. Thus, the term sign $[\sin(2\pi fst)]$ represents a $+1$ for the phase $0 - \pi$ of the subcarrier, whereas a -1 for the phase excursion from π to 2π. fc represents the carrier frequency. Similarly, the cosine-phased BOC may be defined. The subcarrier frequency fs is much lesser than the carrier frequency fc but a few times the chip rate Rc.

Thus, a BOC-modulated navigation signal consists of a sinusoidal carrier, a square wave subcarrier, a PRN spreading or ranging code, and a data sequence. The final signal is produced as the time domain product of these components. The product of the data, PRN code, and carrier essentially makes it a BPSK signal which has its spectrum about the carrier frequency. For this product, when multiplied with a binary valued subcarrier, which is nothing but a square wave, its spectrum again splits and spreads once again. As the centre of its spectrum shifts away from the carrier frequency, i.e., gets offset, and this shift is realized with a binary subcarrier, this is called offset modulation.

For any practical BOC signal, the carrier frequency, subcarrier frequency fsc, and code rate Rc are chosen as multiples of the reference frequency f_0. So

$$1/Ts = fs = m \cdot f_0$$
$$1/Tc = Rc = n \cdot f_0$$
$$(4.69)$$

This refers to the BOC(m, n) modulation. The reference frequency f_0 is generally taken as 1.023 MHz. Furthermore, if we take k to be the total number of subcarrier periods present in one complete code chip period. Then,

$$k = Tc/Ts = fs/Rc = m/n \qquad (4.70)$$

Depending on the relative values of Tc and Ts, the value of k is determined. This value is not always an integer, but it may take semi-integral values, too, as in BOC(5,2). Because a complete period of a square wave has two square pulse elements, one positive and one negative, the double of this value, i.e., $2k = 2m/n$, represents the total number of such square pulses present within a chip. This is denoted by N_{BOC} (Binary Offset Carrier Modulation, 2009) and is used to represent many characteristics of the BOC signal.

Let us try to understand modulation by first taking a simple case in time domain. In a BOC(1,1) modulation there is one complete oscillation of the square wave within the period of one chip. So, a "+1" chip now becomes a "+1 −1" sequence, and a "−1" chip gets converted into a "−1 +1" sequence upon multiplication by the square wave. We will call these finer chips formed on BOC multiplication "BOC chips," to distinguish them from the original subcarrier or encoded data chips. For an arbitrary modulation order, in the BOC (m, n) case, a "1" or +1 value of encoded chips becomes an alternating sequence of "+1 −1 + 1 −1 + 1..." with 2(m/n) elements and a "−1" becomes an alternating "−1 + 1 −1 + 1..." sequence, also with 2(m/n) elements. So, when BOC modulation is applied to PRN-coded navigation binary signals, each original encoded data chip is split into N_{BOC} numbers of finer widths of Ts/2. The case for BOC(5,2) is illustrated in Figure 4.15.

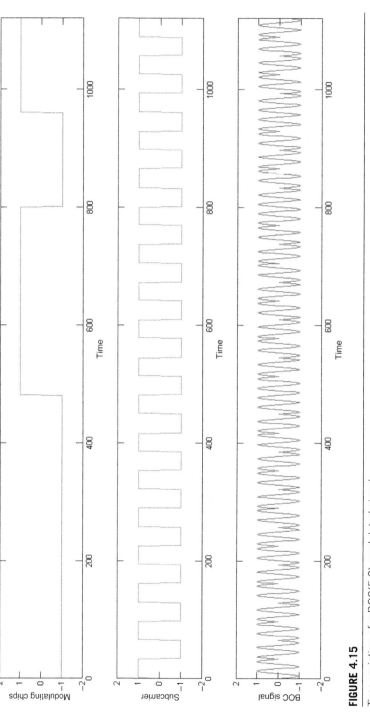

FIGURE 4.15

Time variation of a BOC(5,2)-modulated signal.

There are N_{BOC} subintervals within a chip period, with each subinterval carrying either a $+1$ or a -1 chip alternately. For an integral m/n, the number of BOC modulated chips is even with equal $+1$ and -1s within every chip width. This makes the mean signal level and hence the DC power of the data perfectly zero. When this ratio is a subinteger, the numbers of symbols within a chip period is odd. Consequently, they do not average out to zero in a chip period, and hence manifests some finite DC power.

Let us consider the spectrum of the signal. We know that the effect of multiplying a signal in the time domain with a new tone carrier of frequency f results in a shift of the signal in frequency domain by an amount f. What is a square wave is made of? A square wave signal of repeating frequency fs is the combination of several independent sine and cosine waves, i.e., tone frequencies at $\pm$fs and its integral multiples extending to infinity but with diminishing level.

When this offset carrier is multiplied by the original signal, it is equivalent to multiplying the latter with each of these component tone frequencies of the square wave. It shifts the spread spectrum of the encoded signal by $\pm$nfs, where n is an integer. The resultant spectrum is an offset form of the original signal shifted by amount $\pm$fs and its multiples from the carrier frequency fc and on two sides of it. Any one of these shifted spectrums, which are replicas of the original in shape, can be filtered and use for demodulation.

The nulls of a sinc function representing the spectrum of the encoded data chips, i.e., the baseband nulls, occur at integral multiples of Rc. When, m/n is an integer, making fs an integral multiple of Rc, a shift of fs in the original spectrum as a result of the subcarrier modulation will place a null at the position of the carrier frequency, fc. This is true for both positive and negative shifts of the spectrum. This maintains a naught of the spectral amplitude at fc. When the signal is carrier demodulated, it makes the DC component of the spectrum signify the average power, perfectly zero.

However, if this ratio is a semi-integral number such as 3/2 or 5/2, the splitting and offset movement of the spectrum will place one of the side lobe peaks at the carrier frequency, fc. This is because these peaks occur at semi-integer values of Rc in the original spectrum. Therefore, it cannot eliminate DC power. This corroborates our observation we had regarding the average power from the time signal. The offset spectrum of the BOC for the two cases is shown in the following figure (Figure 4.16). Here the frequency axis is shown relative to the carrier frequency and scaled by the reference frequency f_0.

The expression for the PSD of a BOC (fs, Rc) signal can be derived from of the Fourier transform of the autocorrelation function. When k is an integer, i.e., N_{BOC} is even, it is expressed as

$$G(f) = fc[\sin(\pi f/fs)\sin(\pi f/2fs)/\{\pi f\cos(\pi f/2fs)\}]^2 \qquad (4.71)$$

For a BOC(5, 2) signal, the spread of the original encoded data chips have a spectral spread of 2 MHz on both sides about the carrier. Again, the BOC subcarrier was of 5 MHz. Thus, the spectrum of the original signal, after multiplying by the subcarrier, will appear about the frequency line at $\pm$5 MHz from fc. The separation between the first main peak at the positive and negative side of the carrier will thus

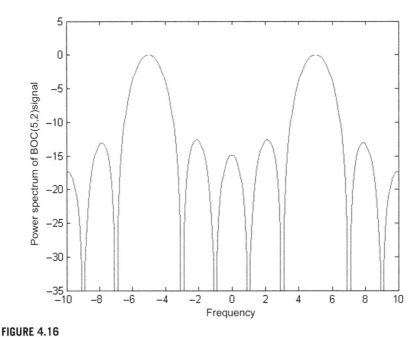

FIGURE 4.16

Spectrum of a BOC(5,2) signal.

be $5 \times 2 = 10$ MHz. Furthermore, each original spectrum had ± 2 MHz bandwidth of the main lobe, i.e., separation of the first nulls from its main peak. Considering the width of these two separated peaks on either side, the bandwidth becomes $2 + 10 + 2 = 14$ MHz. With the BOC modulation, we can shift the spectrum away from the carrier frequency by an amount of our choice by suitably choosing the sub-carrier frequency. However, we have to increase the bandwidth of the receiver accordingly to properly receive them.

An obvious question one may ask at this point is, "What advantages do we really get in doing this?" The main idea behind BOC modulation is that the total spectrum shifts by amount fs, about the original carrier frequency fc upon such modulation. Now, suppose we want two BPSK signals to operate at the same carrier frequency. It is a likely phenomenon for satellite navigation, in which the preferred carrier frequencies are limited. Using another set of BPSK signals with the same carrier frequency will cause interference at the receiver. However, BPSK-modulated signals will have most of their spectral power concentrated around the main lobe that exists about fc. Multiplying one of these signals by a square wave subcarrier to form a BOC will split the spectrum of this signal and will place the two resultant shifted spectral lobes at equal distances of fs on two sides of the center frequency fc, keeping the power low around the center. Consequently, it reduces the interference between the BOC-modulated signal and that in original BPSK, and yet both can use the same primary carrier frequency and same primary modulation.

Autocorrelation of BOC(m, n) can be obtained from its time domain features. As a result of this multiplication of the code chips with the subcarrier, the time domain signal is further subdivided into thinner BOC chips. The width of these finer BOC chips is $T_s/2$, which is N_{BOC} times finer than the original chip width. Thus, if the same BOC signal is multiplied with exact superimposition, i.e. with zero delay, it generates an autocorrelation value of 1 for 0 delay. Whenever a relative shift is applied between them, variation of the product of the chips occurs, but the variation of the subcarrier occurs with N_{BOC} times more rapidity. The product of the chips varies over the time in $+1$ and -1.

The correlation function is given by

$$R_{BOC}(\tau) = 1/T \int C(t)C(t - \tau)\, S(t)S(t - \tau)dt$$
$$= 1/T \int A(t)B(t)dt \qquad (4.72)$$

where $A = C(t)\, C(t - \tau)$ and $B = S(t)\, S(t - \tau)$. Concentrating on the second part of the integral, we see that B is the product of two subcarriers shifted by a delay τ, and is also alternating positive and negative square pulses but not necessarily of equal width. However, in this expression, the term B is a recurrent function of the shift τ and abruptly changes between $+1$ and -1. Because the width of the subcarrier square wave is smaller, the product $B(\tau)$ has a faster rate of variation. Therefore, this correlation also varies at an equally faster rate. This improves the estimation accuracy of correlation during ranging. In addition, this BOC modulation helps in more

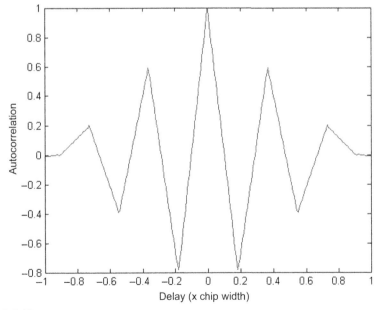

FIGURE 4.17

Autocorrelation of a BOC(5,2) modulated signal.

precise ranging compared with normal BPSK modulation. The autocorrelation of a BOC(5, 2) signal is shown in Figure 4.17 (Box 4.8).

Although BOC modulation typically means BOC (m,n), the BOC modulation can have several variants. A popular type is alternative BOC (Alt-BOC), which we shall discuss next.

4.6.3 **Alt-BOC modulation**

The BOC modulation, as we have seen in Eqn (4.68), is a multiplication of the data, the code, the carrier, and the subcarrier. Among these elements, the carrier and the subcarrier can have orthogonal counterparts. That is, the carrier may be either a sine or a cosine wave, and similarly the subcarrier may be a sine-phased or a cosine-phased one. The inner products of these orthogonal pairs become zero over a complete period. Their temporal variations are shown in Figure 4.18.

From this figure, we can make out that their inner product, over a time that is an integral multiple of their period, is zero. Thus, representing a complete square wave subcarrier with four ranges of phases, $0 - \pi/2$, $\pi/2 - \pi$, $\pi - 3\pi/2$, and $3\pi/2 - 2\pi$, with each range separated by time Ts/4, one may represent the sine- and cosine-phased square wave vectors as $Ss = [1\ 1\ -1\ -1]$ and $Sc = [1\ -1\ -1\ 1]$, respectively. The inner product of these two carriers is zero, and hence they are orthogonal. This can also be shown mathematically as

$$
\begin{aligned}
P &= \int_0^T S_1(t)\,^* S_2(t) dt \\
&= \int_0^{T/4} s1(t)s2(t)dt + \int_{T/4}^{T/2} s1(t)s2(t)dt + \int_{T/2}^{3T/4} s1(t)s2(t)dt + \int_{3T/4}^{T} s1(t)s2(t)dt \\
&= (T/4)(+1)(+1) + (T/4)(+1)(-1) + (T/4)(-1)(-1) \\
&\quad + (T/4)(-1)(+1) \\
&= +T/4 - T/4 + T/4 - T/4 \\
&= 0
\end{aligned}
$$

(4.73)

BOX 4.8 MATLAB BOC MODULATION

The MATLAB program boc_mod was run to generate the time variation of the BOC-modulated signal. In this program, a preselected PRN is multiplied with a sine-phased binary subcarrier with a frequency ratio of $1{:}1$ with the chip rate. The different components of the BOC signal and the autocorrelation of the product signal are shown in Figure M4.8(a) and (b), respectively.

Change the frequency to the chip rate ratio by altering the values of m and n in the program. Observe how the numbers of carrier phase variation change in the signal. Also notice the variation in the autocorrelation within the envelope.

BOX 4.8 MATLAB BOC MODULATION—cont'd

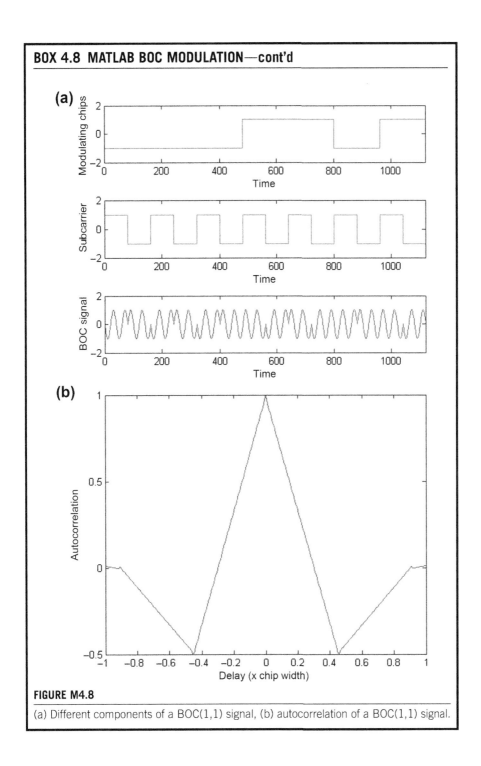

FIGURE M4.8

(a) Different components of a BOC(1,1) signal, (b) autocorrelation of a BOC(1,1) signal.

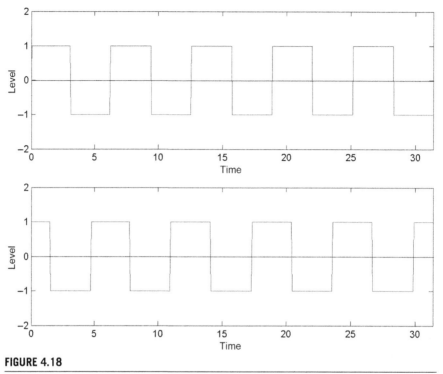

FIGURE 4.18

Sine-phased and cosine-phased binary offset carriers.

This property enables two such BOC to carry different set of signals independently like that of the carrier.

In alt-BOC, the sine-phased binary subcarrier multiplied by the sine and the cosine carriers respectively forms a set of two orthogonal composite carrier components. Similarly, the cosine-phased binary subcarrier on the same sine and cosine carriers forms another set of two separate orthogonal carriers. These two components are also orthogonal to the former two by virtue of the subcarrier orthogonality. Thus, four mutually orthogonal composite carrier—subcarrier components are formed, each carrying a separate set of encoded baseband data exclusively without interference (Margaria et al., 2007; Betz, 2001). Thus, the Alt-BOC signal may be represented as

$$
\begin{aligned}
S(t) = \ & D_1(t)c_1(t)\text{sign}[\sin(2\pi f_s t)]\cos(2\pi f_c t) + \\
& D_2(t)c_2(t)\text{sign}[\sin(2\pi f_s t)]\sin(2\pi f_c t) + \\
& D_3(t)c_3(t)\text{sign}[\cos(2\pi f_s t)]\cos(2\pi f_c t) + \\
& D_4(t)c_4(t)\text{sign}[\cos(2\pi f_s t)]\sin(2\pi f_c t)
\end{aligned}
\tag{4.74}
$$

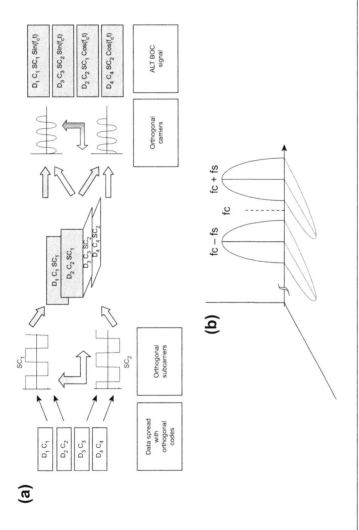

FIGURE 4.19

(a) Alt-BOC modulation scheme, (b) Alt-BOC spectrum.

The four different orthogonal codes for the four channels provide additional isolation over their spectral distinctions. However, the orthogonal channels may use the same data or code as appropriate or necessary. Even some channels may have no data at all, and are called pilot channels. The codes and the subcarriers may be intelligently selected so that unlike simple BOC, the signal spectrum shifts only to any one of upper or lower subcarrier frequencies about the carrier (AltBOC Modulation, 2013). The Alt-BOC modulation scheme is illustrated in Figure 4.19(a); its spectrum is shown in Figure 4.19(b).

Each of these resultant signals will separately have constant amplitude as the operation of subcarrier multiplication to BPSK to form BOC, keep the amplitude invariant. However, when these individual signals are combined to form the composite Alt-BOC signal, this constancy is lost in the resultant phased sum of these components. Hence, the amplitude of an Alt-BOC modulated signal is not constant. This modulation is known as nonconstant envelope Alt-BOC modulation. It has certain negative implications at the receiver, and hence nonconstant envelope Alt-BOC is generally not used in practice.

To solve this problem, a constant envelope modified version of Alt-BOC modulation was introduced. (Betz, 2001). This was achieved by introducing a new signal called an intermodulation product, which is made up of a definite combination of individual components but does not contain usable information (Soellner and Erhard, 2003)and (Ward et al., 2006).

The power spectral density for an Alt-BOC signal for even and odd values of $N_{BOC} = 2$ fs/fc may be shown to be, respectively (Shivaramaiah, Dempster, 2009),

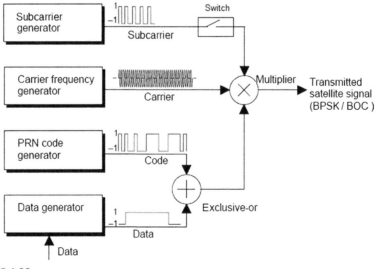

FIGURE 4.20

Schematic arrangements for BPSK and BOC modulation.

$$G(f) = 8Tc[\sin(\pi f\,Tc)/\{(\pi f\,Tc)\cos(\pi f/fs)\}]^2\{1 - \cos(\pi f/fs)\}$$
$$G(f) = 8Tc[\cos(\pi f/fc)/\{(\pi f\,Tc)\cos(\pi f/fs)\}]^2\{1 - \cos(\pi f/fs)\}$$
(4.75)

A schematic arrangement for the BPSK- or BOC-modulated signal generation is shown in Figure 4.20. The generated signal may be made of either of the two by choosing the switch position appropriately. Keeping the switch open makes the output a BPSK-modulated signal, whereas closing the switch makes it a BOC-modulated signal.

4.7 **Typical link calculations**

Very low navigation signal levels around -133 dBm are expected over the targeted region. In addition to the nominal channel noise, these signals may also experience interference from other GNSS signals in these bands. This interference level may be minimized with the help of BOC modulation and CDMA access scheme. The receiver will also receive propagation impairments such as channel attenuation, scintillation, multipath etc. and these effects should be considered to obtain the effective signal condition at the receiver. A typical link calculation in the allocated navigation frequency is shown in the table below.

Types of Signal	Units	Values
Frequency	GHz	1.57542
Wavelength	m	0.1904
EIRP	dBw	23.00
Range	km	20,000.00
Path loss	dB	182.00
Atm. Atn.	dB	1.00
Total attenuation	dB	183.00
Received power	**dBw**	**− 160.00**
Sky temperature	K	250.00
LNA noise temperature	K	80.00
Equivalent system noise temp	K	330.00
Equivalent system noise temp	dBK	25.20
Boltzman constant	dB	−228.60
Rcd. noise power at LNA I/P	dBw/Hz	−203.40
Noise bandwidth	dB	63.00
Noise floor	**dBw**	**− 140.40**

From this table, we can see that the signal power is 20 dB below the noise power. The signal is received by the receiver in this form. When the codes are known to the receivers, it can do the correlation and enhance the power density by dispreading the signal once more.

Conceptual questions

1. Is it possible to raise the signal power from a single satellite above the noise floor at the receiver without performing autocorrelation with the local code using a high-gain antenna only?

2. What problems may occur if the data and the ranging codes are not synchronous at the transmitter?

References

AltBOC Modulation, 2013, ESA Navipaedia, http://www.navipedia.net/index.php/AltBOC_ Modulation (accessed 01.12.13).

Betz, J.W., 2001. Binary offset carrier modulations for radionavigation. Navigation: Journal of The Institute of Navigation 48 (4), 227–246.

Binary Offset Carrier (BOC), 2011, ESA Navipaedia, http://www.navipedia.net/index.php/ Binary_Offset_Carrier_(BOC) (accessed 17.01.14).

Binary Offset Carrier Modulation, 2009, Wikipaedia, http://en.wikipedia.org/wiki/Binary_ offset_carrier (accessed 23.12.13).

Chakrabarty, N.B., Datta, A.K., 2007. An Introduction to the Principles of Digital Communications. New Age International Publishers, New Delhi, India.

Cooper, G.R., McGillem, C.D., 1987. Modern Communications and Spread Spectrum. Mcgraw Hill, USA.

Global Positioning System Directorate, 2012a. Navstar GPS Space Segment/Navigation User Interfaces: IS-GPS-200G. Global Positioning System Directorate, USA.

Global Positioning System Directorate, 2012b. Navstar GPS Space Segment/User Segment L1C Interfaces: IS-GPS-800C. Global Positioning System Directorate, USA.

Gold, R., 1967. Optimal binary sequences for spread spectrum multiplexing. IEEE Transactions on Information Theory IT-13, 619–621.

Katz, J., 2007. Introduction to Modern Cryptography. Chapman & Hall/CRC.

Kovach, K., Haddad, R., Chaudhri, G., 2013. LNAV Vs. CNAV: More than Just NICE Improvements, ION-gnss[+]—2013. Nashville Convention Centre, Nashville, USA.

Lathi, B.P., 1984. Communication Systems. Wiley Eastern Limited, India.

Lin, S., Costello Jr, D.J., 2010. Error Control Coding. Pearson Education, Inc.

Margaria, D., Dovis, F., Mulassano, P., 2007. An innovative data demodulation technique for Galileo AltBOC receivers, journal of global positioning systems. Journal of Global Positioning Systems 6 (1), 89–96.

Mutagi, R.N., 2013. Digital Communication: Theory, Techniques and Applications. Oxford University Press, New Delhi, India.

Proakis, J.G., Salehi, M., 2008. Digital Communications, fifth ed. Mc Graw Hill, Boston, USA.

Shivaramaiah, N.C., Dempster, A.G., 2009. The Galileo E5 AltBOC: Understanding the Signal Structure, IGNSS Symposium 2009-Australia. International Global Navigation Satellite Systems Society.

Soellner, M., Erhard, P., 2003. Comparison of AWGN code tracking accuracy for alternative-BOC, complex-LOC and complex-BOC modulation options in Galileo E5-band. In: Proceedings of the European Navigation Conference. ENC-GNSS, Graz, Austria.

Spilker Jr, J.J., 1996. GPS signal structure and theoretical performance. In: Parkinson, B.W., Spilker Jr, J.J. (Eds.), Global Positioning Systems, Theory and Applications, vol-I. AIAA, Washington DC, USA.

Ward, P.W., Betz, J.W., Hegarty, C.J., 2006. GPS signal characteristics. In: Kaplan, E.D., Hegarty, C.J. (Eds.), Understanding GPS Principles and Applications, second ed. Artech House, Boston, MA, USA.

Navigation Receiver

5

CHAPTER OUTLINE

Understanding Satellite Navigation. http://dx.doi.org/10.1016/B978-0-12-799949-4.00005-1

In Chapter 2, we read that the interface between the navigation signals, offering navigation service, and the user of the service, is the receiver. It accepts the signal and converts it to attain the service goal. In this chapter, we will provide technical details about the receiver. We will start with the generic receiver structure and its different classifications. Then, we will continue with details of the different functional units of the receiver and describe the functionalities of each.

5.1 Navigation receiver

Satellites in the space segment accomplish their objectives by transmitting the necessary signals. It is the task of a receiver to receive these signals and eventually estimate its own position, using information either readily available or derived from the signals. For this, the signals transmitted from different satellites need to be simultaneously received and processed efficiently. This is done by a *navigation receiver*, an instrument the user uses to fix his or her position.

5.1.1 Generic receiver

From the system point of view, a navigation receiver is not very different from any digital communication receiver. It has to do the same job as any communication receiver: that is, receive a signal and derive data from it and use it for the applications. Thus, communications elements are predominant here. However, in terms of processing capability, navigation receivers need to have some added features over a typical communication receiver. This is because of the nature of the signal and the associated criticalities, and also the techniques from which the necessary information is derived. It is also obvious that related applications will be different in accordance with the objective of the receiver.

In this chapter, we will look in detail at how navigation receivers work. We shall stick to the most common implementations of the architectural design of the receiver, and will analyze its theoretical aspects with a focus on its working principles.

To understand the way a receiver works, we have to start from the signals with which the receiver interacts. We first need to understand the condition of the signal the receiver receives. The major receiver characteristics are defined in terms of the features of this signal. In the last chapter, we learned about signal characteristics and how signals are generated and transmitted. Here, we will start with a reiteration of what we learned.

The signal consists of basic navigation information that remains embedded within it in a tiered structure. The job of the receiver is to isolate the individual signal and extract the information from it.

From the previous chapter, we also know that the signal is transmitted in either Code Division Multiplexing Access (CDMA) or Frequency Division Multiplexing Access (FDMA) mode for the purpose of multiple access, so that the receiver can receive signals from different satellites simultaneously without interference. Moreover, irrespective of the technique used for access, there is typically a pseudo random code multiplied with the message for ranging purposes, modulated on a carrier.

In this chapter, we will see how the receiver acquires the signal and then uses it for further processing. Signals from different satellites are received by a common interface and subsequently pass through a common front end. After this, the individual signals are separated, acquired and processed in separate channels of the receiver. These discrete channels thus obtain the signals either in the same frequency and with different codes for CDMA or in different frequencies but with the same ranging code for FDMA.

The signals are carried by a predesignated carrier frequency with known polarization, typically circular. The sinusoidal carriers are modulated by these navigation data, which in turn are already spread by the ranging code. Although the most popular type of modulation for navigation is the Bipolar Phase Shift Keying (BPSK) other types of modulation, including binary offset carrier (BOC), are alternative options. Obviously, the modulation type needs to be known to the receiver so that the demodulation processes for these modulation techniques can take place.

Subsequent to carrier and code demodulation, the processing consists of using the ranging code to find the range from the satellites. These codes are known to the receiver a priori, in terms of the code parameters such as the code sequence, code length, code rate, and so forth. In addition, if any encryption is made to the code, the encryption details including the algorithms and the key, plus other resources to decrypt it, need to be available with the receiver. The range of the satellite is then derived using this code. The carrier and code being removed, now the remaining navigation data are specifically identified from their predefined structure. These parameters are used in turn to find the range, satellite position and finally the user position.

Navigation satellites transmit navigation signals in allotted bands with a given Equivalent Isotropic Radiated Power (EIRP). Yet, a low signal level is expected over the targeted region after it traverses through the intermediate path. In addition, every navigation signal experiences interference from signals in the same bands from other satellites or from the multipath delayed signal from the same satellite. Both effects make the received signal to noise ratio (SNR) poor. This interference levels need to be minimized with appropriate modulation and through proper design. (Global Positioning System Directorate, 2012 a & b).

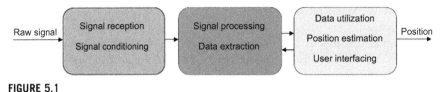

FIGURE 5.1

Functional segments of a generic receiver.

Thus, the basic architecture of the receiver has three main sections composed of

1. Signal reception and conditioning
2. Signal processing and data extraction
3. Data utilization and position estimation

The schematic of a generic receiver is shown in the Figure 5.1.

5.1.2 Types of user receiver

We start our discussion on navigation receivers by classifying different receivers. However, it is not possible to categorically segregate the receivers in a strict sense because the same receiver may fall into different categories according to different perspectives. The basis by which they are categorized may also have identical sense. For instance, a "carrier phase-based ranging receiver" can also be seen as a "precise receiver." These classifications, although not standard, provide an understanding of the divergent features of receivers. The basic categories are as follows.

5.1.2.1 Access technique
Navigation receivers may be classified according to different bases. Because different navigation systems are designated for different access techniques, from the systems point of view, receivers can be divided on this basis into the following.

5.1.2.1.1 CDMA- receivers
CDMA receivers operate in conjunction with a CDMA system in which the satellites transmit the same carrier frequency but with different ranging codes. Here, the combined signal channel is simple and the front-end bandwidth of the receivers is comparatively small. However, this is at the cost of identifying the exact code present in the signal, which thus results in extra processing load and time.

5.1.2.1.2 FDMA receivers
FDMA receivers operate with an FDMA system in which different satellites transmit at different carrier frequencies but with the same ranging codes. Here, the combined signal channel is comparatively of wider bandwidth that accommodate all available signals. Consequently, receiver is required to have a wider band of operation at its input but, unlike CDMA receivers, less overhead time needs to be spent to identify the separate carriers. However, the system front-end bandwidth may be reduced by properly using the frequency reuse plan, as we saw in the previous chapter.

5.1.2.2 Ranging technique

One important task to be accomplished at the receiver is measurement of the satellite's range. Previously, we mentioned that range is calculated from the difference of transmission and reception time. Depending upon which of the signal parameters (code phase) or carrier phase is used for ranging, the receivers may of the following types.

5.1.2.2.1 Code phase-based ranging receivers

When the range is measured in a receiver on the basis of the propagation time of a particular code phase from the satellite to the receiver, it is designated a code phase based ranging receiver. The propagation time is the difference between the receiving time and the transmission time of any definite code phase of the signal. Thus, it is the difference between the current time and the transmission time of the current code phase being received. The ranges measured through this process are noisy and lead to errors in position estimates.

5.1.2.2.2 Carrier phase-based ranging receivers

The range can also be measured by deriving the propagation time of a particular phase of the carrier. Receivers primarily using this mode of measurement to measure the range are referred to as carrier phase-based ranging receivers. Here, the receiver needs measurements of the phase of the carrier signal received, which can be subsequently aided by measurements of the ranging code. This is typically used by receivers needed for precise applications. It adds to the complexity of the receiver and increases computational load, making the receivers costly.

5.1.2.3 Precision

Receivers can also be classified on the basis of the precision they offer. It determines the applications for which they can be used. Although the precision of a receiver is primary determined by the kind of service for which it is meant, there are other factors in receiver construction that help to achieve the required precision. These factors may imply improvements in terms of both computation and hardware used. Thus, on this basis, the receivers can be of similar categories.

5.1.2.3.1 General purpose (standard) receivers

These are receivers that give standard precision in determining the position of the user. They are meant for standard services and are relatively smaller in size and cheaper in cost. This makes these receivers popular for general purpose navigational use.

5.1.2.3.2 Special purpose (precise) receivers

These receivers are used for precision applications such as surveying and other geodetic purposes, and also may be used for strategic applications. Although meant for restricted users, these can use a precision service with decryption capability in order to achieve the goal, the rest depends on relative positioning, processing of the carrier phase of the signal, and many other similar features to improve performance. Such precise receivers also differ by the type of hardware units they use, such as an oven-controlled crystal oscillator or atomic clock, as demanded by the applications for which they are used. There are also primary and differential receivers based upon the service type the receiver is attached to. But we deliberately avoid discussion of this here until we have described the differential positioning in Chapter 8.

5.1.3 Measurements, processing and estimations

The final objective of the receiver is to fix its position. Hence, different requisite parameters are needed to be derived from the signal by the receiver itself. These are done through two major activities in the receiver, viz., *reference positioning* and *ranging*. The following subsections will describe these activities, including all of the related processing necessary to achieve them.

5.1.3.1 Reference positioning

Reference positioning is the process of locating the exact position of navigation satellites used as references. For this, the satellites transmit ephemeredes repeatedly in the signal. Ephemeredes are the Keplerian parameters required for satellite position estimation. We already learned in Chapters 2 that these Keplerian parameters are estimated at the ground segment and are uploaded to satellites, which in turn transmit them to users. It was also clear from Chapter 3 that a minimum of six parameters are required for estimation of the satellite position. There are additional parameters in the signal that can take care of deviations resulting from the perturbations.

The true satellite orbit is different from its designated orbit and it changes with time, resulting in an effective change in these parameters. The most recent ephemeris information, along with the pertinent perturbation parameters, is updated by the satellite in its message, which remains embedded in the navigation signal. Using these, the receiver can calculate the satellite position at any instant, following the equations and methods described in Chapter 3.

5.1.3.2 Ranging

Ranging is a measure of the distance of the user from each reference satellite and is necessary to estimate user position. The distance, i.e. the range, is obtained from the total time taken by the signal to travel from the satellite to the user. Thus, the expression for the measured range is

$$R = c(t_2 - t_1) \qquad (5.1)$$

Here, t_1 is the time of transmission of a definite phase of the signal, t_2 is the time of reception of the same phase at the receiver, and R is the measured range. 'c' is the velocity of light in vacuum. The assumption underlying this method of calculating range is that waves travel at a constant velocity. However, certain errors creep in when assuming this, which we will address in the next chapter. The range is thus derived from the difference of the current time t_2 and the transmission time t_1 of the current received phase. The current time, i.e. the receiving time of the current phase, is obtained from a synchronized clock at the receiver, whereas its transmission time is derived from measuring the current code phase of the received signal and the time stamp on the signal. We will read about these methods in detail later in this chapter, where we will learn how time t_2 and t_1 are identified in a receiver in connection with the appropriate module carrying out the process.

5.1.3.3 Signal processing

It is clear that both prerequisite estimations we have just discussed need to be derived from the information laden signal. Thus, the first basic requirement for this is the signal acquisition.

To acquire the signal and for subsequent information derivation, some preliminary signal processing needs to be done. We now discuss the theoretical essentials of such signal processing.

The combined modulated signal at the input of the receiver manifests itself as a variation of the phase of an electric field. This variation is converted to an equivalent analog variation in electric current or voltage in the receiver, and finally into binary numbers representing the digital levels that represent the values of the parameters. For a normal navigation signal, all information is encoded in the phase variation of the signal. Therefore, first the receiver must follow the phase variation of the incoming signal from these received levels.

This is done by comparing the carrier phase variation of the incoming signal with a local version of the carrier, generated as a reference in the receiver at the same frequency. Then, comparing the phase of the input signal with this reference the difference is determined. This difference is corrected and thus they are aligned in phase. Then any further phase variation in the incoming signal is determined from the subsequent differences generated.

To estimate the relative phase difference and its variation, the frequency of the local reference signal needs to be the same as the incoming one. The frequency of the incoming signal, however, does not strictly remain the same as the transmission frequency owing to the Doppler effects that occur when there is relative velocity between the receiver and the satellite. So, in addition to the initial phase, the initial frequency deviation resulting from the Doppler also needs to be estimated. However, once the phase is set between the two, any small variations in the signal frequency are readily followed by the local reference signal by following its instantaneous phase. Because the width of the encoded data chip also varies proportionately with Doppler, these values obtained by the receiver at the carrier level are also equally useful for proper data demodulation.

Many associated signal processing activities must be done on the signal before ranging and reference positioning. These include analog to digital conversion of the received signal; signal acquisition and tracking; and signal demodulation. However, we will discuss them at appropriate points when describing the different modules of the receiver.

Before we go into more detail, we will discuss some very basic mathematical concepts. Here, we will treat the signals as vectors and analyze them; this will help us understand the total process.

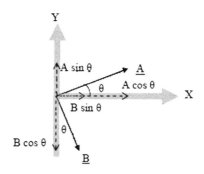

FIGURE 5.2

Vectorial representation of signals.

Let us suppose two orthogonal reference axes are X and Y. Two mutually orthogonal vectors A and B are present here, where A makes a relative angle θ with X, and hence the same angle is made by B with Y. This is shown in Figure 5.2.

So, the component projection of A and B on X, i.e. the dot product of $A + B$ with X, is

$$C_X = A \cos \theta + B \sin \theta \tag{5.2a}$$

and the component projection of A and B on Y, i.e. the dot product of $A + B$ with Y, is

$$C_y = A \sin \theta - B \cos \theta \tag{5.2b}$$

The resultant vector remains invariant as $C_X^2 + C_y^2 = A^2 + B^2$. Again, if $B = 0$, then $C_y/C_X = \tan \theta$ and $C_X^2 + C_y^2 = A^2$.

Now, consider two sinusoidal waves, $S_c = a \cos(\omega t + \phi)$ and $S_s = b \sin(\omega t + \phi) = b \cos\{(\omega t + \phi) - \pi/2\}$. These two waves are mutually orthogonal because they differ by a phase of $\pi/2$. This is ensured by the fact that their inner product is zero. The dot product of two functions is their inner product and is given by

$$\langle S_c, S_s \rangle = 1/T \int_0^{2\pi} S_c(t) S_s(t) dt$$

$$= ab/2T \int_0^{2\pi} \sin(2\omega t + 2\phi) dt \tag{5.3}$$

$$= 0$$

These sinusoidal waves may also be considered two vectors such as A and B. So, when the phase difference between $S_c(t)$ and $S_s(t)$ is maintained at $\pi/2$, they behave like two orthogonal vectors and similarly have zero as their dot product.

Unitary signal $\cos(\omega t)$ and $-\sin(\omega t) = \cos(\omega t + \pi)$ constitute the references. These references are similar to X and Y axes, respectively, and can be combined and represented as $X + jY$. When these signals are multiplied with orthogonal references $X = \cos(\omega t)$ and $Y = -\sin(\omega t)$, the product thus formed is

$$(S_c + S_s) * (X + jY)$$
$$= \{a \cos(\omega t + \varphi) + b \sin(\omega t + \varphi)\}\{\cos(\omega t) - j \sin(\omega t)\}$$
$$= a/2[\{\cos(2\omega t + \varphi) + \cos(\varphi)\} - j[\sin(2\omega t + \varphi) - \sin(\varphi)]$$
$$+b/2[\sin(2\omega t + \varphi) + \sin(\varphi)] - j[-\cos(2\omega t + \varphi) + \cos(\varphi)] \qquad (5.4)$$

So, upon low pass filtering, the components that survive are

$$C = C_i + j\,C_q = [(a/2)\cos(\varphi) + (b/2)\sin(\varphi)] + j[(a/2)\sin(\varphi) - (b/2)\cos(\varphi)] \qquad (5.5)$$

This equation resembles Eqn (5.2a,b), considering that the relative phase angle φ between the considered signals and the references in Eqn (5.5) is equivalent to the angle θ in Eqn (5.2) between vectors. Thus, φ carries the same significance for the signals as θ carries for the vectors in Eqn (5.2a,b).

So, comparing the equivalent equations, the operations of multiplication of orthogonal reference sinusoids with the signals and subsequent low pass filtering may be looked on as an equivalent vector operation of resolving the vectors into the reference axes.

Now, let us derive some special cases out of the general equation we had in Eqn (5.5). Notice that if only the in-phase component S_c exists in the signal, i.e. $b = 0$, and then the products on the cosine component and sine component would respectively be

$$C_i = [(a/2)\cos(\varphi)] \qquad (5.6a)$$
$$C_q = [(a/2)\sin(\varphi)] \qquad (5.6b)$$

It follows from this that the ratio of these components becomes

$$C_q/C_i = \tan(\varphi) \qquad (5.6c)$$

So, the ratio would give us the phase difference between the reference and the signal. Again, if these components are squared and added, then

$$R = \left(C_i^2 + C_q^2\right)^{1/2} = a/2 \qquad (5.6d)$$

So, this will give us the modulus of amplitude of the signal. The same arguments are valid if there is only the S_s component of the signal. The process of multiplying a signal by a reference and subsequently low pass filtering it is called mixing. This operation will be used many times in receiver signal processing.

5.1.4 Noise in a receiver

At this point, it is important to introduce the concept of noise. When the signal passes through different units of the receiver, some unwanted and uncorrelated random electrical variations of spurious nature are added to it and are referred to as *noise*. This results in an error in identifying the correct signal and deriving the correct value of parameters from it as well. Eventually, it affects the measurements done in the receiver. This is analogous to the condition in which we want to hear someone delivering a speech in a crowded room with lots of people talking among themselves. We concentrate on the speaker while the crosstalk creates noise in the background, making it difficult to realize what he or she is actually saying. The more people talk across, the more the noise there is, making understanding the speaker more difficult. At some point it becomes unintelligible.

As the signal propagates through the medium in the form of electromagnetic waves, other random electromagnetic emissions by natural or manmade sources are within the band of the signal. These get added to the actual signal and are picked up as noise when received. Similarly, when the signal is received at the receiver and is transformed into electric parameters of voltage, current, and so forth, in the receiver, the concerned hardware, which is at definite temperatures, also generates similar electrical random variations. These nuisance components adhere to the available voltage signals and add to the noise.

All of these unwanted components are additive and random in nature. So, their value at any certain instant has no relation to the values at any other instant. When these noise values are added up over a considerable time, the sum becomes zero. This means that the noise has a zero mean. However, the square of these values when averaged over time gives the variance σ^2 of the noise and is the index of noise intensity. Furthermore, they generally follow a definite statistics of amplitude probability distribution that is Gaussian about this zero mean, as shown in Figure 5.3(a).

These kinds of noise are spread across all space and at all time. But how much should we consider to be getting added up with our signal? This cannot be done from the time domain analyses of the noise and we need to see its spectral distribution; i.e. we need to see how the noise is distributed across all frequencies.

If we see the frequency spectrum of this noise, we find that within the band of interest, it has equal components across frequency hence it is called white noise. Consequently, a simple model for the thermal noise assumes that its power spectral density Gn(f) is flat for all frequencies, as shown in Figure 5.3(b), and is denoted by a one-sided spectrum as (Maral and Bosquet, 2006)

$$Gn(f) = N_0 \qquad (5.7)$$

So, this thermal noise is additive white, zero mean and Gaussian noise.

We have seen that noise is present at all frequencies, and being white, the spectral amplitude in all frequencies is the same within the band of interest. The baseband signals are distributed about the zero frequency for which the baseband noise is useful. But noise addition takes place mainly when the signal is in a modulated form, and hence bandpass in nature. Thus, under such conditions, the bandpass form of the noise is convenient to deal with. The random noise can be expressed in a bandpass form as (Lathi, 1984)

$$n(t) = n_c(t)\cos(2\pi f_c t) + n_s(t)\sin(2\pi f_c t) \qquad (5.8a)$$

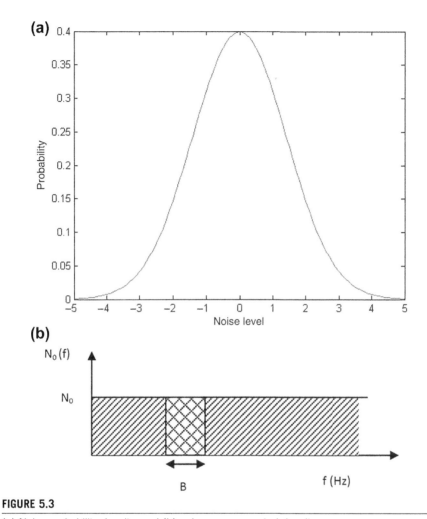

FIGURE 5.3

(a) Noise probability density, and (b) noise power spectral density.

where n_s and n_c are the amplitudes of low-frequency noise components about the DC and bandlimited to f_m when $2f_m$ is the modulated signal bandwidth and f_c is the carrier frequency of the RF signal. Therefore, this expression states that the noise in the bandwidth of the signal is equal to the noise in low pass components of the noise of equal width, shifted to the mentioned band. This is the bandpass representation of the noise where n_c and n_s are the cosine and sine components of the noise amplitude such that the mean noise power in each component is the same, i.e. $\bar{n}_c^2 = \bar{n}_s^2 = \bar{n}^2$. This noise can also be represented by the vectorial sum and the corresponding phase as

$$n(t) = n_r \cos(2\pi ft + \varphi) \tag{5.8b}$$

where $n_r = (n_c{}^2 + n_s{}^2)^{\frac{1}{2}}$ and $\tan \varphi = (n_s/n_c)$. Because both n_c and n_s are slowly varying random components, both $|n_r|$ and φ vary slowly but in a random fashion.

From Parseval's theorem, we know that for power signals the spectral power density at any frequency is proportional to the square of the spectral amplitude, and it is the same across all frequencies for noise. So, the noise power spectral density may be represented by a constant, which we have already seen in Eqn (5.7). It is important to see how we are actually representing the signal. If the signal is represented in a two-sided spectrum, where both positive and negative frequencies are considered, the noise should also be represented likewise with two-sided density $N_0/2$. This is done mainly for radiofrequency (RF) when the signal remains modulated. Once the signals are filtered in a receiver and heterodyned down to baseband, they are conveniently represented in a single-sided spectrum, and hence the noise there should be also represented as a single-sided density of N_0. So, the signal received by a receiver with a bandwidth B has the noise power given by N_0B.

Finally, because the noise is generated from the thermal condition of a body, it can be represented by a linear function of its temperature T as

$$N_0 = kT \qquad (5.9)$$

where k is the proportionality constant called the Boltzmann constant and T is called the noise temperature, which may not be the actual physical temperature of the body. With this introductory note on noise, we return to our main topic and describe the building elements of the receiver.

5.2 Functional units of user receivers

5.2.1 Typical architectures

The architecture of a navigation receiver is typically made up of three distinct functional modules, defined by the objectives mentioned in Section 5.1.1. To achieve these mentioned objectives, each module is again made up of different functional components, which remain more or less the same within a typical receiver, giving an end-to-end solution. These functionalities are:

- Receive RF signal from different satellites
- Further digital processing of the signal
- Acquisition of the desired signal
- Removal of the Doppler, carrier demodulation, and code removal
- Perform measurements of signal transit time (ranging)
- Decode the navigation message to determine satellite position, etc
- Derive other relevant information
- Estimate the position, velocity, and time (PVT)
- Display PVT and provide a suitable interface

The first functional component in a receiver is the analog RF section. This is responsible for picking up the raw electromagnetic signal from the medium through a passive or active antenna and subsequent analog signal conditioning on the resultant electrical parameter. Signal conditioning includes amplification, filtering, and frequency translation, and finally analog-to-digital conversion. The output is the stream of digital binary samples of the signal at an intermediate frequency (IF).

The following section is responsible for high-speed digital signal processing of the data. The discrete digital samples are worked on to demodulate the signal from IF to baseband, find the relative code delay of the signal to find the range, and also wipe the code off to output the stream of navigation data bit.

The last component is a processor-based computational unit responsible for deriving the required navigation parameters and finally fixing position. It also provides feedback to its previous units, the parameters of which are required for their operation. Sometimes this processor looks after the proper display of the data and overall management of the receiver.

In this section, we will describe in detail the different functional units of the receiver with the working principle and the condition of the signal as it passes through them. These functional modules are the building blocks of the receiver to accomplish its final objectives. At every stage, we will also describe how the noise varies and affects the signal at each of these generic modules of the receiver.

5.2.2 RF interface

5.2.2.1 Antenna

The signals transmitted by the satellites are the continuous flow of electromagnetic energy. The first task of the receiver is to receive this energy and convert it into electric parameters of current and voltages that can be used to derive information. This is done by the antenna.

The antenna design is such that it is able to receive the signal over the entire signal bandwidth. It should also have the required sensitivity and gain at the required center frequencies where the signals are present and the required polarization appropriate for the signal. Such an antenna, designed in accordance with the preassigned frequency and polarization and with requisite gain pattern, prevents unwanted components from entering into it. For accommodating signals in multiple frequencies, it may use a single antenna with a large bandwidth or may resort to multiple antennas at different bands when the difference between the frequencies is too large for it.

The antenna also needs to pick up a signal from any part of the sky and for any orientation of the receiver. Therefore, it has to retain the same gain over a large angular range about its bore sight. Consequently, it has low absolute gain across it owing to the low directivity of the antenna. The signal thus remains embedded below the noise when it is reduced by the antenna while it is spread over a wide spectrum.

The atmosphere prevailing in the path of the propagation of the signal also emits energy in an electromagnetic form, creating noise. The noise thus created, unlike the signal, consists of radiation that is incoherent in nature. Noise is radiated at all frequencies, and those in the passband of the receiver are picked up and collected incoherently by the antenna of the receiver. This is called atmospheric noise or sky noise.

As the signal is transmitted to the next unit, i.e. the Low Noise Amplifier (LNA), some noise is added during the process, depending on the loss of the connector. At the same time, the sky noise and the signal are attenuated while they pass through this connector. This composite noise and the received signal are further treated in the next section of the receiver, i.e. the LNA.

5.2.2.2 LNA
As we have mentioned in terms of its reception mode, the navigation receiver is no different from any digital receiver. Its front end is also identical to any typical receiver. This RF section treats the navigation signal the same way a communication signal is treated at the receiver. The received power of the navigation signals is so low, it remains embedded within the noise and the antenna cannot provide additional gain to the signal level. Hence, it is necessary to see that the signal does not get degraded further.

The signal always gets added with the noise as it proceeds through the hardware of different sections of the receiver. So, if the received signal is amplified at the beginning to a large extent without the further addition of noise, the effect of any addition of noise by a subsequent receiver remains proportionately small and hence does not affect the signal much. However, any subsequent amplification amplifies the signal and the noise equally, and hence does not improve the SNR, either. Thus, the SNR obtained at the input of the first amplifier is approximately maintained at all subsequent stages. It is important to see that the first amplifier itself does not add too much noise or it could mar the objective of the amplification. Therefore, a low noise amplifier is used.

The purpose of the LNA is to amplify the RF input signal, adding minimum possible noise. It has a filter preceding it that rejects the out-of-band frequency to reduce the incorporation of unwanted noise. This filter needs to have a large band-width because the data are spread owing to its multiplication with the ranging code. The filter followed by the LNA, in combination, called a preamplifier, is typically placed near the antenna to reduce the effects of noise introduced by the connector and filter present in between.

Let S be the signal power at the input of the LNA and N be the associated noise power, making the SNR as S/N. Now, any pragmatic amplifier will add some noise itself in addition to amplifying the noise already present at its input. This noisy amplifier may be equivalently regarded as a noise-free amplifier with an additional noise at its input so that the output noise after amplification is equal to the combined contribution. The added noise of the amplifier is quantitatively represented by its noise figure F. F is the ratio of the total noise power at the output of the amplifier owing to the combined contribution of the input and the amplifier to the amount of noise present at the output explicitly owing to the input noise only. So, if the input noise is N_0 and the noise added by the amplifier with gain A is N_A, the noise figure of the amplifier is given by (Maral and Bousquet, 2006)

$$F = (N_0 A + N_A)/(N_0 A) \qquad (5.10a)$$

The noise contribution of the amplifier N_A may be equivalently regarded as input noise N_a amplified by the amplifier, such that $N_A = A\, N_a$

$$F = (N_0 + N_a)A/(N_0A)$$
$$= (N_0 + N_a)/N_0$$
$$= 1 + N_a/N_0$$

or

$$F = 1 + T_a/T_0 \qquad (5.10b)$$

where we have used the relation $N = kT$. To set a common reference for all amplifiers, the noise figure F is defined when the amplifier input noise temperature is at the predefined standard value of 290 K. So, T_0 is set as 290 K. It follows from this that the noise contribution of an amplifier equivalently converted as its input noise is

$$T_a - (F - 1)T_0 \qquad (5.10c)$$

So, the total effective noise temperature at the input of the amplifier due to itself is $(F - 1)T_0$, making the corresponding noise $k(F - 1)T_0$. This makes the total effective SNR at the amplifier input, considering the noise added by the amplifier, as

$$S/[kT_i + (F - 1)kT_0] \qquad (5.11)$$

where T_i is the noise temperature at the input of the amplifier because of T_o the noise already present in the signal.

Extending this idea to all subsequent amplifiers in the chain, the effect of the noise added by the next amplifier at its own input is $k(F_1 - 1)T_0$. But to bring all noise contributions to the same reference, we calculate its equivalent noise at the input of the LNA. The equivalent noise owing to the first amplifier following the LNA with noise figure F1 and amplification A_1 at the input of this LNA is $(F_1 - 1)kT_0/A$. So, we see that the effect of the noise added by the next amplifier is drastically reduced by the factor of the LNA gain compared with the signal. Similarly, that of the next amplifier with F_2 and A_2 as the respective noise figure and gain will be $(F_2 - 1)kT_0/(A_1A)$. The effect thus becomes less conspicuous for amplifiers at later stages. Considering all such contributions, when the noise component in the signal at the input of the LNA is N_i, the effective noise becomes

$$N_{eff} = N_i + (F - 1)kT_0 + (F_1 - 1)kT_0/A + (F_2 - 1)kT_0/A_1A$$
$$+ (F_3 - 1)kT_0/A_2A_1A + ... \qquad (5.12)$$

5.2.3 **Front end**

The RF component of the received signal received by the antenna system and passing through the preamplifier proceeds further through the receiver. In this part, the signal is modified and made ready for processing. The activities are commonly known as signal conditioning and are done at the front end of the receiver. It consists of down-converting the signal to IF, filtering, sampling and converting into digital data. All of these steps in this section will be discussed in sequence. However, for

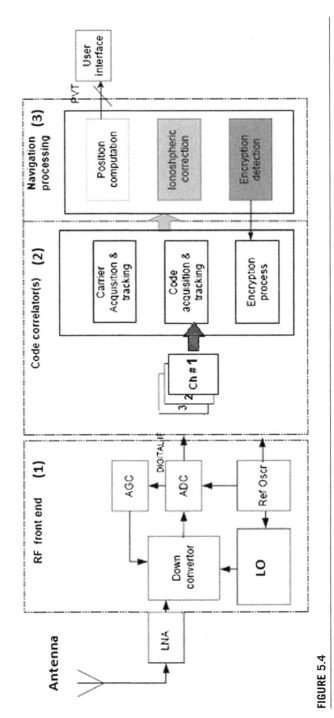

FIGURE 5.4

Architecture of a typical receiver.

all such purposes, we need a timing reference, and hence a timing device is required. It is important to have a precise timing source in navigation, and therefore we will start our discussion with it. The architecture of a typical navigation receiver front end is shown in Figure 5.4.

5.2.3.1 Timing devices

The source of all timing devices inside the receiver is the reference oscillator. It provides time and frequency reference for the receiver. All local signals generated at the receiver are derived from either the local oscillator (LO) if it is analog or from a clock if it is digital. Both the LO and the clock are driven by this reference oscillator.

Because this serves all timing requirements of the receiver, the performance of the receiver depends on its precision. In addition to the short- and long-term stability, the phase noise, power and size of the reference oscillator are also important aspects determining the receiver characteristics. The stability of this reference affects all the time and the frequency units derived from it.

A standard receiver may use a quartz crystal oscillator as the frequency standard. These oscillators are sensitive to the temperature and may vary by $10^{-5}-10^{-6}$ over a typical operating temperature range (Grewal, Weill and Andrews, 2001). Therefore, the frequency excursion must be confined, mainly through temperature control. A temperature-controlled crystal oscillator is typically used for the purpose. A more effective means is to use an oven-controlled crystal oscillator. However, it is commonly restricted to more precise and larger receivers owing to the cost and size.

Where size and cost are not constraints, precise atomic clocks can be used. This is common for the receivers meant to monitor the space segment or for precise applications such as surveying. In addition, recently, chip-scale atomic clocks are being used for the purpose. These have weights of a few tenths of grams and perform much better than crystal oscillators and with much less power consumption.

5.2.3.2 IF down-conversion

The signal received from the antenna is first filtered and amplified at the preamplifier. Then, it is converted to a convenient lower frequency for further filtering and amplification before digitization. This is called the down-converted approach. This allows the receiver to use a lower sampling rate for further processing, which eases the hardware requirement and reduces the processing load. This lowered frequency of suitable value, which is a few orders below the carrier frequency, is called the IF, and the process of reducing the frequency is referred to as down-conversion.

However, because digital signal processing is becoming increasingly powerful, RF sampling or direct digitization is also becoming popular in which digitization and further processing occurs in the RF state only after some limited filtering amplification.

Conventional down-conversion of carrier frequency f_c of the signal to the IF at f_{IF} is conveniently achieved by heterodyning, which is achieved by mixing two separate signals of different frequencies and selecting the lower frequency component of the resultant product. Here, the output of a local oscillator is mixed with the incoming signal.

Mixing, i.e. the multiplication of two signals is implemented by passing their sum through a nonlinear device. The product yields two components, one of sum frequency and the other of difference frequency. The LPF carried out subsequently removes the sum frequency component, allowing the signal with the difference frequency to pass on. Let us consider the input signal to be represented in the phase quadrature form as

$$s(t) = a_i \cos(2\pi f_c t) + a_q \sin(2\pi f_c t)$$
$$= a_r \cos(2\pi f_c t - \varphi) \quad (5.13)$$

where $a_r = \sqrt{a_i^2 + a_q^2}$ is the resultant amplitude and φ is the phase with respect to the cosine reference, given by $\tan \varphi = a_q / a_i$.

In a receiver, down-conversion is carried out by first mixing the RF signal with the frequency of the local oscillator, f_{LO}. This frequency is kept as the difference between the carrier and the required IF frequency, i.e. $f_{LO} = fc - f_{IF}$. So, when the input is multiplied by this signal, we obtain the product as

$$s(t)\cos(2\pi f_{LO}t) = a_i \cos(2\pi f_c t)\cos(2\pi f_{LO}t) + a_q \sin(2\pi f_c t)\cos(2\pi f_{LO}t)$$
$$= a_i/2\left[\cos\{2\pi(f_c - f_{LO})t\} + \cos\{2\pi(f_c + f_{LO})t\}\right]$$
$$+ a_q/2\left[\sin\{2\pi(f_c - f_{LO})t\} + \sin\{2\pi(f_c + f_{LO})t\}\right] \quad (5.14a)$$

Thus, upon multiplying, the resultant beat frequencies produced are at $f_c - f_{LO}$ and $f_c + f_{LO}$, respectively. This signal is then passed through a bandpass filter. The lower frequency of the two, i.e. $f_c - f_{LO}$, remains, whereas the higher frequency components are eliminated as a result. Thus, at the output of the filter, we get

$$s_{IF}(t) = a_i/2\left[\cos\{2\pi(f_c - f_{LO}t\}\right] + a_q/2[\sin\{2\pi(f_c - f_{LO}t)\}]$$
$$= a_i/2[\cos\{2\pi f_{IF}t\}] + a_q/2[\sin\{2\pi f_{IF}t\}] \quad (5.15)$$

The above Eqn (5.15) shows that the down-conversion only reduces the frequency from the RF to the IF, keeping the other features of the signal unaltered. Furthermore, this bandpass filtering limits the out-of-band noise from entering the receiver. In the IF, besides the convenience of signal processing, it is possible to filter out-of-band noise with a sufficient roll-off. The schematic for the IF down-conversion is shown in Figure 5.5.

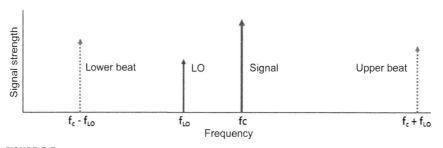

FIGURE 5.5

Signal, LO, upper beat, and lower beat frequencies.

Any signal present at the frequency $f_{LO} - f_{IF} = f_c - 2f_{IF}$ also generates a component at the frequency $f_c - f_{LO}$, if it is multiplied by the LO output. This signal is called the image because this frequency is located at an equal distance from the f_{LO} frequency as the actual signal, but at the opposite side. However, the signal at the image frequency is not our desired signal and should be restricted to contributing to the IF. Thus, care must be taken to remove the image frequency that adds to noise. This is done at the level of the preamplifier even before mixing through proper filtering. Since the frequency separation between the true signal and its image is $2f_{IF}$, a higher value of IF separates them wider and improves the rejection of the image. Conversion to the IF is sometimes done in two successive stages for better image rejection. Another important point is that if any Doppler frequency is present in the RF signal, the same shift is retained in the IF as well. Figure 5.5 shows the position of the signal carrier, LO frequency, image, and beat frequencies along the spectral line in relation to the down-conversion.

Considering the noise components, when the signal from the LO is mixed with the signal, the product with the accompanying noise generates the following:

$$n_{LO} = n_c \cos(2\pi f_c t)\cos(2\pi f_{LO} t) + n_s \sin(2\pi f_c t)\cos(2\pi f_{LO} t) \qquad (5.16)$$

Upon low pass filtering, the components thus generated are

$$n_{IF} = (n_c/2)\cos(2\pi f_{IF} t) + (n_s/2)\sin(2\pi f_{IF} t) \qquad (5.17)$$

So, the effect of the IF down-conversion on the noise is that it produces similar bandpass noise at the IF. Considering the noise before and after mixing, we find the signal to the noise ratio remains unaltered as a result of this process.

Another technique of down-conversion is to achieve it during the sampling. This is called the bandpass sampling (BPS). Here, the sample rate is chosen so that the sampled signal has the signal spectrum in the predefined frequency range. We will discuss this method of BPS when we describe the sampling process of the signal.

5.2.3.3 Analog to digital converter

The signal obtained until now is in analog form with continuous variation over time. It has a bandwidth referred to as the *precorrelation* bandwidth. The analog values of the signal are available at the input of the *analog to digital converter* (ADC), which converts this continuous signal into a discrete form. The analog values are first selected at definite discrete intervals and rejected at all intermediate times. This process is called sampling; the selected values form the samples of the analog signal. These discrete sample values are first characterized and then their levels are converted to corresponding digital numbers through binary coding. The process of sampling and quantization is illustrated in Figure 5.6.

5.2.3.4 Sampling

Today's receivers are digital in nature and can work only on discrete digital values. To enable processing in the digital domain, the continuous signals are sampled generally in the IF before they are converted to base band.

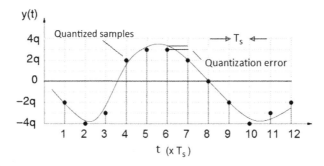

FIGURE 5.6

Sampling and quantization.

Sampling is measuring the amplitude of the continuous signal at discrete regular intervals. Thus, it collects signal information only at these discrete instances and ignore all for the instances in between. However, from sampling theory, it is known that intermediate data and hence the whole signal may be recovered from the sampled data if the sampling is done at a sufficiently fast rate.

The sampling can be done directly at the RF frequency without the need to reduce it to IF, as we mentioned in the previous section. Direct digitization has a major advantage in that, in this type of design, the mixer and local oscillator are not needed. A mixer can generate unwanted spurious signals in nearby frequencies, which can contaminate the output and deteriorate the estimation. A local oscillator can be expensive and any frequency error or impurity produced by the local oscillator will appear in the digitized signal. However, this arrangement does not eliminate the oscillator (or clock) used for Analog to Digital Converter (ADC). The major disadvantage of direct digitization is that the ADC must have a high input bandwidth to accommodate the input frequency of the carrier and a higher sampling rate, unless bandpass sampling is used. In this approach, the sampling frequency must be kept very accurate which is sometimes difficult to achieve practically.

For the preferred IF sampling, the down-converted signal is appropriately amplified to a workable level. The sampling has to be done so that all information in the signal remains contained in the samples and the original signal can be recreated back from it. The important question is, at what rate the signals should be sampled to preserve information? This is governed by criterion set by the Nyquist sampling theorem.

The *Nyquist sampling theorem* states that a band-limited analog signal can be perfectly reconstructed from the complete sequence of its samples if the sampling rate exceeds twice the highest frequency contained in the original signal.

In our case, the sampling is done after down-converting them to IF. It somehow eases the case by eliminating the need for rapid sampling and corresponding processing.

The encoded data is evenly spread on two sides about the IF frequency, f_{IF}. Therefore, from the theorem, the ADC sampling rate must exceed twice the sum of the IF frequency and the signal single-sided bandwidth, to prevent an aliasing effect. The chip period of the encoded data being Tc, the maximum frequency contained in the signal is $f_{IF} + 1/T_c$. This makes the sampling rate $2f_{IF} + 2B$, where $B = 1/T_c$. At the same time, to prevent the spectral fold over, which distorts the signal, the minumum value of the f_{IF} that may be chosen must exceed the single-sided precorrelation bandwidth, i.e. $f_{IF} \geq B$. Thus, combining the two, we see that the minumum sampling rate can be determined from the relation $f_s \geq 2(2B)$. It is necessary to keep the sampling rate more than twice the null to null signal precorrelation bandwidth.

This can also be easily understood from a spectral view of the process. Sampling is multiplying the time variation of the signal by unit impulses located at regular time intervals of sampling, T_s. This train of unit impulses, when transformed to the spectral domain, becomes a train of impulses at interval of $1/T_s = f_s$. So, the spectrum of the sampling impulses is lines at DC and at distances that are integral multiples of f_s (Lathi, 1984). The process of multiplication in the time domain results in convolution in the spectral domain. Convolving the signal spectrum with this impulse train results in an array of the replica of the same signal spectrum with the origin shifted to each of the impulse locations at nf_s but reduced in level. This is shown in Figure 5.7.

This array is passed through a bandpass filter to recover the signal. For the condition in which no two adjacent replica overlap, i.e. no aliasing occurs, separation between the two adjacent impulses of the train must accommodate two single-sided bandwidths of the signal. So,

$$f_s \geq 2(f_{IF} + B)$$
$$\text{Again} \quad f_{IF} \geq B \qquad\qquad (5.18)$$
$$\text{So,} \quad f_s \geq 4B$$

For most compact forms of down-conversion and sampling, avoiding aliasing or spectral fold over, the optimal sampling rate is four times the one-sided bandwidth of the signal. However, this is only a special case with the lowest IF frequencies; for all other f_{IF} values, the required sampling rate is higher.

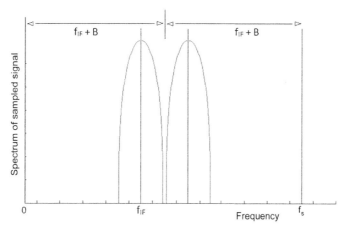

FIGURE 5.7

Spectrum of sampled signal.

According to this, as we have seen, had it been the RF signal, the required condition would be $f_s > 2(f_c + B)$, f_c being the RF carrier frequency. Then, the sampling rate would have become enormously high and difficult to handle. A modulated navigation signal is essentially a bandpass signal with finite bandwidth spread about its center frequency. We can take the advantage of this finite spectral extent to sample the signal at RF directly, but at a reduced rate. In other words, we can reduce the sampling rate using the vacant nature of the spectrum in Figure 5.7 between $f = 0$ and $f = f_c - B$.

Suppose we sample at a rate f_s, such that $nfs = f_c - B$. In this case, the spectral lines of the sampling signal occur at frequencies 0 and multiples of f_s on both positive and negative sides, whereas the signal is around f_c. Figure 5.8(a) illustrates the condition for convenience.

Therefore, from the signal band center, the two adjacent spectral impulses are at distances B on one side and $f_s - B$ on the other. Upon convolution, this will produce a repetitions of the signal spectrum within the interval of every two successive lines of the sampling spectrum. This is shown in Figure 5.8(b).

Now, unlike the previous case, to accommodate both the signal spectrum between the two successive lines and simultaneously to avoid aliasing, the separation between the two must accommodate twice the total bandwidth 2B. So,

$$\text{or}\quad f_s > 4B \tag{5.19}$$

(a)

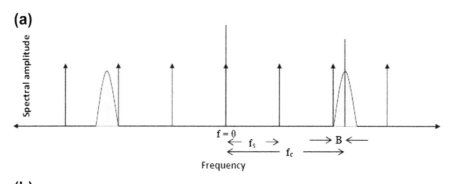

(b)

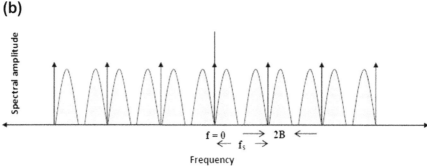

FIGURE 5.8

Spectral condition for reduced rate of sampling.

Thus, the requirement of the sampling rate may be reduced to a minimum, even in the presence of finite IF. This is called BPS, the technique of undersampling a modulated signal to achieve frequency translation using the bandpass characteristics of the signal. Bandpass sampling performs digitization and frequency translation in a single process. In BPS, we can sample the signal at twice the information bandwidth and remain consistent with the Nyquist sampling theorem.

The minimum sampling rate required to avoid aliasing, as we have seen, is twice the one-sided bandwidth of the modulated navigation signal. However, these signals typically carry different information in their I and Q phase components. Sometimes, these two signals in orthogonal phases, and carrying different information, are required to process separately. Hence, they are sampled alternatively and independently from the composite signal. So, for cases in which such sampling is necessary, this requirement needs to be fulfilled for each. This effectively doubles the minimum sampling rate.

The appropriate sampling frequency of the ADC is then determined using either of the two methods, as appropriate. In a real navigation scenario, in which the signals are modulated data bits encoded with pseudo random codes, the sampling rate should be asynchronous with both the IF and chip rate, to ensure that the samples vary in code and carrier phase, collectively encompassing the whole signal waveform (Groves, 2013).

5.2.3.5 Quantization

The samples of the signal still have continuous values of their levels. Quantization is the approximation of the sampled signals in some discrete levels, and then representation of these levels with some known digital codes. The whole range of possible sampled values of the signal is divided into numbers of smaller ranges. A discrete level within each range, called the quantization level, is identified to represent all sample values lying within the corresponding range. Each of these quantum levels is digitally represented by one of the digital codes. These digital codes are basically made up of some finite binary bits, the number of which depends on the number of quantized levels, and hence can take only some finite sets of values. This is called quantization.

In the quantization process, because we represent a whole range of sample levels by a single value, it results in an effective error in the representation of the true value, and is known as a quantization error. This error, which is basically an error resulting from limited representation codes, obviously decreases with an increase in the number of quantization levels for the same range of sampled values. This can be done by increasing the number of binary bits by which the codes are represented.

The errors resulting from quantization may also be treated as an equivalent noise. The difference between the true value of the sample and the value of the quantum level by which it is represented is treated as noise called quantization noise. It can be shown that for linear quantization and evenly distributed signal, the ratio of the signal power, S quantization noise power, N_q can be expressed as (Mutagi, 2013)

$$S/N_q = 1.8 + 6n \quad \text{in dB} \tag{5.20}$$

where n is the number of bits used to represent the levels so that $2^n = N$, the number of levels. It follows that a difference of 1 bit will result in a difference of 6 dB in the SNR.

However, the signals in the conventional satellite navigation system are already embedded in noise. That means the amplitude of noise at any instant is much higher than the amplitude of signal. In other words, the combined amplitude is mainly occupied by the noise. So, large numbers of quantization bits will be used to represent the noise. Instead, only a few bits may be used for quantization, which will sufficiently indicate the levels of polarity of the sampled noisy signal at the instant of sampling. Although reducing a bit results in a deterioration of about 6 dB in SNR by adding quantization noise, this quantization noise does not add much compared with noise already present in the signal.

Increasing the sampling rate or adding quantization bits has implications for processing as well as the cost of the receivers. This is because the increased numbers of these parameters demand increased processing capacity for subsequent signal processing. So, the low-cost receivers may use single bit quantization of sampling. However, this reduces the effective SNR for the reason stated earlier, referred to as the implementation loss. Better performance is obtained when a quantization level of two or more bits is used for the purpose. Although it does not look like a huge improvement at the RF level, it affects precision at final position fixing.

The quantization process is generally improved with automatic gain control (AGC). In a navigation system, the signal is of constant amplitude. Because no information is carried by the amplitude, its variations in the signal, caused by various effects, may be compensated and calibrated without distorting or compromising any information present in it.

Owing to the amplitude variation, the signal may vary between large extremes such that the range crosses the nominal dynamic range of the quantization process. To handle this, amplifiers with a large dynamic range are required, which is difficult to implement. In such situations, alternatively, the AGC can have an effective role by

FOCUS 5.2: QUANTIZATION ERROR

The effective SNR can be represented in terms of signal SNR (SNR_n) and the SNR resulting from the quantization noise (SNR_q) as

$$SNR_{eff} = \left(SNR_n^{-1} + SNR_q^{-1}\right)^{-1}$$

With 2-bit quantization, SNR_q becomes $12 + 1.8 = 13.8 \text{ dB} = 10^{1.38} = 24$.
   whereas $SNR_n = -20 \text{ dB} = 1/100$.
   So, the effective SNR, considering the quantization noise over and above the signal noise becomes

$$SNR_{eff} = (100 + 1/24)^{-1}$$
$$= 1/100.04$$

In logarithmic terms, the effective SNR becomes

$$SNR_{eff} = -10 \log(100.04)$$
$$= -20.01$$

So, the degradation owing to the use of fewer quantization bits is negligible.

varying the amplification of the input to ADC to keep it matched to the dynamic range. However, it only ensures that the composite signal mixed in noise maintains the range, whereas the real signal level embedded in it may vary.

5.2.4 Baseband signal processor

The signal forwarded by the front end of the receiver is a combination of all the signals received from every satellite visible to the receiver. In addition, each signal is composed of all of the signal components: the data, the code, and also the carrier. This composite signal enters into the region where it is treated explicitly for the purpose of identifying individual signals, separating them, and subsequently wiping off the code and the carrier to obtain the navigation data. For this purpose, this sampled and digitized form of the signal, still modulated with IF and added with Doppler, is then processed in the baseband signal processor. The details of this processing will be described in this section.

The baseband signal processor demodulates the sampled signals and recovers the navigation data, removing the carrier and the code added to it. It does so by correlating the signal with the internally generated replicas of the code and the carrier. This section of the receiver is hence called the correlator and data demodulator. Along with the process of code and carrier removal from the signal, it simultaneously performs the process of code or carrier-based ranging, or both.

The architecture of the correlator that receives data from its previous section can be divided into different parallel processing arms that can process separate signals simultaneously and independently. For CDMA, the same composite signal enters as the input to each channel arm, whereas for FDMA receivers, these signals are first separated by filtering and each separated channel is loaded into an arm. Each of these arms segregates and then processes a separate signal channel.

To understand clearly what goes inside this module, let us start with a simple situation. Besides our prior knowledge of the carrier frequency, which is now down-converted to f_{IF}, assume that the carrier phase is known at any arbitrary instant of time. Assume that the code phase at that particular instant is also known. This is a hypothetical situation because it will never happen in an actual scenario. However, this is the condition the receiver attempts to attain. So, starting with this gives us an understanding of what the receiver is trying to achieve in the whole process.

After the removal of the carrier from the signal, if the resultant component is multiplied with a synchronized code, the code is also removed and only the data bit remains. This is done in the section specifically designed for the carrier and code wipe off. The schematic of such a section of the receiver is shown in Figure 5.9.

Two distinct sections are present here; one consisting of the carrier wipe off segment, and the other with the code wipe off segment. The arrangement in each of these sections as appears here is of the open loop type.

Knowing the frequency and the phase of the carrier, the initial job the receiver needs to do is mix a synchronous sinusoid to wipe out the carrier. This can be easily done by using a local carrier. In a carrier wipe off, a local carrier is generated by the

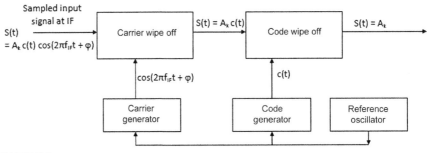

FIGURE 5.9

Schematic of the code and carrier wipe off section.

estimated frequency and phase. This local reference is mixed with the incoming signal to yield the product after low pass filtering as

$$s(t) = A_k c(t) \cos(2\pi(f_{IF} - f_{LO})t + \delta\varphi) \qquad (5.21a)$$

Because the frequency set at the local oscillator is the exact carrier frequency, as per our assumption, and the phase offset $\delta\varphi$ is known and set to the signal, we can put $f_{IF} = f_{LO}$ and $\delta\varphi = 0$ into this equation. Then, the carrier of the incoming signal is eliminated to produce

$$s(t) = A_k c(t) \qquad (5.21b)$$

However, if there is any phase difference owing to the difference in frequency or initial phase, the resultant total phase error is obtained from the LPF product obtained after mixing, as given in Eqn (5.21a).

The local carrier generated and set to the particular frequency and phase for that instant goes on generating the carrier synchronously with the received signal. The local carrier, when mixed with the received signal and low pass filtered, the signal gets carrier demodulated, leaving only the levels, $A_k c(t)$ of the encoded data chips. The code wipe off works in a similar fashion; this can be understood by remembering two things: first, what a complete oscillation for a sinusoidal carrier is a complete code excursion for a code; and second, the low pass filtering, which is a frequency domain concept, is equivalent to integration in the time domain. Thus, the code can be similarly removed by a coherent code generated locally to recover the navigation data.

Recall that during carrier wipe off, we multiplied the local carrier with the incoming signal and then did a low pass filtering of the product. Similarly, here we generate a local replica of the code and multiply it by the code in the incoming signal. The product is then integrated over few integral multiples of the code length. The result thus obtained is given in Eqn (5.22) below, where the local and the signal code phases are separated by delay, τ.

$$P = \int c(t)c(t + \tau)dt \qquad (5.22)$$

Here, for convenience, we are not showing the presence of the data bits because these operations are all carried out within a data bit interval and we assume bit reversal does not occur here. This is equivalent to Eqn (5.21a,b) for carrier tracking. The time over which the integration is to done must be an integral multiple of the code repetition period. When normalized by dividing this integral with the integration time, we get the autocorrelation of the code with delay τ. That is,

$$P/T = 1/T \int_0^T c(t)c(t+\tau)dt \qquad (5.23a)$$

$$= R_{xx}(\tau)$$

So, when the receiver exactly identifies the code phase of the incoming signal, and generates the local code synchronously with it, the above value reduces to

$$P/T = R_{xx}(0) \qquad (5.23b)$$

$$= 1$$

However, this situation is only hypothetical. In a pragmatic case, only the designated value of the carrier frequency is known and not its current phase. Even this frequency changes owing to the Doppler present in the signal, which remains unknown a priori. Moreover, there may be drift in the satellite oscillator that will result in a fixed shift of the signal carrier frequency.

The unique ranging code of an FDMA system is known, but the current phase of the code is still unknown to the receiver. In a CDMA system, the individual code present in the particular signal is also not known. These different issues need to be taken care of.

To know these signal parameters exactly and follow them over time, open loop estimates are not sufficient. These activities are carried out in a closed decision making loop whose generic structure is shown in Figure 5.10.

When the receiver is switched on for the first time or a new satellite comes into view, the condition of the code and the carrier is completely unknown to the receiver. Starting blindly from a completely unknown state of the carrier as well as of the code, the receiver needs to chase these parameters, lock on to the frequency of the incoming signal, and then follow its phase closely.

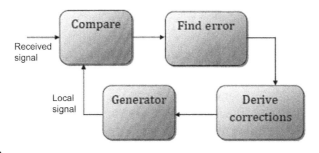

FIGURE 5.10

Schematic for a closed loop arrangement of a signal follower.

Locking on to a signal after identification and being ready to follow it is done by a process called signal acquisition. Thus, the signal acquisition process estimates the code and carrier phase of the signal and the associated Doppler.

5.2.4.1 Signal acquisition

Whatever technique is taken to remove the code and carrier at the receiver, the signal first undergoes a separate procedure in which the receiver approximately adjusts its local carrier and code generators and sets them within close proximity of the signal frequency and phase. This is called signal acquisition.

The receiver may have the oscillator tuned at the designated IF frequency f_{IF}, using which, the IF frequency is removed from it at the carrier wipe off portion. However, the signal still has a slowly varying Doppler frequency as a residual carrier on it and arrives at the code wipe off section in this form.

In this form of the signal, the process of acquisition starts searching for the approximate Doppler frequency remaining and the approximate code phase of a particular individual signal. Moreover, for CDMA systems, identification of the exact unique code present in the 3 signal also needs to be done before code phase identification.

Doppler shifts depend on the relative dynamics of the user and the receiver, and hence vary for different signals. Besides, the ranging codes of the signals although transmitted synchronously from the different satellites, differ in their phase when received, owing to differences in the ranges of the satellites. So, the questions we need to answer through this activity are:

- What is the residual Doppler frequency of this signal?
- What is the ranging code of this signal across the entire range of possibilities?
- What is the code phase of the signal in terms of the chip number and fraction of it?

It requires searching in time and frequency space for the appropriate phase and frequency of the signal carrier and code.

The signal itself remains embedded within the noise, and so the receiver may need to acquire the signal under difficult conditions and within a short period of time. Thus, it is necessary to make the acquisition process simultaneously fast and sensitive. The conventional method of search is through hardware and in the time domain. The codes can be searched serially or in parallel for uncertainties. The most popular methods of code acquisition are

1. Serial search acquisition
2. Parallel code and frequency space search acquisition

5.2.4.1.1 Serial search method

In a serial search, different values of the code phase and Doppler frequencies within the feasible range of these parameters are searched across the uncertainty values in a sequential fashion. Different discrete values of the Doppler frequencies are generated by the receiver in order and mixed with the incoming signal. For each such frequency value generated, the receiver checks for the whole range of code phases.

For any code chosen, every individual code phase is generated and multiplied by the incoming signal, in sequence. The product is then integrated over a finite time to obtain the autocorrelation. If the integral exceeds a certain threshold, it indicates a match and then the selected frequency and the code with its corresponding code phase used are designated as the acquired value. If the integral does not cross the threshold, new combinations are tried. In this way, the receiver checks for locking with all possible code phases in a single code and also over all possible codes in sequence in a CDMA system for the definite Doppler.

Normally, the search is done in every half chip spacing for the code. For every specific combination of frequency and code phase, the autocorrelation value is estimated by integrating over an integral multiple of the code length. Each code phase and Doppler shift searched is called a "bin," whereas each combination of the two is called a "cell." The time spent correlating a signal in each cell is called the "dwelling time" (Groves, 2013). The integration done on the signal samples and passing it for checking for the threshold crossing essentially makes it an integrate-and-dump circuit.

It is the simplest of all the well known algorithms. However, in this algorithm, since each possible combination of Doppler frequency and code phase is tried in sequence until the threshold is exceeded, it requires a large locking time. For FDMA, however, because only a single ranging code is used, it needs to identify only the code phase, which makes the process much faster. Nonetheless, the method works better in noisy conditions.

To see how the signal varies at different parts of the process in quantitative terms, the samples at any instant t will be represented by

$$S(t) = A_k c(t) \cos\{2\pi(f_{IF} + f_d)t + \varphi\} \qquad (5.24a)$$

where A_k is the level of the kth data, $c(t)$ is the code, f_d is the deviation in the frequency owing to the Doppler and other reasons and φ is the phase of the signal. The acquisition value for the Doppler is f_d' with phase φ' and the code phase selected corresponds to delay τ with respect to that in the sampled signal. So, the frequency $f_{IF} + f'$ is loaded onto the oscillator for the carrier wipe off. The output from the carrier wipe off section is thus

$$s(t) = (A_k/2)c(t)\cos\{2\pi(\Delta f)t + \Delta\varphi\}. \qquad (5.24b)$$

$\Delta f = (f_d - f_d')$ is the residual Doppler, i.e. the difference between the true offset and the selected offset of the carrier frequency from the IF and $\Delta\varphi = (\varphi - \varphi')$ is the phase offset. Similarly, because τ is the selected code delay, the code value generated is $c(t + \tau)$. The signal after the code multiplication becomes

$$S_{acq} = (A_k/2)c(t)c(t + \tau)\cos(2\pi\Delta ft + \Delta\varphi) \qquad (5.24c)$$

These samples are accumulated over a time interval that is an integral multiple of the code length, typically about half of the data bit duration. Therefore, these product samples are summed over time T, in which the samples are close enough to consider it as integration, yielding (Van Dierdonck, 1996)

$$S_{int}(t) = A_k R_{xx}(\tau) \, \text{sinc}(\pi\Delta f \, T) \cos(\Delta\varphi) \qquad (5.24d)$$

The value of the expression S_{int} increases as Δf, $\Delta \varphi$ and τ tends to zero. This value should exceed a predetermined threshold before the signal is said to be acquired. This expression is essentially a sine cardinal (sinc) function for the frequency offset; it is evident that the minor peaks appear for certain offset frequencies, as well. Similarly, for CDMA, even for different codes, minor peaks may appear owing to the limitations in the code cross correlation function.

However, because the functional nature of the correlation is known, so is the distribution of its peaks. Therefore, the threshold may be set accordingly, so that the minor peaks remain below it and are not detected even in the presence of expected noise. Furthermore, when the correct peak is acquired, the error for equal offset on either side will theoretically be the same. This feature will not be present for the sidelobes and hence this may be used as a check on the correct acquisition.

The associated uncertainty in frequency that may result from the Doppler or other reasons is used to set the precorrelation filter bandwidth. If the receiver has no a priori knowledge, the initial uncertainty is larger, requiring substantial time to acquire the signal as more frequency bins are required to be searched. Searching for codes starts from the early side to avoid false locking with the multipath component, which always trails the direct signal.

The noise present in the signal deteriorates the autocorrelation results. However, the effect of the noise can be reduced by taking the integral over a longer time interval, i.e. over a large, integral multiple of the code length. But this is limited by the length of the data bit. Any data bit inversion during the correlation integration deviates the result. This is the reason why, in some cases, where low SNR is expected, a pilot channel containing only the product of the code and the carrier but no data bit is sent synchronously but separately from the data channel. The absence of any data bit removes the limitation of integration time while acquiring the signal. So, the signal may be acquired even from a weak condition and poor SNR through longer integration period. Once the pilot channel is acquired, the synchronicity allows the receiver to switch over to the data channel without further acquisition.

If the process is actually executing the reacquisition of a signal it had acquired before but has lost the lock thereafter, it may draw information regarding the time, carrier, and code offsets from what was obtained and stored in the last operating session. This is known as "warm start," and it helps in the acquisition process by drastically reducing the acquisition time. However, the last saved information should be checked for validity across the session before use. Some intelligence may also be obtained from alternative sources of user dynamics or from a satellite almanac. The search efficiency may also be enhanced with exclusive searching of individual channels so that no two channels search the same satellite signals over the entire course of acquisition.

When a complete match is established between participating signal components, the autocorrelation is naturally high. For a CDMA system, the presence of different codes in the composite signal results in the formation of cross-correlation noise during the process. This reestablishes the requirement of proper choice of codes with suitable correlation properties so that the cross-correlation noises do not cumulatively add up to such a considerable amount that it appears like an autocorrelation peak resulting in a false locking condition. Noise will effectually determine the performance of the acquisition process, leading to finite probabilities of a missed detection of the signal and for false locking, as well.

5.2.4.1.2 Parallel search method

The parallel search method is the one in which the sequential search of the Doppler and code and carrier phase done in the previous case is replaced by an equivalent parallel process to increase the searching speed (Scott et al. 2001). Hence, it can be a time-economical alternative to the serial search method. This method is based on the Weiner Khinchen theorem.

The Weiner Khinchen theorem (Proakis and Salehi, 2008) states that for a wide-sense stationary process, x(t), the power spectral density $P_{xx}(f)$ is the Fourier transform of its autocorrelation function $R_{xx}(\tau)$. Similarly, the cross-spectral density $P_{xy}(f)$ of two such wide-sense stationary processes, viz. x(t) and y(t) is the Fourier transform of their cross-correlation function, R_{xy}. So, if F[·] represents a Fourier transform operation,

$$F[R_{xx}(\tau)] = P_{xx}(f)$$
$$= X(f) \cdot X(f) \tag{5.25a}$$

$$F[R_{xy}(\tau)] = P_{xy}(f)$$
$$= X(f) \cdot Y(f) \tag{5.25b}$$

The signal samples of x(t) may be transformed using fast Fourier transform (FFT) to form its amplitude spectrum X(f). Similarly, the FFT of the reference signal y(t) may be obtained to get Y(f). Now, if these two spectra obtained by transforming time signals are multiplied in the frequency domain, we get either the power spectral density when the codes are the same or the cross-spectral density when the codes are different (Proakis and Salehi, 2008). On doing the inverse FFT on this spectral product, we get the correlation values of the input and the reference signal for different delays simultaneously. Thus, the total process of the sequential search for different possible relative code phase delay values is replaced by a single step. The used code is elected if the inverse FFT shows a peak exceeding a threshold and the phase delay is estimated from the relative position of the peak.

However, the cost one has to pay for this is the increased complexity of the computation process at the correlator. The process may be executed only for the code search after complete carrier detection. Alternatively, the frequency domain components of Y(f) may be accordingly adjusted to correspond with the appropriate shift in the frequency of the reference signal. The same process may be repeated for different shifts to obtain the two-dimensional result of the correlation for both frequency and phase shift. So, from the peak obtained in this, both the phase offset and the residual Doppler may be identified. Thus, the parallel frequency space search method provides the correlation values for all possible code shifts in a single go, and hence drastically reduces the search time of the acquisition process. But it is more complex than a serial search and also works poorly in noisy conditions.

Irrespective of which method is used to carry out the process of acquisition, the receiver must know what the possible set of codes is and the expected range of Doppler frequencies of the signal for which it is searching. This knowledge limits the uncertainty bound during correlation for identification and thus determines the precorrelation filter bandwidth. In the same manner, the acquisition accuracy determines the width of the postcorrelation uncertainty.

5.2.4.2 Signal tracking

Acquisition provides only approximate estimates of the frequency and code phase parameters. With these values, the code and carrier demodulation start working just after the acquisition process. Although it approximately wipes off both, it is not exact, and hence some residuals remain in the signal. Besides, it is incorrect to believe that once estimated, the signal frequency and relative carrier and code phase remain the same forever. This is for the obvious reason that the dynamics of the satellite and the receiver will continue to change, thus changing these values owing to the Doppler. Moreover, the presence of the frequency drift and the phase noise in the signal will add to the deviation. So, there will be a residual frequency and phase present in the signal even after acquisition of the signal, which will grow if left unattended.

To handle these effects, the local oscillator used for the carrier wipe off and the clock used for the generation of the code must also have the ability to change the frequency and consequently the phase with finer resolution so that it is able to trace the exact carrier frequency and phase of the input signal and then follow it.

To facilitate this, a numerically controlled oscillator (NCO) replaces the fixed oscillator for signal generation. An NCO is a digital signal generator that generates the discrete values of a complete sinusoidal wave with fixed resolution. These values are generated at regular intervals and at discrete time instants synchronously with clock inputs. With every clock tick, its phase accumulator, which is basically an N bit counter, increases its count by 1. Each count represents the smallest count of phase, which is $2\pi/2^N$ and which defines the resolution of the oscillator. The phase of the accumulator thus increases by this amount at every tick of the clock. As the counter counts modulo 2^N, the phase accumulates from 0 to 2π and resets again. There is a corresponding sinusoidal value stored for every phase. The phase to amplitude converter sends the amplitude corresponding to its accumulated phase, at any instant, to discretely generate the wave. The frequency of the generated wave depends on N and the clock rate. If the clocking rate is R_{clk}, the frequency it generates is $f_N = R_{clk}/2^N$.

Such an NCO is used to drive both the carrier and the code generator. Using these, the input signal is followed exactly in terms of its carrier and code phase through a process known as tracking, which we will now discuss in detail.

5.2.4.2.1 Carrier tracking

At the outset of the carrier tracking, the carrier offset value obtained in the acquisition including the fixed f_{IF} is fed to the carrier NCO to generate the corresponding frequency. Consequently, the signal after acquisition has its nominal carrier wiped off and most of its Doppler removed. To remove the rest and any further variation of it, carrier tracking is done. The main purpose of tracking is to minimize this residual carrier frequency and phase values in the signal and follow its variations with time. This is done by changing the locally generated carrier accordingly so that it follows the temporal phase variation of the input carrier faithfully.

Either the phase or the frequency of the local carrier may be allowed to be attuned to track the corresponding parameters of the incoming signal. Phase tracking may be achieved through a *phase locked loop* (PLL), which is a closed loop arrangement to follow the carrier phase. However, for large deviations in frequency, chasing the signal with PLL is time consuming in addition to its requirement of a large operational range. Moreover, carrier frequency tracking is more robust in poor signal to noise and high dynamics environments because the tracking lock may be maintained here even with larger errors. This is why frequency tracking is preferred in many receivers and they use Frequency Locked Loop (FLL) for the purpose. However, it does not provide the integrated Doppler values which are provided directly by the phase tracking arrangement. In some receivers, FLL is used during the initial part of the tracking as an intermediate process between acquisition and phase tracking.

When the signal remains in a well-acquired state, PLL may work well. Further tracking the carrier phase automatically keeps the frequency following the incoming signal. Optimal tuning of the local source is achieved using the correlation property of the carrier, which maximizes when the two participating waveforms are in phase, i.e. exactly aligned in time. Later, we will see that the same principle works to track codes.

The PLL works by discrete differential adjustment of the phase of the local reference carrier. It is implemented in a closed loop consisting of the carrier wipe off arrangement whose oscillator is controlled by a feedback from the output signal of the wipe off section, passed through an intelligence that resides in a processor. The total control loop is shown in figure 5.11 and performs the following activities:

1. Generate reference signal of certain frequency and phase
2. Compare it with the input and find error in frequency and phase
3. Derive a correction term from the error
4. Generate the corrected reference signal by adjusting its frequency and phase at the oscillator using the correction term

After the generation at the NCO, the reference carrier is mixed with the incoming signal. The code is also removed from this signal by multiplying it with the synchronized local code obtained as the output from the code tracking section. A very narrow band low pass loop filter separates out the sinusoid with the error phase only. Correction terms are derived from this signal inside the processor and fed back to the oscillator, completing the loop. These activities continue in a cyclic fashion over time so that at every instant, the locally generated carrier is exactly in phase with the incoming one. So, it has the following generic blocks arranged in the manner, as shown in Figure 5.10. The exact implementation of the PLL is shown in Figure 5.11.

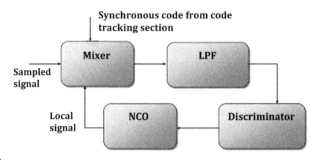

FIGURE 5.11

Schematic of a PLL.

Whenever the sampled signal arrives riding on the IF carrier, its exact phase is not known. So it is multiplied with two orthogonal local reference signals. This reference is initially set to IF but can be attuned to frequencies in the neighborhood. When this multiplied product is passed through low pass loop filter, the incoming signal gets projected onto these references accordingly. We have seen this within mathematical detail in Section 5.1.2 of this chapter. We call these projections of the incoming signal on the cosine and the sine components of the reference signals, respectively, as the in-phase component, I, and the quadrature phase component, Q. From these components, we can derive the relative phase angle between the incoming signal and the local reference.

Let the incoming signal samples at IF be given by the following equation where A_k is the data level.

$$S(t) = A_k cos\{2\pi(f_{IF} + f_d)t + \varphi\} \qquad (5.26)$$

The local reference carrier in two orthogonal phases may be represented by

$$S_c(t) = cos(2\pi f_{acq}t) \qquad (5.27a)$$

$$S_s(t) = sin(2\pi f_{acq}t) \qquad (5.27b)$$

f_{acq} is the nominal frequency set to the oscillator after acquisition and is the sum of the f_{IF} and the estimated Doppler shift $f_{d'}$, obtained through acquisition process, i.e. $f_{acq} = f_{IF} + f_{d'}$. The incoming signal is first mixed with the orthogonal components of S_c and S_s. We do not consider here the code chips assuming that they have been removed synchronously. Passing the product through the loop filter yeilds respectively

$$I(t) = (A_k/2)[cos\{2\pi\delta ft + \varphi\}] \qquad (5.28a)$$

$$Q(t) = (A_k/2)[sin\{2\pi\delta ft + \varphi\}] \qquad (5.28b)$$

These signals thus obtained at the loop filter output have frequency $\delta f = f_d - f_{d'}$ Thus, these two components provide signals with phase and frequency equal to the offset in the phase and frequency, respectively, of the reference with the incoming signal.

The mixed signal after filtering should ideally be a DC if the two signals are perfectly matched. Practically, it is of very low frequency, representing the error frequency only. So, the LPF used here can have narrow bandwidth. However, a very narrow band does not allow large deviations in relative frequency, and therefore the PLL may lose the lock frequently. Such situations may arise owing to unaccounted Doppler shifts. At the same time, wider bandwidth allows larger noise to enter into the loop, thus deteriorating the system's precision.

This signal is then used for generating the correction terms. It is done at the discriminator in this case. The discriminator converts this error signal into a correction term that drives the NCO to generate the correct phase. It is basically a phase to amplitude converter. Thus, accordingly, the clock rate driving the NCO is stepped up or down so that the local reference is synchronized with the input. This process continues over time to keep track of the variations in the signal phase. However, questions that arise here are:

1. How to convert sinusoidal I and Q signals into a correction signal for driving the NCO? The answer to this defines the discriminator characteristics.
2. How the NCO responds to this driving signal to remove the error?

In such a technique for carrier tracking, the problem with using an ordinary PLL is that it is sensitive to the phase reversals, i.e. π phase shifts in the signals. In a navigation receiver, the input signal at this point is multiplied by the encoded navigation data bits. Even if the codes are removed, the data bits, A_k will stay and will shift the phase of the signal by phase π at the instant of transitions. Consequently, the phase difference with the reference changes abruptly.

This problem can be eliminated if either there is no such transition in the binary data chips or if the PLL discriminator is insensitive to such π phase shifts. Here, we will only discuss discriminator algorithms that are insensitive to the data bit variations and intrinsically handle the phase reversal occurring as a result.

This algorithm, which generates a correction value for driving the oscillators by deriving the error terms from I and Q signals, constitutes the discriminator and resides in the processor. They work on the fact that the data bit reversal identically affects both the I and Q components of the signal. The Costas generic discriminator with a differential coder uses the product of I and Q as the discriminator input. This product of two components gives

$$\begin{aligned}
I(t)Q(t) &= \left(A_k^2/4\right)\cos(2\pi\delta ft + \varphi) * \sin(2\pi\delta ft + \varphi) \\
&= \left(A_k^2/8\right)\sin\{2x(2\pi\delta ft + \varphi)\} \qquad\qquad (5.29) \\
&= \left(A_k^2/8\right)\sin 2\Theta
\end{aligned}$$

As A_k gets squared, there is no effect of the bit reversal. From this, it derives the error function $\sin 2\Theta$ as.

Or

$$D_{QXI} = I(t)Q(t)/(A_k^2/8)$$
$$= \sin 2\Theta$$

(5.30)

So, when there is a total phase variation of Θ, the discriminator finds a $\sin 2\Theta$ variation as the error parameter. Although it offers double the nominal sensitivity, as, the sine function repeats in every π radians, it cannot discriminate between φ and $\pi + \varphi$. So, this discriminator has the limitation that it works perfectly for $-\pi/2 < \varphi < \pi/2$ but shows sign error beyond this limit.

A slightly modified version of the algorithm is to get the discriminator function by taking the ratio of the components on I and Q axes. From the ratio of the two components, we can understand the phase difference and can act accordingly to generate the correction term. From Eqn (5.28a,b), the two components can be written as,

$$I(t) = (A_k/2)\cos(2\pi\delta ft + \varphi)$$
$$Q(t) = (A_k/2)\sin(2\pi\delta ft + \varphi)$$

(5.31b)

So, the ratio is expressed by the value of Q over I as

$$D_{Q/I} = Q/I$$
$$= \tan \Theta$$

(5.32)

So, when the error is Θ, the error parameter derived is $\tan \Theta$. However, this equation too, relates the error and the control term in a nonlinear fashion. Therefore, instead of deriving the correction term in this trigonometric form, it can be linearized by taking the inverse tangent function of the ratio. So,

$$D_{atan} = \tan^{-1}Q/I$$
$$= \Theta$$

(5.33)

This is the total phase difference between the incoming and the local carrier at time instant t. Thus, the inverse tangent of the ratio of the signal components available at any instant in I and Q arms, respectively, represents the total phase error.

The discriminator output variation as a function of the total phase error is called the discriminator characteristic. The characteristics of these three methods are shown in Figure 5.12. Note from the figures that for all small phase errors, the discriminator characteristics for both of the nonlinear cases can be approximated to linearity.

Thus, the resultant signals are converted by the discriminator into error terms that become the parameter for driving the NCO. The NCO accepts this term for the correction and adjusts the signal generated accordingly to follow the input.

Because the bit reversal does not affect the operation of carrier tracking, it can be operated without considering the bit transition. When full carrier synchronization is achieved, the sinusoidal variation is completely removed from the the signal reaching the delay locked loop used for code tracking. This eases the code tracking process.

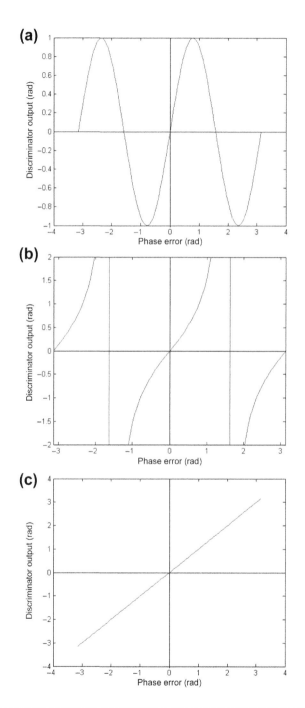

FIGURE 5.12

Discriminator characteristics for (a) $D_{Q \times I}$, (b) $D_{Q/I}$, (c) D_{atan}.

However, here we have considered that there is no noise in the signal. The effect of the noise is discussed in a separate section below.

5.2.4.2.2 Performance in the presence of noise

The noise term in the signal consists of additional variations that will in fact perturb the correction term and hence deviate the loop from accurately tracking the incoming signal. But because this additive noise is zero-mean and random, the corresponding variations in the correction terms will also be random in nature.

The noise introduced within the signal IF bandwidth are all present at the input of this loop, whereas they are down-converted to low pass noise upon mixing. How much of these total noise terms will be passed to the discriminator will define the precision of the loop. This is determined by the bandwidth of the loop filter and is the loop bandwidth.

The loop bandwidth is generally kept narrow. A narrow bandwidth restricts most of the noise to get into the discriminator that could otherwise deteriorate the tuning accuracy of the local oscillator. In the process, performance is improved. However, it cannot be narrowed indefinitely because then it will not be able to accommodate the pragmatic frequency variations in the signal.

The signal plus the noise term at the input may be represented as

$$S(t) = (A_k)\cos(2\pi ft + \varphi) + n(t) \tag{5.34}$$

As we have seen in Eqn (5.8b), the bandpass representation of the noise component at $f = f_1$ is

$$n = n_r \cos[2\pi f_1 t + \varphi_1(t)] \tag{5.35}$$

So, on multiplying this specific component of the noise by the local reference S_c and filtering, the noise term becomes

$$Ni = 1/2[n_r \cos\{2\pi\delta f_1 + \varphi_1\}] \tag{5.36a}$$

Similarly, the noise on the Q component becomes

$$Nq = 1/2[n_r \sin\{2\pi\delta f_1 + \varphi_1\}] \tag{5.36b}$$

For the discriminator, $D_{Q \times I}$, when the I(t) and Q(t) are multiplied, this results in an effective noise of

$$
\begin{aligned}
N(t) &= Ni(t)NQ(t) \\
&= (1/8)[n_r^2 \sin 2\{2\pi\delta f_1 t + \delta\varphi_1(t)\}]
\end{aligned}
\tag{5.37}
$$

From the discriminator characteristics, as given in Eqn (5.30), the error in the correction term that appears as a result of the noise is

$$
\begin{aligned}
\varepsilon(t) &= N(t)/(a^2/8) \\
&= (n_r^2/a^2)\sin 2\{2\pi\delta ft + \delta\varphi(t)\}
\end{aligned}
\tag{5.38}
$$

Therefore, the total amount of resultant error, ε, manifested at the output of the discriminator in the correction signal that drives the local oscillator will be the cumulative sum of such contributions from each of the noise frequency components over the whole loop bandwidth, B_{loop} and will be given by

$$\varepsilon = K \int B_{loop} \varepsilon(f) df$$
$$= \left(K/a^2\right) \left[B_{loop} \times N_0\right] \tag{5.39}$$
$$= K \times \text{Noise power}/\text{Signal power}$$
$$= \left[K/\left(S/N_0\right)\right] \times B_{loop}$$

Because S/N_0 is a signal characteristic, ε depends on only one design parameter, i.e. the loop bandwidth B_{loop} that determines the amount of noise that may come in.

5.2.4.2.3 Code tracking

The acquisition process estimates the approximate carrier as well as the code phase. From this situation, the receiver starts tracking the code, which may vary in its phase and frequency for the same reasons as the carrier that calls for tracking of the signal code phase.

From these sampled data, the exact code phase of the received signal is identified that is required for two different purposes. The first and the most obvious is that it is used to wipe off the code to extract the navigation data bits. The second and equally important use of this estimation of the code shift is that the satellite range is estimated from this value.

Therefore, the open loop design we described in our initial explanation in the code wipe off section with assumed hypothetical conditions needs to be modified with provisions to vary the chipping rate of the local code. Similarly as for the carrier, the code needs to be driven by a code NCO. The code clock is generated in the same way in the code NCO as the reference carrier offset is generated in the carrier NCO. Therefore, there must be some parameter generated that will determine how much faster or slower that the NCO must run to catch up with the input. This in turn demands a closed loop arrangement. Here, too, we will carry the procedure similar to carrier tracking and the basic idea of the tracking system will remain the same as what we discussed referring to the schematic in Figure 5.10.

Code tracking is the process by which the receiver follows the variation in the propagation delay and thus the variation in the received code phase, by estimating the differential shifts in the incoming signal code phase with respect to a reference code generated at the receiver. This is done using a delay locked loop (DLL). Like PLL, DLL is a closed loop arrangement used to detect differential change in the phase of a code. It is implemented with the code removal section with its output going to the integration and dump circuit that we described during the acquisition process. The integration results drive the discriminator to generate the correction signal for controlling the code NCO, which in turn modulates the clock speed accordingly to catch up with the code phase of the input signal.

The carrier recovered during carrier tracking is generally used to remove the modulation before code tracking. However, in some situations, the carrier is not removed and the code is still tracked, and is referred to as the noncoherent mode of tracking.

Coherent code tracking can be done for cases in which the carrier has been exactly identified and wiped off from the signal. Moreover, in coherent tracking, the effects of the Doppler on the code width are also removed using aids from the section that tracks the carrier. This presents unmodulated binary encoded chips to the code tracking system. However, it is difficult to achieve coherency at the beginning, and hence it always starts with the noncoherent mode initially for all practical cases. In the noncoherent mode, the samples integrated at the code wipe off section have a factor for frequency offset δf. The number of samples accumulated is $Me = T/T_s$, where T is the time of integration and T_s is the sampling interval. This number may vary a little because of the code Doppler during the noncoherent operation. But this variation is minimal over 1 ms and its effect is negligible (Van Dierdonck, 1996).

Because acquisition has already found the coarse phase, the task of the tracking system is only to do its fine-tuning and follow any subsequent variation in the code phase with time. To track variations in the code phase, it should first generate an error signal from the integrated samples where the error indicates the current code delay and hence the required shift required in the NCO that will drive the signal generator in such a way that it either moves faster or slower to catch up with the incoming signal.

Recall that during carrier tracking, we multiplied the local carrier by the incoming signal and then performed LPF over the product to get the error in terms of the phase difference using which the corrections were done. Compared with the carrier tracking loop, as shown in Figure 5.11, only two things are required to be remembered: A complete oscillation for a sinusoidal carrier is a complete code excursion for a code. Second, the mixing and subsequent low-pass filtering operation of the carrier is equivalent to autocorrelation of the code. We similarly generate a local replica of the code and multiply it with the encoded signal in both I and Q components. These latter components are obtained by multiplying orthogonal reference carrier signals with the actual incoming signal and subsequently low pass filtering the same. Now integrating the product in both components over a few integral multiples of the code length, we get the result

$$P_I = (A_k) \int c(t)c(t+\tau)\cos(2\pi\delta ft + \delta\varphi)dt \qquad (5.40a)$$

$$P_Q = (A_k) \int c(t)c(t+\tau)\sin(2\pi\delta ft + \delta\varphi)dt \qquad (5.40b)$$

where τ is the offset of the local code with respect to the code of the incoming signal. Here, we are not considering variation of the data bits for convenience because these operations are all carried out within a data bit interval and we assume bit reversal does not occur here.

When these integrals are normalized by dividing this integral by the time length over which the integral has been taken, we get the same expression as we obtained there and which is shown below as

$$P_I/T = (1/T)(A_k) \int c(t)c(t+\delta)\cos(2\pi\delta ft + \delta\varphi)dt$$
$$= A_k R_{xx}(\tau)\text{sinc}(\pi\delta f\,T)\cos(\delta\varphi)$$

(5.41a)

Similarly,

$$P_Q/T = (1/T)(A_k) \int c(t)c(t+\delta)\sin(2\pi\delta ft + \delta\varphi)dt$$
$$= A_k R_{xx}(\tau)\text{sinc}(\pi\delta f\,T)\sin(\delta\varphi)$$

(5.41b)

In our previous chapter, describing basic signal characteristics, we found that the correlation of a pseudo random sequence is highest when there is an exact superposition of the two waves. This value falls off with relative shift of the two signals and grounds to the value of $-1/N$ at a complete one or more chip offset between two. This variation is symmetric on both sides. Thus, the correlations between two such sequences are equal if one of them is equally early or late with respect to the other about the exact match, i.e. exact superimposition condition. Both $+\tau$ and $-\tau$ result in the same autocorrelation values for normal BPSK signals, given by (Cooper and McGillem 1986)

$$R_{xx}(\tau) = (1 - |\tau|/T_c)$$

(5.42)

This depreciation in the correlation value thus obtained indicates the delay between the two signals that is also the shift in the incoming code with respect to the local reference code. Thus, this delay needs to be determined. However, unlike the carrier phase case, because the maximum correlation value is a function of the received signal strength, the actual depreciation cannot be estimated; also, because the autocorrelation is symmetric over the positive and negative values of the delay, δ, the exact delay cannot be readily found.

This problem may be tackled using the symmetric nature of the correlation characteristics. The key is to use two additional separate local codes for the purpose, one early and the other equally late from the reference code, which we call the prompt code.

This forms the Early − Late gate DLL (Cooper and McGillem 1987; Van Dierendonck et al. 1992; Spilker, 1996; Van Dierendonck, 1996; Ward et al. 2006; Groves, 2013). It is widely used in code tracking because it has been an optimum tracking method for fine synchronization of digital signals. As mentioned, this Early − Late DLL consists of an early and late version of the locally generated code. These versions of the codes are used to find the correlation independently but simultaneously with the incoming signal. By comparing the results of the two autocorrelation values, the required shift may be obtained in both magnitude and sign which will align the local prompt code with that of the incoming signal. Most modern receivers use this method for the purpose; its schematic is shown in the Figure 5.13.

Let us see how we estimate the delay using some simple mathematics and our previous knowledge of the correlation. In an Early − Late gate DLL, two codes equally

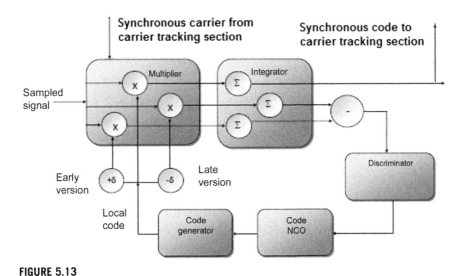

FIGURE 5.13

Schematic for a delay locked loop.

offset by a value δ from a prompt code are generated and used for correlation. One of them, which is phase advanced with respect to the prompt code, is called the *early* code; the other, whose phase trails the prompt code, is called the *late* version of the code. So, for a definite delay τ of the prompt code with respect to the input, the offset of the early code with respect to the input is dE = δ − τ and that of the late code is dL = δ + τ. Because the exact delay cannot be identified from the individual values, the delay is obtained using the difference in the correlation process. The corresponding arrangement in the correlator is shown in Figure 5.13.

In a manner similar to Eqn (5.42), the integrated values for the I and Q signals of the early and late versions over an integration time of T can be represented as (Van Dierendonck, 1996; Groves, 2013)

$$I_e = R_{xx}(\delta - \tau)\mathrm{sinc}(\pi \delta f\, T)\cos(\delta \varphi) \tag{5.43a}$$

$$Q_e = R_{xx}(\delta - \tau)\mathrm{sinc}(\pi \delta f\, T)\sin(\delta \varphi) \tag{5.43b}$$

$$I_L = R_{xx}(\delta + \tau)\mathrm{sinc}(\pi \delta f\, T)\cos(\delta \varphi) \tag{5.44a}$$

$$Q_L = R_{xx}(\delta + \tau)\mathrm{sinc}(\pi \delta f\, T)\sin(\delta \varphi) \tag{5.44b}$$

This equation shows that the integrated sample values depend on the frequency mismatch δf and also the phase mismatch δφ.

5.2.4.2.4 Coherent code tracking

For coherence tracking, the carrier is perfectly synchronized, and hence δf = 0 and also δφ = 0. Under such conditions, only the I components exist while the Q components vanish or carry only the noise. The prompt, early, and late versions of the correlator then give, respectively,

$$P = Ip = R_{xx}(\tau) \tag{5.45a}$$

$$E = Ie = R_{xx}(\delta - \tau) \tag{5.45b}$$

$$L = I_L = R_{xx}(\delta + \tau) \tag{5.45c}$$

For normal BPSK modulation, the expression we obtain for the correlation is given by Eqn (5.42), and so the expressions for these components we obtained in Eqn (5.45) become

$$R_{xx}(\delta - \tau) = [1 - (\delta - \tau)/T_c] \tag{5.46}$$

$$R_{xx}(\delta + \tau) = [1 - (\delta + \tau)/T_c] \tag{5.47}$$

Considering the case, when the delay of the prompt code with respect to the input is within the early—late offset, then defining a discriminator in which the differences of the early and late signals are taken, we find

$$
\begin{aligned}
D_{E-L} &= E - L \\
&= [\{1 - (\delta - \tau)/T_c\} - \{1 - (\delta + \tau)/T_c\}] \\
&= (1/T)[(\delta + \tau) - (\delta - \tau)] \\
&= (2/T_c)\tau
\end{aligned}
\tag{5.48}
$$

The discriminator function is shown in Figure 5.14.

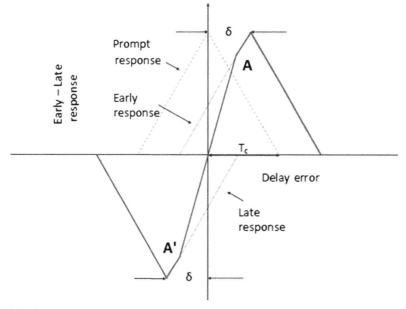

FIGURE 5.14

Discriminator characteristics.

In the working region of the discriminator, the characteristics of the differenced autocorrelation is linear, antisymmetric about zero with slope $2/T_c$. This represents the portion AA' in Figure 5.14 of the early−late discriminator characteristics.

If the code in the incoming signal is aligned with the prompt code, it is equally displaced in phase from these two codes and will produce equal correlation values with both versions and the difference will be zero. If it is skewed toward one of them, the corresponding correlation will be higher than the other. The difference will be positive if it is skewed toward the early version and negative if it is skewed toward the late version. From the magnitude of the difference, the amount of skewness can be determined and the time of arrival can be measured from the corresponding shift.

From this, it is clear that both the positive and negative offsets of the prompt code may be identified using the autocorrelation difference. For obvious reasons, the balanced condition occurs when $\tau = 0$, and then the prompt code may be said to be aligned with the incoming code. For any nonzero values, we can shift the prompt code and its early and late versions to attain the triviality. The early−late correlator output that is proportional to the offset τ acts as a correction factor and drives the code NCO. The latter, in turn, adjusts the shifts in the code phase to align the two codes, which serves our objective.

When the condition $|\tau| > |\delta|$ exists, a balance condition can never be achieved, and hence the DLL is out of its operating range. This is evident from the characteristic curve. Therefore, either the early−late offset needs to be widened or it calls for better acquisition.

It also follows from this that the offset of the early−late increases the sensitivity of the system by increasing the offset response gradient to 2/T from 1/T value of an independent autocorrelation function. But it reduces the linear operating range in the process.

Furthermore, because of its inverse dependence on T, when the bit lengths are small, i.e. the code rates are high, this discriminator has a high sensitivity, and hence its response has a sharp rise as the errors deviate from zero. For a lower slope, the values of the correlation are not different if the delay error is small, whereas they are considerably different when the slope is large. As a result, a receiver that may not resolve the difference in the first case can distinguish it in the latter. A large code rate yields better performance in terms of ranging.

However, this approach, although sensitive to data bit inversion, can be improved by using a coherent dot product method in which the discriminator characteristics are

$$D_{dot} = (E - L)P$$

In this case, the output remains proportional to the relative delay error τ, while owing to the product of the prompt signal in which the inversion takes place simultaneously, the bit inversion effect is nullified.

5.2.4.2.5 Noncoherent code tracking
The previous section assumes that a coherent condition of the signal exists, in which the carriers are exactly removed by matching the reference signal in phase and

frequency. For many practical situations, however, this is not the case and hence noncoherent tracking takes place. For coherent code tracking to take place, accurate carrier tracking is necessary. In fact, there is a considerable interdependence between carrier and code tracking, which makes each of them fragile. So, in noncoherent DLL the signal is treated for code tracking with residual Doppler offset, data modulation, and unknown carrier phase. In such a condition, the incoming signal of the early and late version of the code after integration becomes as given in Eqns (5.43a,b) and (5.44a,b).

There can be different noncoherent discriminator algorithms for code tracking. One of the most popular is the early−late power discriminator. Here, instead of directly differencing the two signals, we square the two and add I and Q to get the total power in these two orthogonal components. So,

$$
\begin{aligned}
P_E &= Ie^2 + Qe^2 \\
&= R_{xx}{}^2(\delta - \tau)\operatorname{sinc}^2(\pi\delta f\tau) \\
P_L &= I_L^2 + Q_L^2 \\
&= R_{xx}{}^2(\delta + \tau)\operatorname{sinc}^2(\pi\delta f\tau)
\end{aligned}
\tag{5.49}
$$

Now, once again differencing the early power from the late power signals we get

$$
\begin{aligned}
P_E - P_L &= \left[\{1 - (\delta - \tau)/T\}^2 - \{1 - (\delta + \tau)/T\}^2\right]\operatorname{sinc}^2(\pi\delta f\, T) \\
&= \left[(4/T)(1 - \delta/T)\tau\right]\operatorname{sinc}^2(\pi\delta f\, T)
\end{aligned}
\tag{5.50}
$$

Thus, for any carrier phase offset, it can be tracked in the same manner as for coherent mode, where the discriminator output remains proportional to the code offset τ and assuming the frequency offset δf remains unaltered during the process.

5.2.4.3 Tracking BOC signals

The BOC (m,n) signals may be seen as normal BPSK signals with the encoded data additionally multiplied by a square wave with a rate m/n times faster than the code. So, tracking a BOC coded data is also done in a fashion almost similar to that of the BPSK. There are two prominent techniques for code tracking:

1. Very Early−Very Late (VEVL) correlator
2. Double estimation

5.2.4.3.1 Very Early − Very Late correlator

We have seen that the ideal autocorrelation characteristic of a BOC signal is a regular bipolar triangular variation with amplitude enveloped within the autocorrelation function of a normal spreading code, as shown in Figure 5.15. The profile has a main peak at the middle for exact superimposition of the signals with diminishing adjacent side peaks at regular delay intervals which corresponds to the shifted matching of the subcarriers.

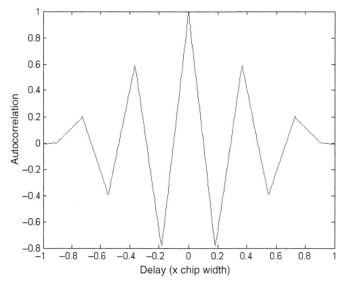

FIGURE 5.15

Autocorrelation of a BOC signal.

The idea in VEVL is that it employs additional correlators-that is, very early (VE) and very late (VL) gates-implemented away from the prompt gate in addition to the nominal Early and Late gates. These correlators monitor the amplitude of adjacent peaks in a correlation function. Because of the symmetric nature of the envelope, adjacent peaks at an equal distance from the center peak at two opposite sides have the same relative peak height. Since the subcarrier period is known, the distances in terms of delays from the center that the adjacent peaks will occur can be predetermined. The VE and VL correlators are placed equally at these delay offsets on two sides of the prompt.

As the side peak heights fall off symmetrically on both sides of the main peak, the VE and VL correlators must get equal values of peak when the prompt is aligned with the incoming code, at the center of the envelope. If the prompt version has coincided with a peak based on the E-L gates, but still a comparison indicates a higher amplitude on either VE or VL, it means the prompt has got latched to a side peak and hence the codes need to be shifted for balancing. However, because the peaks occur at intervals equal to the period of the subcarrier, T_s, the receiver must make the appropriate jump of either $+T_s$ or $-T_s$, in the direction of the correct peak until the autocorrelation values at VE and VL balance.

5.2.4.3.2 Double estimation
A BOC modulation may be seen as the product of the data with the code and the carrier with subcarrier modulation by multiplying it by a square wave. This may be considered a normal BPSK modulation multiplied by a square wave. A BOC (m, n) modulation has m complete square waves in n complete code chips. So,

BOC code tracking can be done by successively tracking the pseudo random noise (PRN) code and the square wave in a noncoherent mode. However, the basic technique for tracking each remains the same as discussed in connection with code tracking in the previous section.

5.2.4.4 Tracking imperfections
5.2.4.4.1 Correlation loss
Correlation loss is defined as the difference between the signal power expected at an ideal correlation receiver from the transmitted signal and the signal power actually recovered in such a receiver, over the same bandwidth. However, the correlation loss in a typical navigation receiver occurs due to the reduction of certain portions of signal band, especially the higher frequency ones. Although we have shown earlier that most signal power remains confined within the first null of the signal, excluding the rest of the spectrum has some effect on the shape of the data bits. Because in a pragmatic channel it is not always possible to retain the whole spectrum, as a result of this limitation, some of the high-frequency components of the complete spectrum get cut off. This removes the sharper edges of the bits in the time domain and gives the rising and falling edges a small but finite curvature. When correlation is done on these bits, the peak that is generated is not sharp but is curved at the top. The reduction in value of the correlation with zero delay from the expected unitary value is the convolution loss (Figure 5.16).

5.2.4.4.2 Jitter and phase noise
Jitter and phase noise are the two different perceptions of the same physical effect. Whereas jitter is described as fluctuations of the exact time variations of a binary bit, particularly its transition instants, phase noise is the appearance of the spurious frequency components when seen in the frequency domain.

We will attempt to understand it, starting from basic yet simple instances that are relevant to navigation. If we have a pure sinusoidal signal and we count the phase and track it continuously with time, it should give us a linearly increasing value modulo

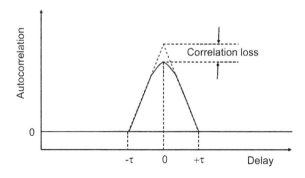

FIGURE 5.16

Correlation loss.

2π, i.e. a saw tooth variation extending from 0 to 2π. However, in a pragmatic case, a signal for certain reasons shows in small and random variations in phase. This leads to broadening of the spectrum of the signal from being a single line, with additional components appearing at nearby frequencies. This does not only result in deviation in phase of the contaminated time signal but also deviates the edges of phase transitions and changes the spectrum of the signal at any instant of time.

The less the amplitude is of these unwanted frequency components generated as a result, lesser the phase variations of the signal. Thus, the potential deviation that may be caused by phase noise is represented by the relative amplitude of these unwanted components appearing at different frequencies over the working range. It is also sometimes represented by the relative power density with unit dBc/Hz.

For a phase shift keying signal, the receivers are sensitive to these variations. This is obvious, because the data are identified by the phase and its variation. When demodulated with a correlated carrier generated at the receiver, these abrupt variations either advance the true phase of the signal or delay it with respect to that expected from the pure input signal. Consequently, they cause phase reversals in the modulated signal to occur at an instant slightly different from the position of true occurrence. Effectually, the receiver thus identifies a bit transition with some time deviation from where it would ideally be. This is the phase jitter in binary data (Figure 5.17).

This problem is serious in the case of navigation because a deviation in the signal phase or bit transition time leads to errors in ranging. This is because range is determined from the true delay of a phase of the code or carrier owing to the propagation only. Because the received and demodulated code bits are deviated owing to jitter, they diverge from the true propagation time. Consequently, the position estimation derived from the ranges suffers in terms of accuracy.

5.2.5 Pseudo ranging

We have learned that one of the prime objectives of the receiver is to carry out satellite ranging. In addition, the ranges may be obtained from the code phase as well as the carrier phase of the signals transmitted from the satellites. Here, we shall see how they are done.

Pseudo range is obtained from satellites that qualify the selection of the best set to estimate the position. We will see the selection criterion in Chapter 7. Pseudo range is the approximate range of the satellite from the receiver as measured by the latter, and in which the measurement errors remain present making it different from the true geometric range.

To understand ranging, let us start with a simple question: How do we measure the traverse time of anything that is moving between two points A and B in space? The answer is simple: We check the time at which it starts at point A, i.e. the start time and also time at which it reaches the point B, i.e. the end time. The traverse time is the difference between these two times. Now that we have the traverse time, we can find the distance it has moved if we know its velocity of movement. However,

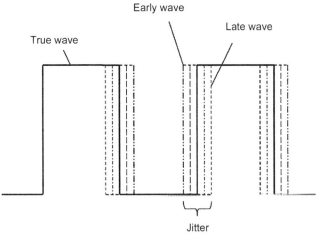

Early wave

Late wave

True wave

Jitter

Time ⟶

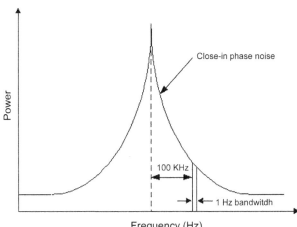

Close-in phase noise

Power

100 KHz

1 Hz bandwitdh

Frequency (Hz)

FIGURE 5.17

Phase noise and jitter.

it is with an assumed condition that it moves with the same constant velocity throughout.

For ranging, we need to identify something that moves from the satellite to the receiver and with a constant velocity. We already learned that the only means by which the system interfaces with the user receiver is through the signal. It is only this signal which moves from the satellite to the receiver, and because it is an electromagnetic wave, it moves with a constant velocity c. Therefore, that *something* must be some specific marked feature on this signal that moves with it so that we can identify when it has started from the satellite and when we have reached the receiver. Our tasks now are

1. To identify this signal parameter
2. To identify the time of its leaving the satellite
3. To identify the time of its reaching the receiver

If we can do these successfully, our purpose is accomplished.

We have read about the signal structure and understand that the three main elements of the signal are its data, code, and carrier. Of these, the data are random, whereas the code and the carrier are structured. It is convenient to use some features of the latter two to find the traverse time. Of all the characteristics, the code and carrier phases are the only features linearly related to time. This means that if the frequencies remain the same, both the code and the carrier phase values increase linearly with time, and hence can be used for the purpose.

Therefore, the pseudo range may be measured from the time of a definite phase of the carrier or code of signal that it takes to travel from the satellite to the receiver. The distance, i.e. the range, is derived from the total time taken by the signal to travel from the satellite to the user. It is obtained by multiplying this traveling time by the velocity of travel. The expression for the measured range is

$$R = c(t_2 - t_1) \tag{5.51}$$

Here, t_1 is the time of transmission of a definite phase of the signal, t_2 is the time of reception of the same phase at the receiver, and R is the measured range. 'c' is the velocity with which the signal phase or code travels through the intermediate space. The velocity of light in vacuum is taken for the purpose. The assumption underlying this method of calculating range is that waves travel with a constant velocity of 'c.' Thus, the measured range can also be regarded as the difference between the current time and the time of transmission of the currently received phase, multiplied by c. Again, to identify the precise transmission time of the currently received phase, we have to identify the current phase with equal precision.

The two main drawbacks in the process are that the phases can be measured only in modulo 2π. Besides, there is no common clock that can measure both the time of transmission at the satellite and the time of receiving at the receiver. Subsequently, in the following two sections, we will see, despite these drawbacks, how the code and the carrier phases are utilized for ranging.

We learned at the beginning of this chapter that receivers can be distinguished on the basis of ranging types, and we mentioned two types of ranging: code based and carrier based. Here, we elaborate on them.

5.2.5.1 Code phase-based ranging

A ranging code remains multiplied by the signal for the purpose of the range measurements. This is true whether it is a CDMA or an FDMA system. This is actually a pseudo random code that remains synchronous with the data as well as the carrier. We already learned this in our previous chapter. In the code based ranging technique, basic measurements made by a navigation receiver are the transit time of a particular code phase from a satellite to the receiver, which is then multiplied by c to get the range.

To define the code phase, recall the previous statement that what is a complete oscillation for a sinusoidal wave is a complete code excursion for a code. Thus, the current code phase is defined by the number of complete chips of the total code and the fraction of the chip that the code state is currently in. The current code phase can also be represented in terms of the fraction of the total code of the current code state.

However, all of the timing measurements need to be with respect to a reference. This reference must be common for the measurement of both the transmission and receiving time. For code based ranging, this reference may be the reference time of the system. Then, recalling that the data and the codes are synchronous, we can take advantage of the time stamp present in the data to obtain the transmission time of the code phase received at the current instant, whereas the current time, i.e., the time of the reception of the code phase can be obtained from the receiver clock in locked condition, which is keeping the system time. The difference between these two provides the transit time, which is used to derive the range.

5.2.5.1.1 Ranging algorithm

The basic assumption in this type of ranging is that the satellite and the receiver clock are synchronous. Although two different clocks are used for the purpose, it is believed that they are completely aligned in time such that using time from one clock is the same as using it from the other. We will discover in the next chapter how we can make the system unconstrained from this restraining assumption, but for now we shall assume it to be there.

If the clocks are assumed to be synchronous, both the satellite transmitter and the receiver will generate the same code phase at the same instant. Thus, the code phase received by the receiver at any instant will be phase lagging from the one currently being generated by the receiver. This is due to the finite time taken by the satellite signal to reach the receiver. Because the code phase changes linearly with time, the propagation time can be derived from this phase lag.

At any instant, the currently received code phase is used as the reference phase from which the satellite range is to be found. The received time of this reference phase, which is the current time kept by the receiver, is readily known by it. The transmission time of this current code phase has to be determined. This is done in the following manner.

The messages of the navigation data are transmitted in a definite structure, say, in frames. The starting of a frame, subframe, etc. can be explicitly identified and hence can act as periodic reference points for time. These points are time marked and the time marks of these points are transmitted in the data. So, the transmission time of such a reference point immediately preceding the current phase is obtained from the time stamp. Now, to know the time of transmission of the code phase in question, we need to know the total time elapsed from the selected reference point to the beginning of the generation of the current received code and the current phase within the code.

Therefore, the first job in this regard is to find the time difference of the currently received code phase relative to this reference point whose time of transmission is obtained.

The receiver hence counts the total number of complete codes that have been received since the receiving of this reference point until the beginning of the current code. In addition to knowing the count of the total codes elapsed since the reference time, we also need to know the fractional part that has elapsed in the code, i.e., the current code phase. This is equivalent to knowing the total number of complete chips that have elapsed in the current code and the fractional part of the chip that the receiver is currently receiving. It is, however, more convenient to obtain the current state in terms of the fractional code phase. The fractional portion of the code phase is obtained by autocorrelating delayed version of a local replica of the same code with the incoming one. The delay that results in maximum autocorrelation indicates the fractional code elapsed.

Autocorrelation is done with a locally generated replica of the code. Since the autocorrelation values have definite linear relation with the relative phase shift, the phase difference between the local and the received code phase may be estimated from their values. As the phase of the local code is known at any instant, the phase of the code received at that instant can then be derived from this estimate, It provides the fractional code part, f. So, if T_0 is the time of transmission of the marked reference point, n is the total number of complete codes elapsed since then, and f is the fractional part of it, transmission time t_1 of the currently received phase of the code may be expressed as

$$
\begin{aligned}
t_1 &= T_0 + n\tau + f\tau \\
&= T_0 + nNT_c + fNT_c
\end{aligned}
\tag{5.52}
$$

where $\tau c = NT_c$, is the code repetition time, N is the numbers of chips in the code, and T_c is the chip interval (Figure 5.18).

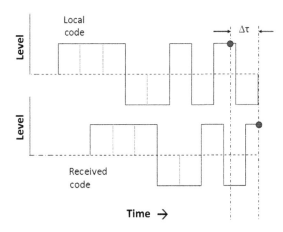

FIGURE 5.18

Estimation of transmission time.

Once the reference instant, i.e. start of the frame, subframe etc. is identified, the receiver can carry on counting the total numbers of complete code it receives up to the current instant. It can keep a separate count 'n' of the complete code elapsed since then. The estimation of the fractional part of the code that the receiver is currently receiving is obtained through alignment of the received code and the locally generated code. The true shift needed to be given to this prompt code, and aligned with the incoming code relative to the total code interval is the required fractional part 'f' of the code being received. We have already seen during code tracking how this phase could be obtained using an Early-Late gate DLL using the autocorrelation property of the codes. These estimations are carried out at every ranging instant.

Once both values of t_1 and t_2 are calculated, the time difference $(t_2 - t_1)$ is multiplied by the speed of light, c, to get the range. However, this estimation carries with it errors owing to propagation and other assumptions. These biased time delay measurements are referred to as pseudo ranges.

As this algorithm utilizes the start of the frame as time reference, its use is only possible once the signal is demodulated and the data bits are retrieved. Code-based ranging can only start after the navigation data frame has been identified to obtain the parameter values.

Furthermore, it is unlikely that the receiver clock will remain synchronous with the transmitting clock at the satellite from the outset. The measurements are contaminated with the clock synchronization error at the beginning. However, once the position is estimated, the clock shift is also determined. This estimation of the clock shift is used to steer the receiver clock to get it aligned with the system time; for subsequent times, the clock is continuously disciplined with this system time.

5.2.5.2 Carrier phase-based ranging

The technique of carrier phase-based ranging is also done on the assumption that the transmitting and the receiving clocks are synchronous. Under this condition, the transmitter at the satellite and the receiver will be generating the same phase of the carrier at the same instant. Also, for the same reason described above, the carrier phase being received at any instant by the receiver will lag the carrier phase that the receiver is itself generating at that instant. This lag is equal to the time of propagation by the signal because the carrier phase has linear variation with time. Hence the carrier phase and the time are proportional and can surrogate the other. Let a particular phase received currently at the receiver be representative of its transmission time, t_1. Similarly, the current phase of the carrier being generated at the receiver at the current instant is representative of the current time t_2. So, the phase difference between the two is the parameter from which the propagation time can be obtained. Multiplying, as usual, this propagation time with the velocity of light, we get the pseudo range. (Figure 5.19).

However, there is a difficulty in measuring the transmission time in this case. Unlike code phase-based ranging, the transmission time cannot be obtained from the navigation data here. As the received phase measurement can only be done modulo 2π, successive carrier phases cannot be distinguished. So, there is no

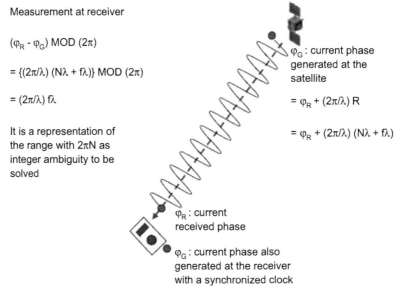

Measurement at receiver

$(\varphi_R - \varphi_G)$ MOD (2π)

$= \{(2\pi/\lambda)\,(N\lambda + f\lambda)\}$ MOD (2π)

$= (2\pi/\lambda)\,f\lambda$

It is a representation of the range with $2\pi N$ as integer ambiguity to be solved

φ_G: current phase generated at the satellite

$= \varphi_R + (2\pi/\lambda)\,R$

$= \varphi_R + (2\pi/\lambda)\,(N\lambda + f\lambda)$

φ_R: current received phase

φ_G: current phase also generated at the receiver with a synchronized clock

FIGURE 5.19

Carrier phase-based ranging.

reference with respect to which the absolute value of the above carrier phase difference may be determined.

Transmission time Tt and received time Tr with respect to any arbitrary start time may be expressed as

$$Tt = k(n_1 2\pi + \varphi_t) \tag{5.53a}$$

$$Tr = k(n_2 2\pi + \varphi_r) \tag{5.53b}$$

Here, $(n_1 2\pi + \varphi_t)$ and $(n_2 2\pi + \varphi_r)$ are those two carrier phases. n_1 and n_2 are the two unknown integers representing the number of complete 2π radians that the respective phases have executed from any arbitrary reference start time. φ_t and φ_r are the corresponding fractional phases. "k" is the constant that converts from phase to the corresponding time with respect to that arbitrary starting instant. It is obvious that factor k is $T/2\pi$, where T is the time period of the carrier. Thus, differencing the two, we get

$$Tr - Tt = k\{(n_2 - n_1)2\pi + (\varphi_r - \varphi_t)\}$$
$$\delta T = kN2\pi + k\delta\varphi(t) \tag{5.54}$$

Here, δT is the propagation time, $\delta\varphi$ is the fractional part of the phase difference measured, and N is the difference between n_2 and n_1. N represents the numbers of complete 2π radians that the phase of the signal advances while traveling from the satellite to the receiver.

Because the measurements are only possible in modulo 2π, only the values of ϕ_r and ϕ_t are known from which we get the differential fractional phase $\delta\phi$. But the integers n_1 and n_2, associated with the two phases cannot be measured directly. Thus, their difference, N, is also not known and is called the "integer ambiguity." This value needs to be solved for proper ranging. This is an involved task and adds the complexity to the receiver.

Multiplying by the velocity of light, c, we get the range, R_{car} on the left hand side (LHS) of the equation and the equation turns into

$$c\delta T = ckN2\pi + ck\delta\phi \qquad (5.55a)$$

but $ck = cT/2\pi = c/\omega = \lambda/2\pi$. So, replacing we get,

$$R_{car} = (\lambda/2\pi)\{N2\pi + \delta\phi(t)\} \qquad (5.55b)$$

Measuring the phases modulo 2π, we can get $\delta\phi$. The value of N in the above equation is unknown and, we need to obtain this value. Searching for N is called ambiguity resolution, which is necessary to get the absolute values of R_{car}. There are different techniques for resolving ambiguity (Cosentino et al. 2006).

Note that the integer ambiguity is a one-time unknown generated at the beginning of the carrier phase measurement process. As time progresses, the equation still holds with the same value of N, but with the updated values of R_{car} and $\delta\phi$ as the currently received $\delta\phi$ changes with time, thus changing the fractional portion of the measured phase. So, at any later time $t + dt$, the range is

$$R_{car}(t + dt) = (\lambda/2\pi)\{N2\pi + \delta\phi(t + dt)\} \qquad (5.56)$$

In this, the receiver can measure only the incremental value of the phase difference $\delta\phi$, i.e., $\delta\phi(t + dt) - \delta\phi(t)$, and add it to the existing value of the fractional phase of Eqn (5.55b) to update it. It can also keep on accumulating the differential phase $\delta\phi$ even above 2π values at any subsequent time from the instant of the start of the measurement. Therefore, the integral part of the equation containing N remains unchanged.

The incremental range from the instant of the start of the measurement can be obtained by differencing the two time separated equations. With this operation, we get

$$R_{car}(t + dt) - R_{car}(t) = (\lambda/2\pi)\{\delta\phi(t + dt) - \delta\phi(t)\} \qquad (5.57a)$$

Or

$$\Delta R_{car} = (\lambda/2\pi)\Delta\delta\phi \qquad (5.57b)$$

The LHS is the differential range with respect to the beginning of the measurement process, and the right hand side (RHS) of the equation is the accumulated differential phase. It is the accumulation of the differential phase difference measured at each instant and accumulated over time from the start of the measurement. Because the temporal change in range results in the Doppler, the differential range ΔR_{car} in the LHS can also be replaced by equivalent Doppler terms as.

So, combining the above equations and representing the the Doppler frequency as f_d, we get

$$(\lambda/2\pi)\Delta\delta\varphi = \Delta R_{car}$$

Or,
$$= \int (dR_{car}/dt)dt \tag{5.58}$$

Or,
$$= (\lambda/2\pi)\int f_d dt$$

Thus, LHS can be regarded as the integrated Doppler that is again derived from the integrated delta range and the RHS is the Integrated Doppler.

In code-based ranging, the maximum error in phase identification that may occur is that corresponding to one code chip. The maximum range error that may happen in such a process of matching is given by

$$\delta R_{cod} = c \times \text{chip width}$$
$$= c/\text{code chip rate} \tag{5.59}$$
$$= c/R_c$$

Assume that if the ranging is done by comparing the carrier phases of the received signal with that generated locally, provided the integer ambiguity is exactly identified, the corresponding range error will be

$$\delta R_c = c \times \text{carrier period}$$
$$= c/\text{carrier frequency} \tag{5.60}$$
$$= c/f_c$$

So, the ratio of the maximum error E is

$$E = \delta R_{Cod}/\delta R_{car}$$
$$= f_c/R_c \tag{5.61}$$

Because the carrier frequency is many times greater than the chip rate of the code, the inaccuracy is reduced manifold by the use of the carrier phase. Thus, the carrier phase measurement is much more precise than that of code phase measurement. The use of a carrier phase, although it improves receiver performance, is computationally more intensive for realization. This includes resolving the issues such as cycle slip and phase scintillations. It also adds to the computation load because of the need to resolve the integer ambiguity.

While continuously measuring the carrier phase of the received signal, the measured phase sometimes shows an abrupt change or a discontinuity in the values. This is called a cycle slip. This causes unwarranted errors in the integrated Doppler term $\Delta\delta\varphi$ appearing in 5.57b or 5.58. This may occur for reasons such as excessively low received signal strength, receiver anomaly, interference, or ionospheric disturbances. The receiver processing algorithm must be robust enough that it can identify any cycle slip occurring during the carrier phase measurement and mend any damaging effect resulting from it. A simple way to handle it is to restart the phase-based ranging again, resetting all accumulated values once the cycle slip occurs.

5.2.6 Navigation processor

A navigation processor is one of the important subsystems of the user receiver. It is marked as Section (3) in Figure 5.4. The navigation signal carries the navigation data with it, the format of which is known a priori to the receiver. The navigation processor receives from its previous modules the decoded data separately from all visible satellites simultaneously with the respective estimated range of each. It then processes them to calculate satellite position. Finally, using this information, the navigation processor calculates the user's position. Although the ranging is done in the processor of the receiver, it is categorically separated from the navigation processor because they have distinct differences in functionality.

To achieve this end, the demodulated data stream from all different channels are required to be received by the processor. It checks and stores the data and utilizes the necessary algorithm for position fixing.

The data are used by the algorithm to obtain the navigation solutions for the final position fix. Associated tasks such as controlling the display and other user interfaces and interfacing with external devices are executed by this processor. In some receivers, a separate logical module is superimposed on the nominal processing tasks, which oversee the activities and perform the management of receiver functionalities. We will now describe the sequence of procedures done on the derived data until the position is estimated and displayed.

5.2.6.1 Handling navigation data

The first job of the navigation processor is to retrieve and store the navigation data. At this stage, the processor is provided with only an uninterrupted stream of data. It needs to identify the frames or the specific message structures and then select the values of the definite parameters from their respective positions in these messages, as per the defined format. Thus, it is important that the receiver identifies when the message frame has actually started in the stream of bits. All subsequent data bits can then be relatively located with respect to this initial reference. However, decoding the channel codes are necessary at this point.

Identification of the data structures is generally carried out using a preamble. Frame identification and synchronization are achieved by utilizing a specific pattern of bits at the beginning of the frame transmitted with the signal. When this pattern is located in the bit stream at the receiver, it is identified as the beginning of a frame. The definite pattern is chosen such that there is a little chance that the same pattern appears in the data in between. Otherwise, false detection may occur. This probability decreases with an increase in the preamble code length and increases with the length of the frame. However, the true preamble is repeated after every frame width, but a similar repetitive pattern appearing probabilistically within the message body is highly unlikely.

Like false locking, there can be missed detection. This condition arises when preamble bits are incorrectly received owing to error in communication. The receiver cannot identify the preamble and goes on searching until it finds the next.

To avoid false detection and counter missed detection, the repeating nature of the preambles is verified during locking, considering that they must reappear after a regular interval of bits. Some appropriate algorithm must be defined to ensure that the identification of proper preamble has been achieved. For example, the frame may be said to be locked when four successive correct preambles are detected at an interval of the frame width.

There are a few intricacies related to this, such as variation of the bit width owing to jitter, etc., that may hinder the synchronization process and need to be taken care of in the receiver.

Data are now checked for corrections and identification of the frames have been done. Each of the constituent parameters of the navigation data are now needed to be extracted and stored along with the time of data and its period of validity. This is because the same navigation data will remain valid up to only a finite interval, and whether this validity has expired or not needs to be checked with every use. However, the data are typically updated well within this period and hence the validity naturally gets extended further. The process of decryption or descrambling needed to be done on the classified data present is performed before this. These stored data are then further processed to get the user's position.

5.2.6.2 Ephemeris extraction and reference positioning

The navigation signal contains ephemeris data, which are required for finding the positions of the respective satellites. These ephemeris data are selected from the dataset already stored. Because the same ephemeris remains valid over a range of time, this activity is repeated only when a new set of updated ephemeris data appears in the transmitted data. In deriving the values of the ephemeris, and for any other parameters as well from this set of navigation data, appropriate scale factors may need to be applied. When the ephemeris is known, current satellite positions are calculated from it using relations as described in Chapter 3.

5.2.6.3 Selection of satellites

We will learn in Chapter 6 that at least four satellites are required to find the user's position and time in an ordinary receiver, and that the accuracy of position estimation depends on the relative geometry of the user and these four satellites. It is represented by a term called geometric dilution of precision (GDOP). The next job of the navigation processor is to select the best four satellites based on the GDOP. However, an approximate position of the user receiver and the satellites needs to be known a priori to obtain the GDOP. If the approximate user position is not known to the receiver, an alternative approach may be to use any four satellites at first and get the first position estimate, although this may lead to a compromise in the position accuracy. Using this approximate position, one may obtain the DOP values and select the most suitable four satellites to further obtain the precise position.

These values must be updated at frequent intervals if not at every estimate, because, owing to the dynamics of the user and the satellite, this value goes on changing with time.

5.2.6.4 Range corrections

Although the range estimation is done in the processor, it may be considered an activity separate from those done by the navigation processor with the objective of position fixing. According to the perspective of the navigation processor, the range is obtained from an external ranging processor.

The atmosphere around the earth affects the traveling speed of RF signals and causes propagation delay errors. These errors may either be derived from the correction parameters sent in the data or may be obtained directly from dual frequency range measurements and the range may be corrected. Similar corrections may be done for the range errors occurring for reasons other than those for propagation.

5.2.6.5 Calculate user position and time

Using the corrected pseudo range measurements and the calculated satellite positions, the user position coordinates and the time offset of the receiver clock from that of the satellite can be calculated by solving the following range equation:

$$R_i = \sqrt{\left\{(x_i - x_u)^2 + (y_i - y_u)^2 + (z_i - z_u)^2\right\}} + b_u \qquad (5.62)$$

where (x, y, z) is the user position coordinates; the satellite coordinates (x_i, y_i, z_i) are known from ephemeris data where i is the satellite number, i.e. 1, 2, 3, 4. The pseudo range R_i for four satellites can be estimated by measuring signal propagation delay. We will learn about the necessary algorithms for position estimation in the next chapter. To improve the computed position accuracy, a Kalman filter may be used. Kalman filters are briefly described in Chapter 9.

5.2.6.6 Convert position

The user position calculated from simultaneous equations is in a Cartesian coordinate system. It is usually desirable to convert positions to geodetic coordinates and represents the position in terms of latitude, longitude, and altitude. Assuming the shape parameters of the earth, that can be obtained from the datum these conversions may be easily done.

5.2.6.7 User interface

The position and time thus obtained need to be presented to the user through a suitable interface. A convenient display interface is popularly used for the purpose. Typically, other information such as the signal strength, satellite positions, and expected precision are displayed along with the position. Graphical displays such as sky plots showing the satellite positions and geographical grids for user positions are also common.

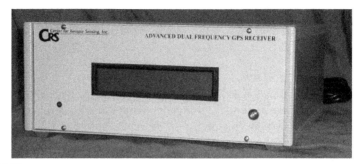

FIGURE 5.20

Complete navigation receiver.

Courtesy: CFRSI

The user interface is also needed for purposes other than the display of positions, such as configuring the receiver, exchanging data with other devices through suitable interfaces for application, and storage. The necessary conversion of the data format is also done here. The user interface also needs to support encryption key input.

Finally, the whole receiver is packed in a form suitable for the user. Receivers developed for any particular application need to have some appropriate casing suitable for that specific application. Figure 5.20 shows a type of complete navigation receiver in its final shape.

Conceptual questions

1. Do you expect the accuracy of the ranging to change if the early—late correlator difference is changed?
2. How can the navigation process handle a situation when it is reading from a stored navigation file and a new set of data arrives?
3. How does it help to have the code-based ranging simultaneously when carrier phase-based ranging is done?
4. What are the advantages of using early—late power rather than early—late during code tracking?
5. Which of the mentioned PLL discriminators would you prefer to use when there is a finite probability of large deviations?

References

Cooper, G.R., McGillem, C.D., 1986. Modern Communications and Spread Spectrum. Mcgraw Hill, USA.

Cosentino, R.J., Diggle, D.W., de Haag, M.U., Hegarty, C.J., Milbert, D., Nagle, J., 2006. Differential GPS. In: Kaplan, E.D., Hegarty, C.J. (Eds.), Understanding GPS Principles and Applications, second ed. Artech House, Boston, MA, USA.

Global Positioning System Directorate, 2012a. Navstar GPS Space Segment/Navigation User Interfaces. IS-GPS-200G. Global Positioning System Directorate, USA.

Global Positioning System Directorate, 2012b. Navstar GPS Space Segment/User Segment L1C Interfaces. IS-GPS-800C. Global Positioning System Directorate, USA.

Groves, P.D., 2013. Principles of GNSS, Inertial and Multisensor Integrated Navigation System. Artech House, London.

Lathi, B.P., 1984. Communication Systems. Wiley Eastern Limited, India.

Maral, G., Bosquet, M., 2006. Satellite Communications Systems, fourth ed. John Wiley & Sons Ltd., U.K.

Mutagi, R.N., 2013. Digital Communication: Theory, Techniques and Applications. Oxford University Press, New Delhi, India.

Proakis, J.G., Salehi, M., 2008. Digital Communications, fifth ed. Mc Graw Hill, Boston, USA.

Scott, L., Iovancevic, A., Ganguly, S., 2001. Rapid signal acquisition techniques for civilian and military user equipments using DSP based FFT processing. In: Proceedings of 14th International Technical Meeting of the Satellite division of The Institute of Navigation, Salt Lake City, UT, pp. 2418–2427.

Spilker Jr., J.J., 1996. Fundamentals of signal tracking theory. In: Parkinson, B.W., Spilker Jr., J.J. (Eds.), Global Positioning Systems, Theory and Applications, vol. I. AIAA, Washington DC, USA.

Van Dierendonck, A.J., 1996. GPS receivers. In: Parkinson, B.W., Spilker Jr., J.J. (Eds.), Global Positioning Systems, Theory and Applications, vol. I. AIAA, Washington DC, USA.

Van Dierendonck, A.J., Fenton, P., Ford, T., 1992. Theory and performance of narrow correlator spacing in a GPS receiver. Navigation 39 (3), 265–283.

Ward, P.W., Betz, J.W., Hegarty, C.J., 2006. Satellite signal acquisition, tracking and data demodulation. In: Kaplan, E.D., Hegarty, C.J. (Eds.), Understanding GPS Principles and Applications, second ed. Artech House, Boston, MA, USA.

Navigation Solutions

6

In this chapter, we will learn to perform the most important activity in the whole satellite navigation process, estimating the position. The whole purpose of the navigation is to achieve this objective. Here, we will first understand the approach to obtaining the solution intuitively. Then, we shall see how the inputs required to form the mathematical equations needed to solve for the position are obtained. Finally, we will obtain the solution mathematically using the linearization technique, and will discuss a few intricacies of the process. As a continuation, we will also learn other processes to obtain the same solution.

6.1 Fundamental concepts

In a real-life scenario, all distances are required to be measured in a three-dimensional (3D) space. In Chapter 1, we mentioned that to represent the position of a point with respect to a reference, we need to fix the position of the reference and get the distance of the point from it along three orthogonal directions. Now, let us see in detail the geometrical aspects of setting distances along three orthogonal

Understanding Satellite Navigation. http://dx.doi.org/10.1016/B978-0-12-799949-4.00006-3

directions. We know that a point in a 3D space is the result of the intersection of three nonparallel planes. So, our task to define the point reduces to defining these three planes. Once the reference system is set, the reference point is consequently defined and the directions of the three orthogonal axes are fixed. The origin of the coordinate system (i.e. the reference point coordinates) becomes $(0, 0, 0)$. To define the position of point $P (x_1, y_1, z_1)$, we first fix a distance x_1 along axis x. This defines a plane $x = x_1$ that is parallel to the yz plane at this point. This is at a distance x_1 from the origin along the x axis. Thus, moving this distance along the x axis, our position is reduced to some point $(x_1, 0, 0)$ on this plane. The job is to define the other two planes orthogonal to the one already defined. At this stage, we define another plane, $y = y_1$, which is parallel to the xz plane with the normal distance, y_1, from the origin. This further restricts the locus of our point on a straight line parallel to the z axis formed by the intersection of these two planes defined. Moving a distance, y_1, from the current position along the y axis puts us on some point $(x_1, y_1, 0)$ on this line. Then, as the final plane $z = z_1$, perpendicular to the z axis is defined, it intersects the mentioned line of locus at $z = z_1$. Again moving a distance, z_1, along the z axis, we reach the intersection of the three planes at the position (x_1, y_1, z_1). Thus, these three mutually perpendicular movements completely fix our point at the point $P (x_1, y_1, z_1)$. This is shown in Figure 6.1.

Because the task is only to define the planes, it can be done with respect to any arbitrary reference point in the frame using the relative distances. Even then, the point in question may be fixed with respect to the original reference point, if we know the position of this new secondary reference in the original frame. Defining planes with respect to any reference is done by representing the vectorial distance of these planes from this new reference point to the point of interest in terms of three independent bases or coordinates. Remember, we said the vectorial distance and not simple range, because it requires both a sense of direction and magnitude to define the point.

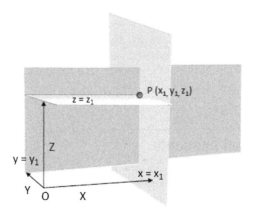

FIGURE 6.1

Defining a position in three dimensions.

But, what if we know only the radial range from this new reference point to our point of interest instead of the vectorial distance? In a 3D space, any distance vector should have three independent components. In a spherical system, the three coordinates are the radial range and two angular deviations from some fixed planes. So, when we mention only the range, it means we lose two sets of information out of three. In such cases, the exact position of the point of interest may still be derived by adding some more independent information, which effectually compensates the loss. This may be done by adding independent range measurements from other new reference points whose positions are known. The basic idea is to form definite intersecting surfaces to reduce the possible intersection to a point.

This is what is done during satellite navigation. Here, the new planes are relative to different reference points, which are the point of locations of three satellites whose absolute positions are known in the chosen reference frame. The ranges from these satellites to the point form the independent information from which three independent spherical surfaces may be obtained. Using these, the position of the point may be derived locating the point of intersection in terms of absolute reference. However, because the surfaces thus created are nonlinear, it requires an adequate number of such surfaces to explicitly indicate the position.

In a 3D space, two planes intersect to form a straight line that is a linear function of the coordinates. On the contrary, two spherical surfaces intersect to form a circle that is a quadratic function. So, unlike the former case, in which an additional plane surface sufficiently defines a point, an additional spherical surface intersects with the circle at two different points, and hence is not enough to explicitly fix a point.

How much information, therefore, is sufficient for this need? This may be elucidated by considering an example. First, let us understand in a scenario where we attempt to represent the position of a point, P, located on the xy plane, i.e. in a two dimensional (2D) space.

In 2D space there are two unknowns for the location of a point, P. In Cartesian coordinates, these are "x" and "y" with respect to the absolute reference point O at the origin of the axes. The point may also be represented in terms of its range from the origin and its direction given by the angle it makes with the axes. The range may be obtained from the individual Cartesian coordinates as $r = \sqrt{x^2 + y^2}$ and the angle, $\theta = \tan^{-1}(y/x)$. If we know only the radial range of the point from this reference, we can only represent the distance as

$$r^2 = x^2 + y^2 \tag{6.1a}$$

This is the equation about the origin with radius r. In squaring the values, we actually lose the information about its sign and cannot find θ from it. Thus, we have a deficit of information, its exact direction.

Furthermore, let there be more such information available that states that the radial range of the same point from a new reference point at (x_1, y_1) is r_1. So,

$$(x - x_1)^2 + (y - y_1)^2 = r_1^2 \tag{6.1b}$$

The above equation represents another circle centered at (x_1, y_1) with radius r_1. Replacing Eqn (6.1a) in Eqn (6.1b), we get

$$x_1^2 + y_1^2 - 2xx_1 - 2yy_1 = r_1^2 - r^2 \tag{6.2}$$

This is a linear equation of the form $ax + by = c$, where $a = 2x_1$, $b = 2y_1$, and $c = r^2 - \{r_1^2 - (x_1^2 + y_1^2)\}$. Thus, two quadratic equations formed out of two observations in a 2D space thus represents two circles which intersect at two different points that lie on a straight line represented by a linear function of x and y. The values of the coordinates are yet to be determined. The solution can be obtained with an additional equation. Stated mathematically, this establishes the fact that the quadratic nature of the range equation results in two possible roots of the unknowns are formed from two equations; and we need an additional equation to get the exact solution.

The same fact is also evident from the corresponding geometry of the equations as shown in Figure 6.2. In 2D space, using only the range information R_1 and R_2 from two relative reference points, A and B, respectively, leads to ambiguity with probable positions at P_1 and P_2, where the conditions set by both equations satisfy. In addition, if point P is known to be positioned at a distance R_3 from another relative reference point C, it must lie on the perimeter of the circle around C with radius R_3. This cuts only the point P_1 among the probable two. Only then can we unambiguously find the position of P.

We can generalize this observation for higher-order spaces with more numbers of unknowns. For "k" numbers of unknowns, k nonlinear equations of order 2 leave us with two equiprobable solutions. Thus, we need 1 more equations to explicitly solve for the unknowns. Therefore, a total of $[k + 1]$ equations are required. So, in real-life conditions, where the observation equations are quadratic in a 3D space, it requires $3 + 1 = 4$ equations to solve for positions explicitly using them.

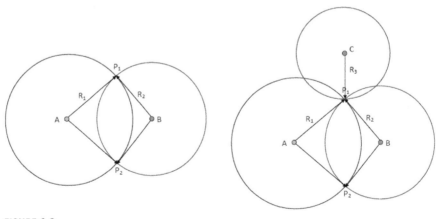

FIGURE 6.2

Two-dimensional case of position fixing.

FOCUS 6.1 REQUIREMENT OF EQUATIONS FOR SOLUTION

Here, we explain with simple examples the requirement of numbers of equations in the case of spherical nonlinearity. We treat the problem for both cases of solving with a constraint equation and for an additional observation equation of spherical nature.

Suppose we have a second-order equation such as:

$$x^2 + y^2 = 9 \qquad \text{(i)}$$

To solve for variables x, y, and z, we need more equations like this. Let another equation be

$$x^2 + y^2 - 10x + 9 = 0 \qquad \text{(ii)}$$

Combining Eqns (i) and (ii), we get the value of x as

$$x - 18/10$$
$$= 1.8 \qquad \text{(iii)}$$

Thus, the point at which these two equations simultaneously satisfy has the value of x as 18/10. However, taking this expression for x derived in Eqn (iii) and putting in any of the equation of Eqn (i) or Eqn (ii) leaves us with a quadratic in y. Putting it in Eqn (i), we get,

$$(18/10)^2 + y^2 = 9$$
$$\text{or,} \quad 100\, y^2 = 576$$
$$\text{or,} \quad y = \pm 2.4$$

Thus, from these two Eqns (i) and (ii), we cannot obtain an explicit solution of (x,y) even in a 2D case. Now, we need to add a new equation.

First, let us suppose that we add a constraint equation of the form

$$3x + 4y = 15 \qquad \text{(iv)}$$

This is a constraint equation that states that the solutions should also simultaneously lie on the line defined by it. Putting the value of x = 1.8 into this equation, we get the exact value of y as

$$y = (15 - 3 \times 1.8)/4$$
$$= 2.4$$

Thus, we get the exact solution for x and y.

Second, assume that we have a separate equation of the form

$$5x^2 + 5y^2 - 18x - 44y + 93 = 0 \qquad \text{(v)}$$

Notice that all of the nonlinear equations that we have used are spherical in nature. Again, using the derived value of x = 1.8 here, we get

$$16.2 + 5y^2 - 32.4 - 44y + 93 = 0$$
$$\text{or,}\ 5y^2 - 44y + 76.8 = 0 \qquad \text{(vi)}$$

So, the possible solutions of y are

$$y = 6.4$$
$$\text{and } y = 2.4$$

Therefore, the actual solution of (x, y) is (1.8, 2.4), which we get from three independent equations.

Let us consolidate what we have learned in this discussion:

1. First, the position of any unknown point may be represented in a reference frame, by knowing the position of a new reference point in the frame and vectorial distance from this relative reference point to the point of interest.
2. If, instead of the vectorial distance, only the radial distance (range) is known, the lack of adequate information may be compensated for by adding enough numbers of similar range of information from other such points of known location to explicitly define the position of the point.

6.2 Generation of observation equation

This general observation may be extended to the practical scenario of 3D space, in accordance with the requirements of satellite navigation.

In satellite navigation systems, the preferred absolute reference system is the geocentric earth-centered earth-fixed (ECEF). However, it may be transformed to any other frame according to the requirements. The representation of the position of any point in this frame may be done using the secondary references, which are the navigation satellites in the sky. For this, we obviously need the absolute position of the satellite and the vectorial distance of the point from the satellite. However, because it is not feasible to obtain vectorial distances, positioning of the point can still be done, as we have just seen, if we know the ranges from other satellites placed at known locations. This is known as *trilateration*, the estimation of the position of a point unambiguously based on the measurements of distances from three or more known reference locations.

To estimate a position of a point using a satellite navigation system, it is thus necessary to have two things ready: the position of the satellites and the distance of the point from these satellites. Range observation equations are generated from this information and the position is estimated in 3D space by solving these equations. Therefore, two aspects are important at this point:

1. To obtain this required range information
2. To solve the equations

We have discussed the first aspect in the previous chapter, and hence shall concentrate only on the second here.

Ranges of satellites are measured at the receiver. If these ranges measured from three satellites, S1, S2, and S3 are R_1, R_2, and R_3, respectively, we get three quadratic equations, formed using three reference positions and the corresponding ranges. These equations are

$$R_1 = \sqrt{(x_{S1} - x)^2 + (y_{S1} - y)^2 + (z_{S1} - z)^2} \tag{6.3a}$$

$$R_2 = \sqrt{(x_{S2} - x)^2 + (y_{S2} - y)^2 + (z_{S2} - z)^2} \tag{6.3b}$$

$$R_3 = \sqrt{(x_{S3} - x)^2 + (y_{S3} - y)^2 + (z_{S3} - z)^2} \tag{6.3c}$$

where, x_{si}, y_{si}, and z_{si} are the coordinates of the ith satellite (i = 1, 2, 3) and x, y, z are the unknown user position. This constitutes the set of observation equations.

Geometrically, the ith of the above equations state that the user is located anywhere on a spherical surface formed with a radius R_i with the satellite located at (x_i, y_i, z_i) placed at the center of the sphere. Simultaneous measurements from two such references put the possible location of the user on the intersection of two spheres. Using two of the equations, of (6.3), the information may be reduced to a linear equation in (x, y, z), signifying that the intersection of the corresponding two spheres leads to planar geometry. The possible locus of the positions actually forms a circle that lies on this plane represented by the obtained linear equation in 3D space.

A third reference satellite and the measured range of its user generate the third Eqn (6.3c). This equation combines with the effective circular intersection obtained previously to reduce the positions of the point into two solutions at two equally possible locations. Therefore, either an additional measurement from a fourth satellite is required or there must be some separate independent relationship existing between the user's coordinates. The latter is called the constraint equation. This additional information augmenting the three observation equations suffices to obtain the particular solution for the position.

Thus, the nonlinearity of range equations demands four navigational satellites to solve the unknown position coordinates. However, this issue of solving the nonlinearity problem can be resolved in a different manner. Nevertheless, we shall see that even without nonlinearity, we still require four satellites to solve the position; the reason for this is different and will be clear in the next section. First, let us see how this nonlinearity problem is resolved.

6.3 **Linearization**

From our discussion in the last section, the problem of position fixing is reduced to solving the problem of simultaneous quadratic equations. At the same time, we know that the three simultaneous quadratic equations in (6.3) cannot be solved to get the unknowns, x, y, and z unambiguously, because these quadratic equations give two possible solutions for their unknown variables. We therefore need more information to solve the unknowns.

There are three different methods to attack this problem. One is to take an additional independent observation equation and solve four simultaneous quadratic equations. The second option is to use a constraint equation instead, to resolve the ambiguities that arise on solving three quadratics. A constraint equation is the definite relation that the unknown variables always maintain between them. The final possibility, as chosen in most cases of navigational receivers, is that these quadratic equations are linearized to form three linear differential equations. We will now see how it is done.

Linearization (Kaplan et al. 2006), defined in this context, is the technique of converting quadratic observation equations into linear differential equations about a fixed nominal approximated position, which may be an initial guess of solution. These linear equations are then solved by standard methods to get the differential values of the true position coordinates with respect to those of the initial guess. Once these differential values are estimated, they can be added to the initial guess to obtain the absolute position solution.

To elaborate on this, we first start from Taylor's theorem, which says that if the value of a multivariate function $f(X)$ is known at a point X_0, its value at a nearby point X is

$$f(X) = f(X_0) + f'(X_0) \, dX + \tfrac{1}{2} f''(X_0) \, dX^2 + \dots \qquad (6.4)$$

where f' and f'' are the first- and second-order derivatives of function f, respectively, with respect to X and obtained at the known position X_0.

In our case, the function is the range, R, of the satellite from the user position, P, which is a function of both user position $P = (x, y, z)$ and that of the satellite, $P_s = (x_s, y_s, z_s)$. Thus, we can write function $R = R(x_s, y_s, z_s \, x, y, z)$ as

$$R = \sqrt{(x_s - x)^2 + (y_s - y)^2 + (z_s - z)^2} \qquad (6.5)$$

For any instant (supposing we have frozen the time at that instant), the known satellite positions are fixed. Effectively, the range remains a function of unknown user position variables (x, y, z). At this instant, let us consider an approximate (practically close enough) user position, $P_0 = (x_0, y_0, z_0)$. Now, expressing the range at true position P by expanding the range function about the point and in terms of the range at position P_0 according to Taylor's theorem, and considering only up to the first-order derivatives, it follows from Eqn (6.4),

$$R(x, y, z) = R(x_0, y_0, z_0) + \partial R/\partial x|_{P0} \, \Delta x + \partial R/\partial y|_{P0} \, \Delta y + \partial R/\partial z|_{P0} \, \Delta z$$

$$\text{or,} \quad R(x, y, z) - R(x_0, y_0, z_0) = \partial R/\partial x|_{P0} \, \Delta x + \partial R/\partial y|_{P0} \, \Delta y + \partial R/\partial z|_{P0} \, \Delta z$$

$$(6.6)$$

The higher-order derivatives can be neglected owing to our assumption of the close proximity of our approximated position from the true values. Here, the finite differences along the coordinates between the two points are represented as

$$\Delta x = x - x_0$$
$$\Delta y = y - y_0 \qquad (6.7)$$
$$\Delta z = z - z_0$$

Calling the geometrically calculated range from the P_0 to the satellite as R_0, so that

$$R_0 = \sqrt{(x_s - x_0)^2 + (y_s - y_0)^2 + (z_s - z_0)^2} \qquad (6.8)$$

Equation (6.6) can be rewritten as

$$R - R_0 = (\partial R/\partial x \, \Delta x + \partial R/\partial y \, \Delta y + \partial R/\partial z \, \Delta z)\big|_{P_0}$$

$$\text{or,} \qquad \Delta R = (\partial R/\partial x \, \Delta x + \partial R/\partial y \, \Delta y + \partial R/\partial z \, \Delta z)\big|_{P_0} \qquad (6.9)$$

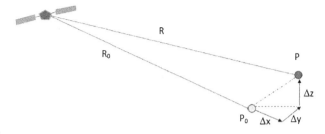

FIGURE 6.3

Elements of linearization process.

where $\Delta R = R - R_0$, is the finite differential range between the true and the approximated position. This is illustrated in Figure 6.3.

Partially differentiating the expression of the range equation with respect to x, y, and z at the approximated position of $P_0(x_0, y_0, z_0)$, and putting the values into Eqn (6.9), we get

$$\Delta R = -(x_s - x_0)/R_0\, \Delta x - (y_s - y_0)/R_0 \Delta y - (z_s - z_0)/R_0\, \Delta z$$
$$= g_\alpha\, \Delta x + g_\beta\, \Delta y + g_\gamma\, \Delta z \tag{6.10}$$

g_α, g_β, and g_γ are the partial derivatives of the range at the approximated point with respect to x, y, and z, respectively. Because the position coordinates of this nominal point (x_0, y_0, z_0) are known, the values of g_α, g_β, and g_γ can easily be determined. They also represent the direction cosines of the vector joining the satellite from the nominal position. The negative sign shows that an increment in Δx will result in a decrement in the range error with the sense that the parameters are chosen.

Thus, the nonlinear range equations with unknown coordinates have turned into a linear differential equation about the nominal position, assumed a priori, with relative errors in position as unknowns. Thus, the set of such linearized observation equations can be solved to obtain these position errors for each coordinate. Because the nominal position is known, the true position can be determined from it by adding the estimated relative errors to it.

Nevertheless, we still need four satellites, even after linearization to find the position and time. To understand that, let us recall how the user receiver measures the ranges of each reference satellite. Broadly, it measures the range from the time delay with which a definite phase of the signal, transmitted from the satellite, is received at the receiver. The fundamental question here is, how does the receiver know when any signal is being transmitted by the satellite? Here comes the utility of the ranging codes. We learned in Chapter 4 that in a satellite navigation system, the ranging codes are transmitted synchronously with the navigation data and carrier phase of the signal such that the transmission instant of each code bit or the partial phase thereof can be exactly derived from the time stamp present in the signal and the knowledge of the code chip rate. The received time is derived from the clock present in the receiver itself. Thus, the propagation time is measured from the difference of time at which a message was transmitted and the time when the message was received at the receiver. This time interval is multiplied by the velocity of light, c, to obtain the range.

The transmission and the received times are obtained from two different clocks: the former from the satellite clock and the latter from the receiver. But the issue here is that the receiver clock does not have as much accuracy and precision as the satellite clock. The satellite clock is an atomic clock of very high stability, about $\sim 10^{-13}$, and thus keeps time very accurately. The receiver clock, on the other hand, is cheap and consequently its stability is low, of the order of about $\sim 10^{-6}$ to 10^{-9}. Thus, the clock at the receiver is not synchronous to the satellite atomic clock and drifts with respect to the latter, leading to a relative clock offset. There remains an intrinsic time delay (or advancement) between the satellite and the receiver time. This affects the ranging process and hence the measured range. The error in ranging is equal to the product of the offset between the two clocks and the velocity of light.

Let a definite phase of the signal be transmitted at true time T_t, which is also the satellite clock time. Also, let this phase be received at true time T_r after a traverse time of $\Delta t = T_r - T_t$. At the instant of receiving, if the receiver clock is shifted by an amount, $+\delta t_u$, with respect to the satellite time, the time that the receiver registers for to receive the phase is $T_r + \delta t_u$. Thus, to the receiver the time of propagation is

$$\begin{aligned} \Delta t_u &= (T_r + \delta t_u) - T_t \\ &= (T_r - T_t) + \delta t_u \\ &= \Delta t + \delta t_u \end{aligned} \tag{6.11a}$$

The range thus obtained at the receiver is

$$\begin{aligned} R &= c\Delta t_u \\ &= c\Delta t + c\delta t_u \\ &= \rho + c\delta t_u \end{aligned} \tag{6.11b}$$

An error of $c\delta t_u$ incurs in range in the process owing to its clock shift. It must be appreciated that an error of $1\mu s$ leads to an error of 300 m. This resultant offset in range remains added to the measured range. Therefore, this unknown receiver clock offset also needs to be determined for range correction.

This resultant offset in range is taken as an unknown variable in solving for the position. So, even for the linear equations, one more unknown is added in addition to the three unknown position coordinate variables. Thus, to get the solution for these four unknown variables, four linearized observation equations are required. One more reference satellite and its distance are therefore necessary. This results in a requirement of four satellites for position and time estimation.

6.4 Solving for position

With the introduction of a new unknown (i.e. the receiver clock shift with respect to the satellite clock), the basic observation equation changes as given in Eqn (6.11b) and below as

$$R_i = \rho i + c\delta t_u$$

where R_i is pseudorange measurement to the ith satellite, ρ_i is the corresponding geometric range, and δt_u is the receiver clock offset with respect to the satellite time. Because all satellite clocks are synchronous, this shift is the same for all observations. We are considering the effective additional length that gets erroneously added to the range as a result of to this shift, so we multiply the clock bias, δt_u with c, the velocity of light in vacuum, to convert the same into an effective range error.

Four observation equations for four satellites can be written by expanding the geometric range as a function of coordinates as

$$R_1 = \sqrt{(x_{S1} - x)^2 + (y_{S1} - y)^2 + (z_{S1} - z)^2} + c\delta t_u$$
$$R_2 = \sqrt{(x_{S2} - x)^2 + (y_{S2} - y)^2 + (z_{S2} - z)^2} + c\delta t_u$$
$$R_3 = \sqrt{(x_{S3} - x)^2 + (y_{S3} - y)^2 + (z_{S3} - z)^2} + c\delta t_u \qquad (6.12)$$
$$R_4 = \sqrt{(x_{S4} - x)^2 + (y_{S4} - y)^2 + (z_{S4} - z)^2} + c\delta t_u$$

These simultaneous observation equations are linearized about the approximate solution of $X_a = (x_0, y_0, z_0, c\delta t_{u0})$, and the resultant linearized equations become

$$\Delta R_1 = -G_{\alpha 1} \cdot \Delta x - G_{\beta 1} \cdot \Delta y - G_{\gamma 1} \cdot \Delta z + \Delta b$$
$$\Delta R_2 = -G_{\alpha 2} \cdot \Delta x - G_{\beta 2} \cdot \Delta y - G_{\gamma 2} \cdot \Delta z + \Delta b$$
$$\Delta R_3 = -G_{\alpha 3} \cdot \Delta x - G_{\beta 3} \cdot \Delta y - G_{\gamma 3} \cdot \Delta z + \Delta b \qquad (6.13a)$$
$$\Delta R_4 = -G_{\alpha 4} \cdot \Delta x - G_{\beta 4} \cdot \Delta y - G_{\gamma 4} \cdot \Delta z + \Delta b$$

where all terms are in accordance with the terms defined in Eqns (6.6)–(6.10) and have the ordered index for the four different satellites. Δb is the new term added and represents the finite difference of the true effective range error owing to the receiver clock bias from the same assumed in the initial guess of the same (i.e. $\Delta b = c [\delta t_u - \delta t_{u0}]$). The set of equations in (6.13a) may be written in matrix form as

$$\Delta R = G_a * \Delta X_a \qquad (6.13b)$$

where

$$G_a = \begin{pmatrix} -G_{\alpha 1} & -G_{\beta 1} & -G_{\gamma 1} & 1 \\ -G_{\alpha 2} & -G_{\beta 2} & -G_{\gamma 2} & 1 \\ -G_{\alpha 3} & -G_{\beta 3} & -G_{\gamma 3} & 1 \\ -G_{\alpha 4} & -G_{\beta 4} & -G_{\gamma 4} & 1 \end{pmatrix}$$

$$\Delta R = [\Delta R_1 \quad \Delta R_2 \quad \Delta R_3 \quad \Delta R_4]^T \quad \text{and} \quad \Delta X_a = [\Delta x \quad \Delta y \quad \Delta z \quad \Delta b].$$

Now ΔX_a may be obtained by standard techniques such as iteration, simple least-squares solution, weighted least-squares solutions, and so forth (Axelrad and Brown, 1996; Strang, 2003). The least-squares solution becomes

$$\Delta X_a = \left(G_a^T G_a\right)^{-1} G_a^T \Delta R \qquad (6.14)$$

When this derived ΔX_a is added to the initially assumed approximate position, X_a, about which the equations have been linearized, the true position solution is obtained as $X = X_a + \Delta X_a$.

Recall that we assumed at the beginning that the approximated point is near the true position such that the linearity condition in the error holds good and higher-order differentials of Taylor's theorem may be omitted. However, it is not always possible to approximate such a position because the receiver may have no idea about the location. In such cases (which are more likely), the estimation may start with any arbitrary approximate position where the above conditions are not fulfilled and hence higher order terms in the Teylor's series exist. As a result of neglecting these higher-order terms in the estimation process, the solution arrived at will definitely carry some error. However, it will be near the real position. So, this solved position of the first estimation may now be taken as the initial guess and another iteration of solving for the position may be carried out with the same set of data to reach a further nearer position. This way, after a few iterations for a single definite point, the solution will converge to the true position. These situations may thus be handled through the iteration process, in which the previous steps are required to be repeated until solution is obtained. Box 6.1 explains the estimation process where the solution obtained after few iterations.

FOCUS 6.2 SOLVING FOR POSITION

We assume the following constant values for radius of earth, Re, and radial distance of satellites, Rs, from the earth's center expressed in km is as below:

$$Re = 6.3781 \times 10^3$$
$$Rs = 2.6056 \times 10^4$$

Let the true position of the user be:

$$\text{Latitude} = 22° \text{ N}$$
$$\text{Longitude} = 88° \text{ E}$$

The coordinates in ECEF frame are:

$$x_t = 206.3853 \text{ in km}$$
$$y_t = 5.9101 \times 10^3 \text{ in km}$$
$$z_t = 2.3893 \times 10^3 \text{ in km}$$

Let the clock offset of the user receiver be such that $c*\Delta t = b = 15$ in kilometers. These are unknown values initially. Only the satellite positions and measured ranges are known. The solution of the position estimation, obtained from measured ranges and approximate position, must converge to these values.

Let the Cartesian coordinates of satellites S1, S2, S3, and S4 in ECEF frame in kilometers, as obtained from the satellite ephemeris transmitted by the satellites in their message, be:

Sat:S1	Sat:S2	Sat:S3	Sat:S4
$x_{S1} = 2.1339 \times 10^3$	$x_{S2} = 0.0$	$x_{S3} = 4.1006 \times 10^3$	$x_{S4} = 2.0581 \times 10^3$
$y_{S1} = 2.4391 \times 10^4$	$y_{S2} = 2.4484 \times 10^4$	$y_{S3} = 2.3256 \times 10^4$	$y_{S4} = 2.3525 \times 10^4$
$z_{S1} = 8.9115 \times 10^3$	$z_{S2} = 8.9115 \times 10^3$	$z_{S3} = 1.1012 \times 10^4$	$z_{S4} = 1.1012 \times 10^4$

The measured range for satellite S1 to S4, expressed in km are:

$$R_{t1} = 1.9708 \times 10^4$$
$$R_{t2} = 1.9702 \times 10^4$$
$$R_{t3} = 1.9773 \times 10^4$$
$$R_{t4} = 1.9714 \times 10^4$$

Initially, the true position of the user is not known. Thus, let us take an initial approximate position in Cartesian coordinates in ECEF frame as:

$$x_{a0} = 458.9177 \text{ km}$$
$$y_{a0} = 5.8311 \times 10^3 \text{ km}$$
$$z_{a0} = 2.5433 \times 10^3 \text{ km}$$
$$b_{a0} = 10 \text{ in km}$$

and the ranges calculated in km from the initial approximate position to four satellites are:

$$R_{a01} = 1.9703 \times 10^4$$
$$R_{a02} = 1.9726 \times 10^4$$
$$R_{a03} = 1.9723 \times 10^4$$
$$R_{a04} = 1.9691 \times 10^4$$
$$dR_{a0} = [4.2000; -23.5262; 50.2975; 23.1277]$$

The directional cosines with respect to the approximate position with satellite S1 are:

$$G_{\alpha 1} = 0.0851 \quad G_{\alpha 2} = -0.0233 \quad G_{\alpha 3} = 0.1847 \quad G_{\alpha 4} = 0.0813$$
$$G_{\beta 1} = 0.9424 \quad G_{\beta 2} = 0.9461 \quad G_{\beta 3} = 0.8839 \quad G_{\beta 4} = 0.8990$$
$$G_{\gamma 1} = 0.3234 \quad G_{\gamma 2} = 0.3230 \quad G_{\gamma 3} = 0.4296 \quad G_{\gamma 4} = 0.4303$$

The observation equations after linearization turns into:

$$04.2 \quad = -0.0851 \, dx - 0.9424 \, dy - 0.3234 \, dz + db$$
$$-23.5262 = +0.0233 \, dx - 0.9461 \, dy - 0.3230 \, dz + db$$
$$50.2975 = -0.1847 \, dx - 0.8839 \, dy - 0.4296 \, dz + db$$
$$23.1277 = -0.0813 \, dx - 0.8990 \, dy - 0.4303 \, dz + db$$

This can be expressed in the form: $dR = G_a * dX_a$

$$\begin{pmatrix} 04.20 \\ -23.5262 \\ 50.2975 \\ 23.1277 \end{pmatrix} = \begin{pmatrix} 0.085055 & 0.94244 & 0.32337 & 1 \\ -0.023277 & 0.94611 & 0.32301 & 1 \\ 0.184740 & 0.88393 & 0.42959 & 1 \\ 0.081258 & 0.89903 & 0.43029 & 1 \end{pmatrix} \begin{pmatrix} dx \\ dy \\ dz \\ b \end{pmatrix}$$

The dX_a can be simply solved as: $dX_a = G_a^{-1} * dR$

$$dX_a = \begin{pmatrix} -252.9364 \\ 73.1880 \\ -156.2947 \\ 1.1212 \end{pmatrix}$$

Hence, the new positions after the first iteration become $X_{a1} = X_{a0} + dX_a$:

$$x_{a1} = 458.9177 - 252.9364 = 205.9813 \text{ km}$$
$$y_{a1} = 5.8311 \times 10^3 + 073.1880 = 5.9043 \times 10^3 \text{ km}$$
$$z_{a1} = 2.5433 \times 10^3 - 156.2947 = 2.3870 \times 10^3 \text{ km}$$
$$b_{a1} = 10 + 001.1212 = 11.1212 \text{ km}$$

After the first iteration, we come closer to the actual solution of position than the first approximation.

Now, with the new values of x, y, z, and b, we repeat the steps. From the position obtained after the first iteration, the ranges calculated from the initial approximate position to the four satellites and expressed in km are:

$$R_{a11} = 1.9710 \times 10^4$$
$$R_{a12} = 1.9704 \times 10^4$$
$$R_{a13} = 1.9775 \times 10^4$$
$$R_{a14} = 1.9716 \times 10^4$$
$$dR_{10} = [-2.3798, -2.3653, -2.3110, -2.3670]$$

Note that the absolute values of differential ranges have reduced compared with the first iteration. The directional cosines with respect to the approximate position with satellite S1 are:

$$G_{\alpha 1} = 0.0979 \qquad G_{\alpha 2} = -0.0105 \qquad G_{\alpha 3} = 0.1971 \qquad G_{\alpha 4} = 0.0940$$
$$G_{\beta 1} = 0.9385 \qquad G_{\beta 2} = 0.9435 \qquad G_{\beta 3} = 0.8779 \qquad G_{\beta 4} = 0.8942$$
$$G_{\gamma 1} = 0.3312 \qquad G_{\gamma 2} = 0.3313 \qquad G_{\gamma 3} = 0.4364 \qquad G_{\gamma 4} = 0.4377$$

The observation equations turn into the form: $dR = G_a * dX_a$

$$\begin{pmatrix} -2.3798 \\ -2.3653 \\ -2.3110 \\ -2.3670 \end{pmatrix} = \begin{pmatrix} -0.0979 & -0.9385 & -0.3312 & 1 \\ 0.0105 & -0.9435 & -0.3313 & 1 \\ -0.1971 & 0.8779 & -0.4364 & 1 \\ -0.0940 & -0.8942 & -0.4377 & 1 \end{pmatrix} \begin{pmatrix} dx \\ dy \\ dz \\ b \end{pmatrix}$$

The dX_a can be simply solved as: $dX_a = G_a^{-1} * dR$

$$dX_a = \begin{pmatrix} 0.4042 \\ 5.8132 \\ 2.3114 \\ 3.8808 \end{pmatrix}$$

Thus, the updated positions after the second iteration become $X_{a2} = X_{a1} + dX_a$:

$$x_{a2} = 205.9813 + 000.4042 = 206.3855 \text{ km}$$
$$y_{a2} = 5.9043 \times 10^3 + 005.8132 = 5.9101 \times 10^3 \text{ km}$$
$$z_{a2} = 2.3870 \times 10^3 + 002.3114 = 2.3893 \times 10^3 \text{ km}$$
$$b_{a2} = 11.1212 + 3.8808 = 15.0020 \text{ km}$$

After the second iteration, the solution converges to the true position.

BOX 6.1 MATLAB FOR POSITION FIXING

The MATLAB code position_main.m was run to obtain the position solution from known values of satellite positions and corresponding ranges. The input to the program is the navigation and observation information preloaded in text files. From the navigation data present in the former file, the positions of the visible satellites at any instant are obtained, whereas the

BOX 6.1 MATLAB FOR POSITION FIXING—cont'd

corresponding range information was obtained from the latter. On running the program, the following information is provided in sequence:
   The position and range of satellites.

$$x = 22657881.0793 \quad \text{in meters}$$
$$y = 13092933.5636 \quad \text{in meters}$$
$$z = 5887881.983 \quad \text{in meters}$$
$$R = 22889484.2157 \quad \text{in meters}$$

The program then assumes an approximate position and displays it as:
Approximate position assumed:

$$x\_apx = 302536.5663 \text{ in meters}$$
$$y\_apx = 5772741.575 \text{ in meters}$$
$$z\_apx = 2695567.787 \text{ in meters}$$
$$b\_apx = 10 \text{ in meters}$$

   From these data, the best set of four satellites is selected by obtaining the minimum dilution of precision (DOP), which the program flashes as output as:
   The minimum DOP obtained is 0.094484.
   The program starts the iterative calculation of the position following the user input for the numbers of iteration required.
   The program then displays the results it derives sequentially, as in the following.
   Iteration # n
   Approximated ranges to the selected four satellites are:

$$1.0e + 007^*[2.2673 \quad 2.3734 \quad 2.1757 \quad 2.0181] \quad \text{in m}$$

Differences in ranges are: dR

$$1.0e + 004^*[-3.9911 \quad -4.5629 \quad -4.5205 \quad -4.5642] \quad \text{in m}$$

Linearized observation equation is: dR = G * dX.

$$-39910.5553 = 0.4043 \ *dx + -0.74338 \ *dy + 0.53285 \ *dz + 1 \ *db$$
$$-45,629.4328 = -0.5439 \ *dx + -0.04069 \ *dy + -0.83816 \ *dz + 1 \ *db$$
$$-45,204.8555 = 0.0474 \ *dx + -0.58482 \ *dy + -0.80978 \ *dz + 1 \ *db$$
$$-45,641.5591 = -0.2551 \ *dx + -0.95848 \ *dy + -0.12731 \ *dz + 1 \ *db$$

Solution for dX is: GT * inv(GT*G):

$$1.0e + 004^*[0.4120; \quad 0.3869; \quad 0.3305; \quad -4.0461] \quad \text{in m}$$

Solution after n iterations is:

$$1.0e + 006^*[1.1965 \quad 6.2759 \quad 1.5954 \quad 0.1143] \quad \text{in m}$$

Finally, after the requisite numbers of iterations are over, it displays the final solution as:
Final solution of coordinates of user are:

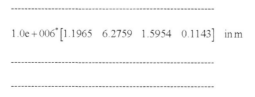

$$1.0e + 006^*[1.1965 \quad 6.2759 \quad 1.5954 \quad 0.1143] \quad \text{in m}$$

6.5 Other methods for position fixing
6.5.1 Solving range equations without linearization

Here, we discuss how to fix the positions of the user by using the same range measurements, but with different methods to get the solution. A number of methods have been put forward by researchers in this field that employ the analytical method to find the solution. However, we will describe only two methods that use the original quadratic observation equation, but will handle it with two different approaches. The first will be done by using an additional linear equation called a constraint equation; the other will logically select the true position from two alternatives.

6.5.1.1 Using a constraint equation

As stated before, it is possible to solve for four unknowns from quadratic equations without linearization. For this, we need a constraint equation defining the fixed relationship that the coordinates follow, in addition to the four quadratic equations. This makes a total of five equations, and is in accordance with our previous observation for the requirement of the equations: four unknowns and one more constraint equation to remove the quadratic ambiguity.

One such constraint may be the assumption that the difference between the square of the radius of the user position and the square of the error resulting from clock shift is a constant (Grewal et al. 2001). This constraint may be mathematically represented as

$$(x^2 + y^2 + z^2) - b^2 = k^2 \tag{6.15}$$

We have chosen this constraint to make our computations simple when we describe the process. However, other constraints relating the position coordinates and bias will do, as well, provided the constraint equation is independent.

The first observation equation was

$$R_1 = \sqrt{(x_{S1} - x)^2 + (y_{S1} - y)^2 + (z_{S1} - z)^2} + c\delta t_u \tag{6.16}$$

Expanding the square terms, and denoting $c\delta t_u$ as b, we get the equation

$$R_1 = \sqrt{x_{S1}^2 + x^2 + y_{S1}^2 + y^2 + z_{S1}^2 + z^2 - 2x_{S1}x - 2y_{S1}y - 2z_{S1}z} + b \tag{6.17a}$$

With a few manipulations, it follows from the above equation

$$R_1^2 = x_{S1}^2 + x^2 + y_{S1}^2 + y^2 + z_{S1}^2 + z^2 - 2x_{S1}x - 2y_{S1}y - 2z_{S1}z + 2R_1b - b^2 \tag{6.17b}$$

Replacing the constraint Eqn (6.15) in Eqn (6.17b), we get

$$R_1^2 - k^2 - R_s^2 = -2xx_{S1} - 2yy_{S1} - 2zz_{S1} + 2R_1b \tag{6.18a}$$

$$\text{or,} \quad A_1x + B_1y + C_1z + D_1b = k_1 \tag{6.18b}$$

where $A_1 = 2x_{S1}$, $B_1 = 2y_{S1}$, $C_1 = 2z_{S1}$, $D_1 = -2R_1$, and $K_1 = R_s^2 + k^2 - R_1^2$. We can construct four such linear equations with each observation equation and the constraint, and solve for the four unknowns. The simultaneous equations thus formed are

$$A_1x + B_1y + C_1z + D_1b = k_1$$
$$A_2x + B_2y + C_2z + D_2b = k_2$$
$$A_3x + B_3y + C_3z + D_3b = k_3 \qquad (6.19a)$$
$$A_4x + B_4y + C_4z + D_4b = k_4$$

This can be written in matrix form as

$$GX = K \qquad (6.19b)$$

where $G = \begin{pmatrix} A_1 & B_1 & C_1 & D_1 \\ A_2 & B_2 & C_2 & D_2 \\ A_3 & B_3 & C_3 & D_3 \\ A_4 & B_4 & C_4 & D_4 \end{pmatrix}$

$$X = \begin{bmatrix} x & y & z & b \end{bmatrix}^T \quad \text{and} \quad k = \begin{bmatrix} k_1 & k_2 & k_3 & k_4 \end{bmatrix}^T$$

Using standard least-squares method, the solution for X becomes

$$X = (G^TG)^{-1}G^TK \qquad (6.20)$$

6.5.1.2 Bancroft's method

In our general discussion about finding the solution, we said that it becomes convenient to find a solution if the original quadratic observation equation is turned into a linear one. The equation was then linearized to serve the purpose. In the last subsection, we put an additional constraint on the positional coordinates to get the solution. In Bancroft's method, the equation is kept quadratic and some algebraic manipulations are carried out using the given relation to reduce the equations to a least-squares problem. Then, from the two possible solutions of this quadratic equation, the required solution is logically chosen. This method of solution is algebraic and noniterative in nature, computationally efficient, and numerically stable, and admits extended batch processing (Bancroft, 1985). It is a classic example of efficiently handling quadratic conditions and was further analyzed by Abel and Chaffee (1991) and Chafee and Abel (1994).

The observation equation is written in terms of satellite position, user positions and receiver bias as

$$R = \sqrt{(x_s - x)^2 + (y_s - y)^2 + (z_s - z)^2} + b \qquad (6.21a)$$

Expanding the terms in the observation equation as functions of the unknown terms of the user and the satellite, we get

$$x_s^2 - 2x_sx + x^2 + y_s^2 - 2y_sy + y^2 + z_s^2 - 2z_sz + z^2 = R^2 - 2Rb + b^2 \qquad (6.21b)$$

Rearranging the terms, we get

$$(x^2 + y^2 + z^2 - b^2) - 2(x_s x + y_s y + z_s z - Rb) + (x_s^2 + y_s^2 + z_s^2 - R^2) = 0$$

(6.21c)

Notice that this may be recognized as a common quadratic equation of the form $X^2 + kX + c = 0$, where $X = [x\ y\ z\ b]^T$ is a multivariate vector of dimension 4.

However, the intelligent part comes at this point. Instead of solving directly for X, notice that the quadratic unknown terms, i.e. $(x^2 + y^2 + z^2 - b^2)$, remain in the form of a scalar. Those who are aware of the Special Theory of Relativity can identify that this form is similar to that of the Lorentz equation. Thus, this composite term is called the Lorentz inner product of X. This function may be defined as

$$\lambda = <X * X> \\ = x^2 + y^2 + z^2 - b^2$$

(6.22)

Remember that this term λ is a scalar function of the unknown variables in X. Similarly, defining the vector $S = [x_s\ y_s\ z_s\ R]^T$, we get

$$<S * S> = (x_s^2 + y_s^2 + z_s^2 - R^2) \\ = \alpha$$

(6.23)

Using these definitions, the equation becomes

$$\lambda - 2\beta X + \alpha = 0$$

(6.24a)

where α and β are known and λ and X are unknown quantities. Also, $\beta = [x_s\ y_s\ z_s\ - R]$. This can equivalently be written as

$$\beta X = \tfrac{1}{2}\lambda + \tfrac{1}{2}\alpha$$

(6.24b)

Because this equation holds well for each of the satellites, a similar equation may be formed from "n" different satellites to form the matrix equation

$$BX = \tfrac{1}{2}\Lambda + \tfrac{1}{2}A.$$

(6.24c)

λ, which is a scalar function of the user position, remains the same for all satellites. Hence, here $\Lambda = \lambda\ U$ and $U = [1\ 1\ 1\ 1\ ....]^T$ and are both $[n \times 1]$ matrices. "α" is a constant but it is different for different satellites, thus forming the matrix A. So,

$$B = \begin{pmatrix} x_{S1} & y_{S1} & z_{S1} & -\rho_{S1} \\ x_{S2} & y_{S2} & z_{S2} & -\rho_{S2} \\ x_{S3} & y_{S3} & z_{S3} & -\rho_{S3} \\ & \vdots & & \\ x_{Sn} & y_{Sn} & z_{Sn} & -\rho_{Sn} \end{pmatrix}$$

and $A = [\alpha_1\ \alpha_2\ \alpha_3\ ...\ \alpha_n]^T$.

If we have n satellites, B is an $[n \times 4]$ matrix, U is $[n \times 1]$, and A is an $[n \times 1]$ vector. If we have enough satellites, a least-squares solution solves the normal equation. From Eqn (6.24c), we can derive the least-squares solution X^* of X as

$$X^* = (B^TB)^{-1}B^T\left(\tfrac{1}{2}\Lambda + \tfrac{1}{2}A\right)$$

$$= K\left(\tfrac{1}{2}\Lambda + \tfrac{1}{2}A\right) \tag{6.25}$$

where $K = (B^TB)^{-1}B^T$ is a $[4 \times n]$ matrix. However, our solution X^* involves λ, which again is a function of X. Substituting X^* into the definition of the scalar λ, we get

$$\lambda = <\tfrac{1}{2}K(\Lambda + A) * \tfrac{1}{2}K(\Lambda + A) >$$

$$= \tfrac{1}{4}\lambda2 < KU * KU > +\tfrac{1}{2}\lambda < KU * KA > +\tfrac{1}{4} < KA * KA > \tag{6.26a}$$

or, $\quad \lambda^2 < KU * KU > +2\lambda(< KU * KA > -1) + < KA * KA >= 0 \quad$ (6.26b)

or, $\quad \lambda^2 C1 + 2\lambda\, C2 + C3 = 0 \tag{6.26c}$

This is a scalar quadratic equation in λ. You may verify by comparing the dimensions of each constituent matrix of the given operations that the equation contains all scalar coefficients. All three of these values can be computed because all components in them are known. Hence, the two possible solutions for λ (λ_1 and λ_2) can be obtained. Each of these two solutions is valid, but these are scalar functions of X and not X itself. Thus, each value of λ can be put into Eqn (6.25) to obtain the final closed value of X, as

$$X_1 = K\left(\tfrac{1}{2}\lambda_1 U + \tfrac{1}{2}A\right)$$

$$X_2 = K\left(\tfrac{1}{2}\lambda_2 U + \tfrac{1}{2}A\right) \tag{6.27}$$

One solution for this will only give a logical result that makes sense. For example, for ground-based users, one solution for X will be on the surface of the earth with a radius of R, and one will not. Thus, one of the two is selected through rational reasoning, and the other is rejected. The mathematical equivalence of this, however, is only a constraint equation. Box 6.2 describes the method for solving for position using Bancroft's method.

FOCUS 6.3 SOLVING WITH BANCROFT'S METHOD

We illustrate here an example of solving the position of a point using the measured ranges and the satellite positions as input. For the Cartesian coordinates of satellites S1, S2, S3, and S4 in an ECEF frame in kilometers, as obtained from the satellite ephemeris, transmitted by the satellites in their message:

Sat:S1	Sat:S2	Sat:S3	Sat:S4
$x_{S1} = 2.1339 \times 10^3$	$x_{S2} = 0.0$	$x_{S3} = 4.1006 \times 10^3$	$x_{S4} = 2.0581 \times 10^3$
$y_{S1} = 2.4391 \times 10^4$	$y_{S2} = 2.4484 \times 10^4$	$y_{S3} = 2.3256 \times 10^4$	$y_{S4} = 2.3525 \times 10^4$
$z_{S1} = 8.9115 \times 10^3$	$z_{S2} = 8.9115 \times 10^3$	$z_{S3} = 1.1012 \times 10^4$	$z_{S4} = 1.1012 \times 10^4$

and the measured range from a definite position P on the earth after all corrections are made on that for satellites is

$$R_{t1} = 1.9708 \times 10^4$$
$$R_{t2} = 1.9702 \times 10^4$$
$$R_{t3} = 1.9773 \times 10^4$$
$$R_{t4} = 1.9714 \times 10^4$$

We need to find the position of P. The first job is to use the position and range information to generate the B matrix. For our case, the B matrix becomes

$$B = \begin{bmatrix} 0.2134 & 2.4391 & 0.8912 & -1.9708 \\ 0 & 2.4484 & 0.8912 & -1.9702 \\ 0.4101 & 2.3256 & 1.1012 & -1.9773 \\ 0.2058 & 2.3525 & 1.1012 & -1.9714 \end{bmatrix} \times 10^{04}$$

From this value of B, the K matrix is derived as

$$K = \begin{bmatrix} 0.06 & -0.06 & -0.02 & 0.02 \\ 0.38 & -0.37 & -0.39 & 0.39 \\ 0.11 & -0.16 & -0.16 & 0.21 \\ 0.52 & -0.54 & -0.56 & 0.57 \end{bmatrix} \times 10^{-2}$$

Now, because both KU and KA values are known and formed using known parameters, these can be used to generate Eqn (6.26b) with coefficients c1, c2, and c3, respectively, where

$$c1 = -1.1445 \times 10^{-09}, c2 = -0.810, c3 = 1.4786 \times 10^{08}$$

The solutions for the equation thus formed in Eqn (6.26c) can be obtained using standard methods of solving quadratic equations, and are

$$\lambda1 = -8.5821 \times 10^{08} \text{ and } \lambda2 = 1.5053 \times 10^{08}$$

Putting these values into Eqn (6.27), we get vectors X1 and X2 as

$$X1 = [0.041; \quad 1.137; \quad 0.468; \quad 16.026] \times 10^{03}$$

$$X2 = [0.223; \quad 6.464; \quad 2.619; \quad -1.979] \times 10^{03}$$

To validate the feasibility of the two results thus obtained, the radius of the two solution points is determined. These radii turn out to be R1 and R2 for solutions X1 and X2, respectively, where

$$R1 = 1.2299 \times 10^3 \text{ km}$$
$$R2 = 6.9787 \times 10^3 \text{ km}$$

Considering that the point of interest was on the earth's surface, the first solution does not satisfy the case because its radius is too short, whereas the second solution does, and looks more probable. So, X2 is our solution.

BOX 6.2 MATLAB FOR BANCROFT'S METHOD

The MATLAB program Bancroft.m was run to obtain the solution, as shown above. Information regarding the satellite position and the measured ranges was obtained from an external file. This file, "sat_pos.txt," was read by the program through in-line commands.

Run the program and use different sets of data to check for the following:

1. How the condition of matrix B changes as the satellite passes close by

2. What happens when the measured range is exact (i.e. $x_s^2 + y_s^2 + z_s^2 - R^2 = 0$) so that A = 0

6.5.2 **Other methods**

6.5.2.1 Doppler-based positioning

Among other different methods, the Doppler-based position fixing is important. It was the technique first used by many initial satellite navigation systems. This technique was used for the first time with satellite Sputnik when the position of the satellite was determined using the Doppler frequency from known receiver positions. Here, we describe the fundamentals of position estimation using Doppler (Axelrad and Brown, 1996).

We first define Doppler frequency as a shift in the frequency of the received signal from what is transmitted as a result of the relative radial motion between the transmitter and the receiver. Thus, if v_{rs} is the radial velocity of the receiver relative to the transmitter, the shift in the received frequency of the signal of wavelength λ owing to the Doppler is given by

$$\Delta f = -v_{rs}/\lambda \tag{6.28}$$

We have used the convention that when the relative velocity is such that it increases the intermediate distance, it has a positive sense. Therefore, when the radial distance decreases as the transmitter and the receiver approach each other, the frequency increases, resulting in a positive Doppler frequency; whereas whenever they relatively recede with a corresponding increase in radial range, the received frequency decreases, causing a negative Doppler frequency. Furthermore, because λ is fixed for a signal, the relative velocity v_{rs} at any instant can readily surrogate the Doppler shift. In this section, we will use the terms "Doppler shift" and "relative radial velocity" synonymously.

Doppler and integrated Doppler can be used to determine the position of the receiver when the position and the velocity of the transmitter are precisely known. We have seen that the range measured by the receiver can be expressed as

$$R_1 = \sqrt{(x_{S1} - x)^2 + (y_{S1} - y)^2 + (z_{S1} - z)^2} + b \tag{6.29}$$

where the notations carry their usual meanings. Because the relative velocity is the rate of change in the range, considering the rate of change of this range, we get

$$\begin{aligned} v_{rs} &= dR/dt \\ &= (\alpha_s v_{sx} - \alpha_r v_{rx}) + (\beta_s v_{sy} - \beta_r v_{ry}) + (\gamma_s v_{sz} - \gamma_r v_{rz}) + db/dt \end{aligned} \tag{6.30}$$

where $\alpha_s = \partial R/\partial x_s$, $\beta_s = \partial R/\partial y_s$, and $\gamma_s = \partial R/\partial z_s$. Similarly, $\alpha_r = -\partial R/\partial x_r$, $\beta_r = -\partial R/\partial y_r$ and $\gamma_r = -\partial R/\partial z_r$. Here, v_{sx}, v_{sy}, and v_{sz} are components of the satellite velocities and v_{rx}, v_{ry}, and v_{rz} are the components of the receiver velocities along the X, Y, and Z axes, respectively. $d(\delta t_u)/dt$ is the drift of the receiver clock. These velocity components of the satellite and user velocities, multiplied with their respective projection factors α, β and γ, contribute to the total relative radial velocity along the line joining the user receiver to the satellite.

Now, to calculate position, the receiver must remove the effect of his own velocity. Thus, an effective choice is that the receiver remain stationary during the estimation then, under the condition of static receiver location, v_{rx}, v_{ry}, and v_{rz} become zero. The previous equation reduces to

$$v_{rs} = (\alpha_s v_{sx}) + (\beta_s v_{sy}) + (\gamma_s v_{sz}) + db/dt \tag{6.31}$$

We have already seen that derivatives α_s, β_s, and γ_s are the direction cosines of the vector joining the satellite and the receiver. These are also the components of the unit vector **e** along the direction of the satellite onto the Cartesian axes. We can also write Eqn (6.31) as

$$
\begin{aligned}
v_{rs} &= \begin{bmatrix} v_{sx} & v_{sy} & v_{sz} & 1 \end{bmatrix} \begin{bmatrix} \alpha_s & \beta_s & \gamma_s & db/dt \end{bmatrix}^T \\
&= \begin{bmatrix} v_s & 1 \end{bmatrix} \cdot G
\end{aligned}
\tag{6.32}
$$

This gives a relation between the Doppler shift, alternatively represented by v_{rs}, and the unknown positions present in parameter G through absolute satellite velocity v_s. Therefore, if the satellite position can be derived from the transmitted data along with its velocity, then from the Doppler-derived relative velocity, v_{rs}, the receiver can use Eqn (6.32) to deduce his own position.

In this equation, parameters, α_s, β_s, and γ_s depend on the relative position of the transmitter and the receiver. It is a nonlinear function of the receiver position of x_r, y_r, and z_r along with the satellite position.

For a definite satellite position, any point on the line joining the satellite and the true receiver position will have the same value of G. This ambiguity of position, however, is removed to reduce the solution to a single point when many such equations are simultaneously considered. Then, to solve the position few distinct measurements are done for the same receiver position. The batch of observation equation, thus created, is then solved through nonlinear least square methods. A relative approach may also be used to avoid the nonlinearity while solving. Here, to get the fix, an approximate solution $X_0 = (x_0, y_0, z_0, db_0/dt)$ is first assumed. Because the Doppler variations will be different for different positions, when it is put in the batch of observation equation, this assumed position will yield different value of v_{rs}, say v_{rs1} for each of the observation. Because these two Doppler values are attained for the same value of satellite velocity v_s, the corresponding differential expression in a matrix form, considering all the observation equations in the batch, will be

$$
\begin{aligned}
\Delta v_{rs} &= v_{rs} - v_{rs1} \\
&= \begin{bmatrix} v_s & 1 \end{bmatrix} \cdot \begin{bmatrix} G_{true} - G_{approx} \end{bmatrix} \\
&= \begin{bmatrix} v_s & 1 \end{bmatrix} \cdot \Delta G \\
&= \begin{bmatrix} v_s & 1 \end{bmatrix} \cdot \operatorname{grad}(G)|_{x_0} \Delta X
\end{aligned}
\tag{6.33}
$$

where ΔX is the differential values of the unknowns with respect to the approximate values.

Thus, we get a relation between the Doppler error with the position estimation error. Thus, the problem reduces to finding the differential values from these linear equations. There are different standard algebraic methods to find them. Once these differential values are solved, they can be added to X_0 and the values of x, y, and z may be obtained in turn.

With a single measurement at any definite processing instant, and no a priori knowledge of the position, it is difficult to estimate accurate values. However, with enough data collected over time for a specific position of the user, the explicit position, satisfying all sets of equations over time in a least-squares sense can be obtained.

In summary, here the position fix begins with an approximated position and determines the differential positioning error, by solving for the shift in that position required to best match calculated slant range rate with those measured using by Doppler.

6.6 **Velocity estimation**

The receiver estimates the position, P, and time, T, at every instant from the updated measurements. Thus, it is easy to find the velocity from these estimates as the instantaneous ratio of the incremental position to the incremental time. This ratio, derived along any definite axis, will give the velocity along that axis. So,

$$V = \Delta p / \Delta t \qquad (6.35a)$$

This can be resolved into an estimation along the axis components as

$$
\begin{aligned}
v_x &= \Delta x / \Delta t \\
v_y &= \Delta y / \Delta t \\
v_z &= \Delta z / \Delta t
\end{aligned}
\qquad (6.35b)
$$

However, this numeric approach for velocity estimation has drawbacks. A better alternative may be devised from the fact that the Doppler shift in the receiver frequency is a function of the relative radial velocity of the user with respect to the satellite (Kaplan et al. 2006). We saw this in Eqn (6.28). We can write it and expand the term into its components so that the equation turns into

$$
\begin{aligned}
\Delta f &= -v_{rs}/\lambda \\
&= -\left[(\alpha_s v_{sx} - \alpha_r v_{rx}) + (\beta_s v_{sy} - \beta_r v_{ry}) + (\gamma_s v_{sz} - \gamma_r v_{rz}) \right]/\lambda
\end{aligned}
\qquad (6.36)
$$
or, $\quad \lambda \Delta f + v_s \cdot G_s = v_r \cdot G_r$

where $G_r = (\alpha_r \hat{i} + \beta_r \hat{j} + \gamma_r \hat{k})$ and $G_s = (\alpha_s \hat{i} + \beta_s \hat{j} + \gamma_s \hat{k})$ are the unit vectors along the direction of the receiver and the satellite, respectively. λ is the wavelength of the transmitted signal and is equal to c/f_t, where c is the velocity of light and f_t is the frequency of the transmitted signal. Δf is the measured Doppler shift. v_s and v_r are the velocity of the satellite and the user, respectively.

If the user is able to receive the signal and measure the Doppler frequency Δf, and can estimate the velocity of the satellite from the received ephemeris, he can also derive the velocity using this relation.

However, there are still practical considerations. How do we measure Doppler frequency? The answer is, simply by differencing our known signal frequency from the measured frequency of the signal. Again, we measure the incoming frequency of the signal by counting the total number of complete oscillations of the received signal occurring in a second. Finally, count of a second is derived from a local clock. Thus, if the local clock in the receiver is in error, the frequency measured becomes erroneous, and also the Doppler shift.

Besides receiver clock error, the transmitted frequency is not exactly a designed frequency known to the receiver. It has errors owing to satellite oscillator drift. However, this value is typically estimated by the ground segment of the system and the required correction is sent through the navigation message to the user. Therefore, we may consider the error resulting from this effect to have been corrected.

Considering that the receiver clock error is the only source of error in this estimation, we recall that receiver clock errors are clock bias and clock drift. Receiver's fixed-clock bias (i.e. deviation from the true time) creates no difference in frequency measurement. This is because the difference between the start and the stop times in counting the oscillations remain equally shifted from the true time, and hence the error is nullified in the difference.

If t' is the clock drift, during the interval Δt, between the start and the end time of Δt, the clock shifts in this interval by

$$\delta t = (t' \text{x} \Delta t) \tag{6.37}$$

This term δt becomes the error in timing. Therefore, if n is the total count of oscillations in the time period Δt, the true received frequency of the signal is $f = n/\Delta t$. But, owing to the error in measuring the time period, the frequency actually measured is

$$\begin{aligned} f_m &= f + \Delta f \\ &= f + \partial f/\partial \Delta t\, \delta t \end{aligned} \tag{6.38}$$

The error in measuring the frequency is

$$\begin{aligned} \delta f &= \partial f/\partial \Delta t\, \delta t \\ &= \partial/\partial \Delta t (n/\Delta t)\delta t \\ &= -n/\Delta t^2 \delta t \\ &= -n/\Delta t^2 (\Delta t\, t') \\ &= -n/\Delta t\, t' \\ &= -f\, t' \end{aligned} \tag{6.39}$$

The negative sign, as usual, indicates that if the drift is positive, it leads to a decrement in the measured frequency. So, the measured Doppler frequency becomes equal to the true Doppler and the Doppler estimation error due to clock drift. For each satellite, we then get the expression from the Doppler observation as

$$\begin{aligned} v_r\, G_r + \lambda f t' &= v_s\, G_s + \lambda\, \Delta f_m \\ [G_r\ \lambda f][v_r\ t'] &= v_s\, G_s + \lambda\, \Delta f_m \\ G_r'\, [v_r\ t'] &= v_s\, G_s + \lambda\, \Delta f_m \end{aligned} \tag{6.40}$$

or,

or,

where $G_r' = [G_r\ \lambda f]$.

We can now construct a similar matrix relation to solve for the receiver velocity. The solution becomes

$$[v_r \ t'] = (G'_r)^{-1} G_s v_s + (G'_r)^{-1}(\lambda \Delta f) \tag{6.41}$$

This equation has an implicit assumption that the line of sight joining the satellite and the user is known before the estimate of the velocity, which enables us to find G_r. Thus, velocity estimation with this method can be done only after the position of the user is estimated. Like the position, these velocity values are derived from the instantaneous measurements and are updated after every measurement.

However, errors come in this estimation, too. We will learn about the errors in the chapter specifically on this topic in Chapter 7. Also, other techniques can be used to estimate the position and velocity. One of these is by the Kalman filter, in which position, time, and velocity are determined simultaneously, considering them to be the state variables of the receiver. An introduction to the Kalman filter and associated estimation process is discussed in Chapter 9.

Conceptual questions

1. Is it possible to find the position of a flying aircraft by measuring its range from a known position on the earth? If yes, how many such receivers will be required to find it?
2. Instead of using the measured range, if we take the range difference values, the common term of clock bias cancels out and the equations are left with three unknowns. Is it possible to use only three satellites and the corresponding three difference equations to derive the position coordinates?
3. What advantages do we obtain by using an atomic clock while determining the position and velocity?
4. Do you expect the accuracy of the navigation solution to improve if more than four satellites are used for the purpose?

References

Abel, J.S., Chaffee, J.W., 1991. Existence and uniqueness of GPS solutions. IEEE Transactions on Aerospace and Electronic Systems 27 (6), 952–956.

Axelrad, P., Brown, R.G., 1996. GPS navigation algorithms. In: Parkinson, B.W., Spilker Jr., J.J. (Eds.), Global Positioning Systems, Theory and Applications, vol. I. AIAA, Washington, DC, USA.

Bancroft, S., 1985. An algebraic solution of the GPS equations. IEEE Transactions on Aerospace and Electronic Systems 21, 56–59.

Chaffee, J.W., Abel, J.S., 1994. On the exact solutions of pseudorange equations. IEEE Transactions on Aerospace and Electronic Systems 30 (4), 1021–1030.

Grewal, M.S., Weill, L., Andrews, A.P., 2001. Global Positioning Systems, Inertial Navigation and Integration. John Wiley and Sons, New York, USA.

Kaplan, E.D., Leva, J.L., Milbet, D., Pavloff, M.S., 2006. Fundamentals of satellite navigation. In: Kaplan, E.D., Hegarty, C.J. (Eds.), Understanding GPS Principles and Applications, second ed. Artech House, Boston, MA, USA.

Strang, G., 2003. Introduction to Linear Algebra, third ed. Wellesley-Cambridge Press, Wellesley, MA, USA.

Errors and Error Corrections

Understanding Satellite Navigation. http://dx.doi.org/10.1016/B978-0-12-799949-4.00007-5
Copyright © 2014 Elsevier Inc. All rights reserved.

In all our previous chapters, we have talked about the theoretical ways of fixing positions. But the conditions in which we derived these parameters were very hypothetical, since we assumed that all the measurements that we are doing at the receiver during ranging are perfect and the positions of satellites were determined without any error. However, in all practical scenarios, we cannot reduce these without making errors. These errors creep into our measurements from almost everywhere, corrupt them, and make our position estimates wrong. We cannot get rid of this menace unless we categorically know how the errors are actually occurring. In this chapter, first we identify these errors coming from different sources and quantify their relative effects on position estimations; and we will also briefly discuss how to negotiate with these errors.

7.1 Scope of errors

The only means through which the system information reaches the user receiver is the navigation signal, loaded with significant information. It is practical to realize that some errors must be already present in this information content. Errors also occur while the signal flows to the user and while the receiver derives the information from it.

It is important to note here that we are not much concerned about the communication errors. Communication errors result in incorrect identification of the navigation data bits. However, these data bits are either coded with forward error correction or have provisions for cyclic redundancy check to identify and restore their correctness. So, whenever any error in the bit occurs, they are either corrected or discarded. So, these errors that may lead to erroneous information in the receiver are not used in the process.

We now know that the only relationship that is used between the measurements and our required parameter is the observation equation of the form:

$$R = \sqrt{(x_s - x_u)^2 + (y_s - y_u)^2 + (z_s - z_u)^2} \qquad (7.1a)$$

To this equation, we assumed that the receiver clock bias is offset by an amount δt_u with respect to the system reference time and added the compensation term. This modified the observation equation to

$$
\begin{aligned}
R &= \sqrt{(x_s - x_u)^2 + (y_s - y_u)^2 + (z_s - z_u)^2} + c\delta t_u \\
&= \sqrt{(x_s - x_u)^2 + (y_s - y_u)^2 + (z_s - z_u)^2} + b
\end{aligned}
\qquad (7.1b)
$$

where b is the effective change in range due to the receiver clock bias. Both the reference positioning parameters (x_s, y_s, z_s) and the ranging parameter R have their respective positions in this equation. These known values are used in the equation and from the definite relationship of these parameters the unknowns $X = (x_u, y_u, z_u, b)$ are derived. Use of linearization method is nothing but differentiating this relation into a linear form and solving of simultaneous linear equations of the form

$$\underline{G} \ \Delta\underline{X} = \Delta\underline{R}$$
$$\Delta\underline{X} = \underline{G}^{-1}\Delta\underline{R} \tag{7.2a}$$

where $G = \partial R/\partial X$, ΔX is the difference of the true receiver position X with respect to a known reference position X_0, and ΔR is the difference in measured range of the receiver R with respect to the range R_0, geometrically calculated from the point X_0. So, replacing $\Delta\underline{X} = (\underline{X} - \underline{X}_0)$ and $\Delta\underline{R} = (\underline{R} - R_0)$, we get

$$\underline{X} = X_0 + \underline{G}^{-1}(\underline{R} - R_0) \tag{7.2b}$$

Now, if the measurements are erroneous so that R turns to $R + dR$ due to error, the solution also becomes $\underline{X} \mid d\underline{X}$, where

or,
$$\underline{X} + d\underline{X} = X_0 + \underline{G}^{-1}(\underline{R} + dR) - \underline{G}^{-1}R_0$$
$$d\underline{X} = \underline{G}^{-1}dR \tag{7.2c}$$

So, the error in solution dX is directly related to the measurement error dR through the inverse of G. So any error in the measured range values will ultimately lead to an error in position estimates and the magnitude of the effect will be dependent upon the relative geometry represented by G. Therefore, to reduce the error in the position estimation solution, on one hand the error in measurements, i.e. dR, has to be small. On the other hand, G^{-1} should be such that it does not inflate the errors in dR. Now, as for any matrix A, $A^{-1} = C_A^T/|A|$, where C_A is the cofactor matrix of A. For the last condition to be true $|G|$ should not be singular or near singular. In other words, the matrix G has to be well conditioned and to achieve the latter the rows of A must be linearly independent.

Thus we have seen that the requirements to make the estimation error dX small are to minimize dR and to make G well conditioned. But before going into any further detail, we shall see in Box 7.1 how the estimated position, fixed using observation equations, assuming no clock bias, will vary with the error in the range.

BOX 7.1 MATLAB

The MATLAB program posi_err.m was run to find the positional error for a definite set of satellites with their respective positions as given below.

Satellite positions:

[32.2° N; 67.0° E]; [0.5° N; 73.8° E]; [−15.3° N; 91.4° E]; [5.6° N; 117.2° E].

The corresponding ranging errors were

[5.2; 4.1; 3.7; 9.2] in meters.

The error in Cartesian coordinates thus obtained was

dx = −6.832 m; dy = 16.259 m; dz = 7.887 m

This makes the effective radial error 19.749 m. The corresponding condition number of the G matrix was 57.382, which is quite high compared to the desirable value near 1.

Change the satellite positions, keeping the range errors the same, and see the variation in position errors. Check the condition number of the matrix G every time. Arrange the errors ordered by this parameter of G.

Similarly, keep the satellite positions fixed and change the range errors and see the result.

7.1.1 **Sources of errors**

You must be willing to know about the factors responsible for introducing errors in the measurements. That is what we are going to address in this section. First we shall discuss here the formal classification of these errors. Then we shall elaborate on each of these kinds. Errors in the known parameters of the observation equation come from various sources, and all three architectural segments of a typical navigation system contribute to these errors. So, we generally categorize the errors on the basis of the segment from which the errors are coming and under each segment there are different reasons for the errors to be generated.

The primary sources of error at the various segments are (Parkinson, 1996):

1. Control segment errors: ephemeris error,
2. Space segment errors: satellite clock bias, satellite code bias
3. Propagation errors: ionospheric delay, tropospheric delay, multipath
4. User segment errors: receiver noise, receiver bias

Sometimes it is convenient to classify errors by their characteristics. On this basis the errors can be divided into

1. Constant errors: These are the errors that are constant over the geographic location of the users and remain the same for all users irrespective of the receiver position. These errors come into the signal before they start traversing toward the users and hence are not influenced by the position of the receiver. This category includes errors like the satellite clock bias, satellite hardware delay, etc.
2. Correlated errors: These are the errors that are dependent upon the location of the users and are geographically correlated at the same time. The errors introduced by the medium as the signal propagated through them are of this type. Errors like the ephemeris errors, ionospheric and tropospheric errors fall under this category. The amount of these errors varies from place to place as the receiver relocates. However, there remains a large correlation between the values at one place and those at its vicinity. This makes it possible to derive the error value at one place if the error values at adjacent places are known.
3. Uncorrelated errors: These are the errors that also are dependent upon the location of the users but are not geographically correlated. These errors are independent and mutually exclusive, such that one cannot estimate the error magnitude at one place from any other related information of the same at another place. Errors due to multipath fall under this category.

We shall once again review these error categories in our next chapter, while the individual errors will be described here.

7.2 **Control segment errors**

In our previous discussion, we have seen that the ephemeris and clock correction parameters are computed at the control segment. These values are updated at regular intervals and transmitted to the users through the navigation message. During this process of computing some errors are generated, which are discussed below.

7.2.1 **Ephemeris errors**

While estimating the ephemeris of the satellite, the control segment uses different models that predict its future orbital locus. In this process there remain some modeling errors that lead to errors in ephemeris issued by the control segment. This is called ephemeris error.

How does the ephemeris error affect the range? The ephemerides are the parameters from which the satellite positions are derived. These satellite positions in turn are the parts of the observation equation, as given in Eqn (7.1b), which is again given below

$$R = \sqrt{(x_s - x_u)^2 + (y_s - y_u)^2 + (z_s - z_u)^2} + c\delta t_u$$

Notice that here we equate the measured range on the left-hand side to the theoretical Euclidean range obtained from the satellite and user positional coordinates on the right-hand side. The measurement is done from the true position of the satellite. However, the deviation of the satellite position from the true values due to the ephemeris error will result in the calculation of a theoretical range of the satellite with respect to a position where the satellite is not actually located. So, this can no longer be equated with the measured range R, as it is in Eqn (7.1), and will result in an unbalanced equation of range. This will eventually lead to errors in user position solution.

Despite this fact, we really have no means to correct the values of these coordinates in the expression. So, instead we keep the satellite coordinates as derived from the transmitted ephemeris and then add a compensatory term in the theoretical expression of the equation to pay off the act. This extra term compensates the range error that occurs due to putting of the incorrect satellite coordinates and balances the equation. The two sides are now ready to be equated through the expression

$$R = \sqrt{(x_s - x_u)^2 + (y_s - y_u)^2 + (z_s - z_u)^2} + c\delta t_u + \delta r_{eph} \qquad (7.3)$$

In this expression, the values of the satellite coordinates x_s, y_s and z_s, are actually those derived from the ephemeris and thus contain error and δr_{eph} is the compensatory term that balances the resultant range error.

Figure 7.1 describes the deviation of the estimated position of the satellite from its true position. Here, the apparent position of the satellite is at A, whereas its true position is at T. The deviation $S = \varepsilon$ can be resolved into ε_θ and ε_R. ε_θ is normal to

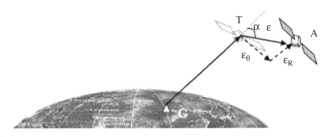

FIGURE 7.1

Ephemeris error.

the radial distance GT and does not affect the range when measured from the ground receiver, G. The radial error ε_R is along the radial distance and adds an extra magnitude to the actual range of the satellite. Thus, it is this effective range error component that affects ranging and is of concern here. It is this term ε_R that is used in Eqn (7.3) as δr_{eph}, considering the sign accordingly.

It is obvious that, for a particular observation, the compensatory term δr_{eph} is dependent upon the position of the receiver. When we separate the total ε into its components, the radial component of the error that actually affects the range is ε_R. Now $\varepsilon_R = \varepsilon \cos \alpha$, where α is the angle made between the range and the error vector. The error vector being fixed for a particular ephemeris error, the angle α is dependent upon the relative position of the satellite and the receiver. So, for different location G of the receiver, α will be different and so will the value of ε_R for the same value of ε. Consequently, different amounts will be added to the actual radial range due to the same ephemeris error.

To understand this effect a little more mathematically, we shall first see how a very small error in x_s, y_s, and z_s due to ephemeris error will effectually change the range.

Let there be a radial range deviation of dR in addition to a cross-radial error, when there is a difference in the true and transmitted value of the satellite position coordinates by $dS = [dx_s \ dy_s \ dz_s]$. Since we know the expression of the absolute range R in terms of the satellite coordinates x, y, and z, we can derive the differential as well. So, the differential error in radial length due to these errors in coordinates may be given by

$$dR = (\partial R/\partial x_s)|_{xs}dx_s + (\partial R/\partial y_s)|_{ys}dy_s + (\partial R/\partial z_s)|_{zs}dz_s \qquad (7.4a)$$

Note that here the derivatives are measured at those satellite coordinates derived from the transmitted ephemeris. From Eqn (7.1), the expression for the derivatives can be obtained as

$$\partial R/\partial x_s|_{xs, \ ys, \ zs} = (x_s - x_u)/\sqrt{(x_s - x_u)^2 + (y_s - y_u)^2 + (z_s - z_u)^2}$$

$$\partial R/\partial y_s|_{xs, \ ys, \ zs} = (y_s - y_u)/\sqrt{(x_s - x_u)^2 + (y_s - y_u)^2 + (z_s - z_u)^2} \qquad (7.4b)$$

$$\partial R/\partial z_s|_{xs, \ ys, \ zs} = (z_s - z_u)/\sqrt{(x_s - x_u)^2 + (y_s - y_u)^2 + (z_s - z_u)^2}$$

Note that the derivatives are the cosine of the angle that the direction of the radial range makes with the respective axes. So these derivatives are also the components of the unit range vector along the axes of the coordinates. Therefore, Eqn (7.4) can be written as

$$dR = (\partial R/\partial x_s)dx_s + (\partial R/\partial y_s)dy_s + (\partial R/\partial z_s)dz_s|_{xs,\ ys,\ zs}$$
$$= \cos\alpha\ dx_s + \cos\beta\ dy_s + \cos\gamma\ dz_s$$
$$= e_x\ dx_s + e_y\ dy_s + e_z\ dz_s \tag{7.4c}$$
$$= \underline{e} \cdot d\underline{S}$$

where $\underline{e}$ is the unit vector along the range and e_x, e_y, and e_z are its components along the three coordinate axes. Thus, the effective radial error is the dot product of the satellite position error and the unit radial vector, making it the projection of the total errors onto the radial direction. Hence, this effective radial error is dependent upon the mutual angle between them. It is also evident from the fact that the expressions of each component are dependent upon the receiver position coordinates and hence on receiver–satellite geometry. Referring to Figure 7.2, if T and A are the true and apparent positions of the satellite due to the ephemeris error, there is a substantial change in range for the receiver Rx1 while for Rx2 the difference is trivial.

Though the ephemeris error values change with the position of the users, there exists a definite function that the variation follows. So, these errors are correlated over space. However, this change is so small for nearby distances that for small variations, the errors may be considered to be constant. To establish this, we shall start by referring to Figure 7.2.

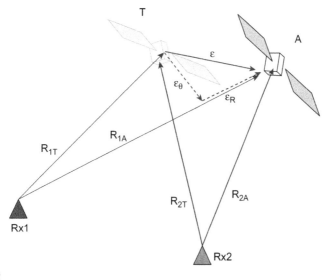

FIGURE 7.2

Variation of ephemeris error.

Using Eqn (7.4c) we can say that the difference in the ephemeris error at the two locations is

$$\Delta dR = (e_1 - e_2) \cdot dS \tag{7.5}$$

Here, e_1 and e_2 are the unit radial vectors from the two receivers Rx1 and Rx2, respectively, located at nearby positions $[x_{u1} \ y_{u1} \ z_{u1}]$ and $[x_{u2} \ y_{u2} \ z_{u2}]$. This can be written by expanding the terms e_1 and e_2 in terms of the position coordinates as

$$\Delta dR = \left[(x_s - x_{u1})/\rho_1 - (x_s - x_{u2})/\rho_2\right]dx_s + \left[(y_s - y_{u1})/\rho_1 - (y_s - y_{u2})/\rho_2\right]dy_s$$
$$+ \left[(z_s - z_{u1})/\rho_1 - (z_s - z_{u2})/\rho_2\right]dz_s$$

$$\tag{7.6a}$$

For nearby stations, we can always put the condition that the range remains almost equal. This makes $\rho_1 \approx \rho_2 = \rho$. This modifies the above equation to

$$\Delta dR = (x_{u2} - x_{u1})/\rho \ dx_s + (y_{u2} - y_{u1})/\rho \ dy_s + (z_{u2} - z_{u1})/\rho \ dz_s \tag{7.6b}$$

Again, each of the coefficients in the above equation are the ratio of the coordinate differences between the receivers to the range to the satellite. So, each of these ratios is negligibly small, making the total equation approach zero.

So, the triviality of this equation indicates that for two closely placed receivers the ephemeris errors are almost the same. In Box 7.2 we shall see how this ephemeris

---

BOX 7.2 MATLAB

The MATLAB program eph_err.m was run to obtain the ephemeris errors for a given variation of the receiver position and for a given position of the satellite. Plots are obtained for a specific deviation of the satellite. It is shown in Figure M7.2.

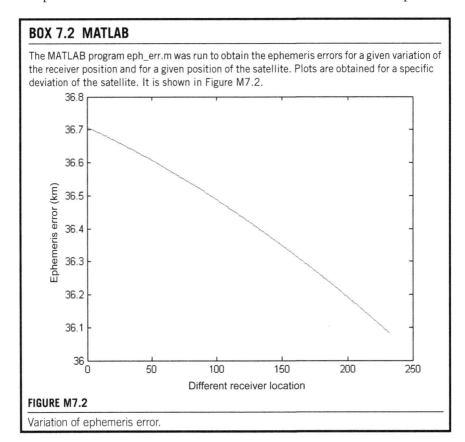

FIGURE M7.2

Variation of ephemeris error.

BOX 7.2 MATLAB—cont'd

Note that even for large ephemeris errors resulting in large deviation of the satellite from the true satellite position, the corresponding error observed over a large departure of the receiver is quite small.

Run the program again with different departures of the receiver and also for larger deviations of the satellite and observe the major differences in this result.

---

error affects the range for different positions of the receiver and for a given deviation of the satellite.

7.3 Space segment errors

The space segment errors include all those errors in ranging that occur due to the elements in the satellites. These include primarily the satellite clock error and the satellite code bias.

7.3.1 Satellite clock error

The satellites carry atomic clocks, which provide the reference time and frequency for all the activities done in the satellite, including the time stamping of the signal. The satellite clocks, despite being atomic clocks, have their respective biases and drifts. These drifts generally come from the associated electronics of the clock that determine the transition resonance condition of the atoms in such a clock. This error is very small, as the precision remains $\sim 10^{-13}$ or better. But, since this error grows cumulatively and we measure the range by multiplying the time by the velocity of light c (3×10^8 m/sec), this small variation also results in considerable range errors. The effective range deviation due to the clock error is given by

$$b_s = -c\ \delta t_s \tag{7.7}$$

where c is the velocity of light and δt_s is the satellite clock error. The negative sign is due to the fact that if δt_s is positive, making the satellite clock advance with the system time, the signal transmission time from the satellite will be marked later than what it actually is. So it would appear that the traveling time is less by δt_s, leading to a reduction in the effective calculated range.

However, as we have read in the previous section, these errors are estimated by the control segment and the corrections are transmitted along with the signal so that the user can correct the effects, using them during position fixing. Still, there remain some residual errors that occur due to the estimation of this error at the control segment of the system. These residual errors cannot be corrected; hence some effective range errors get added in the range measurement due to this fact. So, effectively what remains in the range is the residual error due to inaccuracy in the ground segment estimation of the satellite clock bias. As a result, the observation equation in Eqn (7.3) changes to

$$R = \sqrt{(x_s - x_u)^2 + (y_s - y_u)^2 + (z_s - z_u)^2} + c\delta t_u - c\delta t_s + \delta r_{eph} \qquad (7.8)$$

7.3.2 Code bias

The ranging is done by estimating the delay incurred by the signal in propagating through the intermediate path from the satellite to reach the receiver. It is assumed that the ranging codes are transmitted by all the satellites synchronously, i.e. the start of the transmission of a code or any specific code bit within it occurs at the same instant in all satellites. Moreover, where the same encoded data bits are transmitted using multiple frequencies, precisely the same code phase is assumed to exist for the two different frequencies. But pragmatically, this is not the case. Some small time differences always remain between them. A part of this is contributed by the hardware delay in transmitting the signals from the marked time in the time stamp to the actual instant it is transmitted from the antenna phase center of the satellite. These differences are of the order of nanoseconds, which are of the order of typical hardware delays of the devices, but even that amounts to about a meter of difference in the range. Also, the differential value of this hardware bias adds to the differential code bias between two same codes transmitted by the same satellite but on different carrier frequencies. These errors are effective when calculations are done by comparing two different signals.

7.4 Propagation and user segment errors
7.4.1 Propagation errors

The propagation errors in a navigation signal are those which occur during the propagation of the signal through various atmospheric layers. Each of these layers contributes in a different way to the error in the received signal. The signals pass through the ionosphere and then through the troposphere. In both these layers the signals experience some impairment causing errors in measured range.

7.4.1.1 Ionospheric effects

The radio waves of the navigation signal experience an additional delay as the signal traverses through the ionosphere, a part of the atmosphere of the earth. This delay gets converted into range error at the receiver, since the receiver derives the range from the traverse time. This delay contributes the most to the total range error, and so it is mandatory that we understand this in greater detail. But the ionosphere is a vast subject and excessively interesting. So, we should be cautious not to lose track of our main discussion while exploring this wonderful element of nature, the ionosphere.

The ionosphere is the part of the atmosphere that extends from 50 km to above 1000 km from the earth's surface. This layer of the atmosphere has large numbers of free electrons that are generated from the abundant ozone and oxygen molecules present there. These molecules have loosely bound electrons, which get separated from their parent molecule by absorption of solar radiation. These free electrons get recombined with the parent molecule, diminishing the free electron density. These two processes, viz. generation and recombination, occur simultaneously and with equal rates at equilibrium to produce and sustain a definite density of free electrons. However, this density, being a function of the solar flux, varies with time of day, season, solar activity periods, and various other factors. The path integral of the ionospheric electron density is known as the total electron content (TEC). So

$$TEC = \int N_e(s)ds \qquad (7.9)$$

If we consider a straight line passing through the ionosphere connecting the satellite to the receiver, with unit cross-section area, then it forms a cylinder of volume V where $V = 1 \times s$; s is the length that this line crosses through the ionosphere. So, we may also define TEC as the total number of the ionospheric free electrons present in this volume. For obvious reasons, the length of the line varies depending upon the elevation of the look angle. So, the part of the ionosphere that the path will move through will be different, thus containing different electron contents. So, the TEC for slant and vertical look angles will be different. It will also be different for different times of day and for different seasons, as the density of the ionospheric electrons changes with these factors. In Figure 7.3 the total number of electrons present in the cylinder S is the total electron content for the shown path. The unit for TEC is electrons/m^2. However, the pragmatic values of the parameter when expressed in this unit are very large. Hence a more convenient unit for the same is TECU, where 1 TECU $= 10^{16}$ electrons/m^2.

The ionosphere has a different refractive index than free space, which is denoted by n. Thus the velocity with which the signal codes move through this region is c/n, where c is the velocity of light in free space. Therefore, if D_1 and D_3 are the length traversed above and below this region and D_2 within it, then the time taken

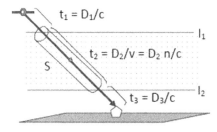

FIGURE 7.3

Ionospheric total electron content and traverse time.

respectively to traverse through these lengths will be $t_1 = D_1/c$, $t_2 = D_2n/c$ and $t_3 = D_3/c$. It makes the excess time taken in traversing the ionospheric section to

$$\Delta t = (D_1 + D_2 n + D_3)/c - (D_1 + D_2 + D_3)/c$$
$$= (n - 1) D_2/c \tag{7.10a}$$

This causes an additional error in the range of equivalent path length given by

$$c\Delta t = (n - 1) D_2 \tag{7.10b}$$

The deviation of the refractive index, $(n - 1)$ is a function of TEC which we shall derive in the next subsection. Figure 7.4 shows the typical temporal variation of the vertical TEC at a station near the magnetic equator.

7.4.1.1.1 Ionospheric delay

As the radio waves traverse through the ionosphere, there is an interaction between the wave and the electrons. The electric field of the wave, which is oscillatory in nature, forces the free electrons in the medium to oscillate with it. Since the electrons are free, there is no other force acting on these. So, the force equation on these electrons can be written as:

$$mf = e E \exp (j\omega t) \tag{7.11}$$

where m is the mass of the electron, e is the charge, and f is the acceleration produced. E is the incident electric field with the angular frequency ω. So solving for

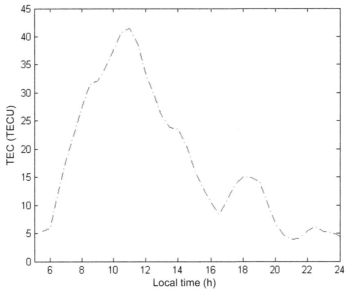

FIGURE 7.4

Variation of ionospheric total electron content with local time.

the electron velocity, v, and its displacement, x, by integrating the above equation once and twice, respectively, with respect to time, we get

$$v = e \, E \, \exp{(j\omega t)}/mj\omega \qquad (7.12a)$$

$$x = -e \, E \, \exp{(j\omega t)}/m\omega^2 \qquad (7.12b)$$

So, both the velocity and the amplitude of the motion of the electrons decreases as the frequency ω of the forcing signal increases. The velocity containing the imaginary term 'j' indicates that the velocity is in phase quadrature with the forcing field. Further, the negative sign in x implies that the displacements of the electrons are just opposite to the implied field.

Now, using the above expression of v, the resultant current density J, generated due to the motion of the electrons originating as a result of the imposed electric field is given by:

$$\begin{aligned} J &= N \, e \, v \\ &= -j \left(Ne^2/m\omega\right) \cdot E \end{aligned} \qquad (7.13a)$$

where N is the electron density in the region. So, the resultant conductivity σ can be expressed as

$$\sigma = J/E = -j \left(Ne^2/m\omega\right) \qquad (7.13b)$$

As these electrons flow through the medium generating conduction current and change with time, similar electric fields are produced with proportionate magnitude and with a phase that is 90 degrees apart from the forcing field. As a result, the effective electric field becomes the vector sum of the original one and the one due to the forced oscillation of the electrons in the medium.

Not only does the resultant field, at every point on the medium, have a phase difference with the original impressed field, but the combined phase also moves with a different velocity due to the superimposition. The phase of the electromagnetic waves travels with somewhat higher velocity than the velocity c in free space. This velocity v_p is called the phase velocity. However, the group velocity v_g with which the energy moves is slower than that in vacuum or free space. For the ionosphere, these two velocities are related such that $v_p \times v_g = c^2$. Hence, using free space velocity for ranging calculations, for the whole transmission path, causes error, which is one of the major effects of the ionosphere on the navigation signal.

This modified phase velocity of the wave in the medium can also be described as being due to a difference in refractive index of the medium compared to that of free space due to the free electrons present in it. The fact that finite conductivity of a medium modifies the refractive index of a medium can also be derived (Feynman et al. 1992; Reitz et al. 1990) from the Maxwell equation. The ionosphere, due to the existence of the free electrons, has a finite conductivity, which in turn influences the effective relative permittivity of the medium and the refractive index, in turn. So, the corresponding Maxwell equation can be written as

$$\begin{aligned} \nabla \times B &= \mu_0 \, J + \mu_0 \, \varepsilon \, dE/dt \\ &= \mu_0 \, \sigma \, E + \mu_0 \, \varepsilon_0 \, dE/dt \end{aligned} \qquad (7.14)$$

Here, σ is the conductivity of the ionosphere. μ_0 and ε_0 are the permeability and permittivity, respectively, in this region, which can otherwise be treated almost as free space in the absence of conducting electrons. For the medium with propagating field strength given by E, the current density J is expressed as $J = \sigma \cdot E$ and B is the resultant magnetic field. Again, considering the field generated by the propagating signal to be sinusoidal with angular frequency ω,

$$E(t) = E_0 \exp (j\omega t)$$

$$\text{or,} \quad dE/dt = j\omega \, E \, (t) \tag{7.15}$$

$$\text{So,} \quad E(t) = -(j/\omega) \, dE/dt$$

Replacing this expression in Eqn (7.14), we get

$$\nabla \times B = \mu \, \varepsilon_0 \, (1 - j\sigma/\omega\varepsilon_0) \; dE/dt \tag{7.16}$$

Thus, due to the partial conductance of the medium an imaginary term appears in the expression and the effective relative permittivity gets transformed into a complex parameter

$$\varepsilon' = \varepsilon_0 \, (1 - j\sigma/\omega\varepsilon_0) \tag{7.17}$$

The complex refractive index "n" of the medium is thus formed utilizing the standard relation and we get:

$$n^2 = (\varepsilon'/\varepsilon_0)$$
$$= (1 - j\sigma/\omega\varepsilon_0) \tag{7.18a}$$

$$= \left(1 - Ne^2/m\varepsilon_0\omega^2\right) \tag{7.18b}$$

$$= \left(1 - \omega_p^2/\omega^2\right) \tag{7.18c}$$

Here, in Eqn (7.18b), we have replaced the expression for the conductivity as derived in Eqn (7.13b) to get ω_p in Eqn (7.18c). This ω_p is called the plasma frequency. The final expression for the ionospheric refractive index "n" is hence obtained by putting values of the constants e, m, and ε_0 and we get

$$n = \sqrt{1 - 80.6 \, N_e/f^2} \tag{7.19a}$$

where Ne is the density of free electrons responsible for conductivity σ and $f = \omega/2\pi$ is the frequency of the wave in Hertz. For satellite signals, the frequency being very high, Eqn (7.19) can be approximated to

$$n = 1 - 40.3 \, N_e/f^2 \tag{7.19b}$$

Consequently, the velocity with which the phase of the wave travels is

$$v_p = c/\left(1 - 40.3 \, Ne/f^2\right) > c \tag{7.20a}$$

So whenever electromagnetic waves pass through this region, their phase moves faster compared to that in vacuum, resulting in an effective phase advancement at the receiver. Consequently, the group velocity with which the codes move becomes

$$v_g = c^2 \Big/ \Big\{ c \Big/ \Big(1 - 40.3 \, Ne/f^2 \Big) \Big\}$$
$$= c \Big(1 - 40.3 \, N_e/f^2 \Big) < c \qquad (7.20b)$$

Therefore, the energy contained in the signal and carried by the constituent codes and data traverses slower than in vacuum, causing an additional code delay.

Now, let the path through ionosphere be l. Following Eqn (7.8), the excess time taken by the codes to traverse a path length of dl is equal to

$$\Delta t(dl) = dl/v_g - dl/c$$
$$= dl \left(v_p/c^2 - 1/c \right)$$
$$= (dl/c) \left(v_p/c - 1 \right)$$
$$= (dl/c) \left(1/n - 1 \right) \qquad (7.21a)$$
$$= (dl/c) \Big\{ \Big(1 - 80.6 \, N_e/f^2 \Big)^{-\frac{1}{2}} - 1 \Big\}$$
$$= (dl/c) \Big(1 + 40.3 \, N_e/f^2 - 1 \Big)$$
$$= 40.3 / \big(c \, f^2 \big) \, N_e \, dl$$

So, integrating it over the length of S that the signal traverses through the ionosphere, we get the total excess time of traverse as

$$\Delta t(s) = (1/c) \left(40.3/f^2 \right) \int_S N_e \, dl \qquad (7.21b)$$

This additional delay is thus proportional to the integral of the electron density along the path of the wave. The quantity $[\int Ne \, dl]$, which is the path integral of the electron density, is the TEC of the ionospheric path through which the wave passes. It is expressed in number of electrons/m^2 or alternatively in TECU, as defined.

Consequently, the excess path that gets added up with the range due to the assumption of uniform velocity is thus

$$\delta l = c \, \delta t(l) = \left(40.3/f^2 \right) \int N_e \, dl$$
$$= \left(40.3/f^2 \right) TEC \qquad (7.21c)$$

This delay is corrected for obtaining improved performance. Focus 7.1 and the following Box 7.3 demonstrate how the TEC is derived from the measured ranges. The correction methods are discussed in the next section.

FOCUS 7.1 IONOSPHERIC TEC MEASUREMENTS

The ranges measured at two frequencies, 1575.42 and 1227.6 MHz, of a specific navigation satellite are 20,217,324.41 m and 20,217,331.67 m, respectively. To find the true range, assuming that the difference is due to the ionospheric delay only, we write

$$D1 = D + \delta_1 = D + (40.3/f_1^2)\text{TEC}$$

and

$$D2 = D + \delta_2 = D + (40.3/f_1^2)\text{TEC}$$

So,

$$D1 - D2 = \delta_1 - \delta_2 = 40.3(1/f_1^2 - 1/f_2^2)\text{TEC}$$

or,

$$\text{TEC} = (D1 - D2)/40.3 \times (1/f_1^2 - 1/f_2^2)^{-1}$$

Putting in the values, we get

$$\text{TEC} = 7.26/40.3 \times \{1/(1227.60)^2 - 1/(1575.42)^2\}^{-1} \times 10^{12}$$

$$= 691126.10 \times 10^{12}$$

$$= 6.9 \times 10^{17}$$

$$= 69 \text{ TECU}$$

Hence the correction for D1 is $\delta 1 = 40.3 \times 6.91126 \times 10^{17}/(1575.42 \times 10^6)^2 = 11.22$.

So, the corrected range is $20,217,324.41 - 11.22$ m $= 20,217,313.19$ m.

7.4.1.1.2 Faraday rotation

Another effect that takes place on the propagating navigation signals as they pass through the ionosphere is the Faraday rotation. As the electric fields of the radio waves with a definite polarization move across the magnetic fields of the earth, the field direction undergoes a rotation that is proportional to the TEC and also the total path in addition to the magnetic field strength.

7.4.1.1.3 Ionospheric Doppler

Doppler effect is the change in frequency of the received signal due to the relative velocity between transmitter and receiver. When the transmitter and the receiver approach each other, the phases of the transmitted signal thus received at the receiver will change relatively faster due to this relative velocity compared to the phase change observed when the distance between them is constant. This consequently results in effective increase or decrease in frequency of the signal as experienced by the receiver. This phenomenon occurs whenever the effective optical path between

---

BOX 7.3 MATLAB

The MATLAB program TEC_est.m was run with a given file containing measured range values in frequencies 1575.42 and 1227.6 MHz for a definite day. The variation observed in TEC for a station located near the geomagnetic equator is as shown in the following figure. Observe the high TEC value manifested (Figure M7.3).

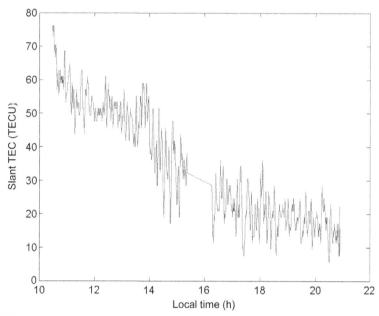

FIGURE M7.3

Variation of TEC.

Go through the program and identify how the TEC is derived from the measured ranges. Use the program to find the equivalent excess path and add necessary scripts to correct the range. Plot the variation of the measured and the corrected ranges. Verify that corrected ranges are equal for both the frequencies.

the transmitter and the receiver changes with time. So, if we represent the received signal phase φ_r as

$$\varphi_r = \varphi_t - 2\pi/\lambda \, R \tag{7.22}$$

where φ_t is the current phase at the transmitter, R is the range between the transmitter and the receiver and λ is the wavelength, then the corresponding change in the received phase due to the change in R with time will be

$$d\varphi_r/dt = -(2\pi/\lambda) \, dR/dt \tag{7.23a}$$

This change in phase is solely due to the change in the intermediate path of the signal and is in addition to the true change in the signal phase by the signal generator at the transmitter. So, it causes an additional frequency deviation δf and is equal to

$$\delta f = 1/2\pi \; d\varphi/dt$$
$$= -1/\lambda \; dR/dt \tag{7.23b}$$
$$= -v/\lambda$$

In the ionosphere, the increase in the effective path length due to the ionospheric TEC is given in Eqn (7.21). So, as the TEC varies with time, so does the effective path. This in turn results in additional phase variation, causing a change in the received frequency even if the satellite-receiver distance remains constant. So, the total effective path variation is the sum of the geometric path variation and is due to the change in TEC along the path. This effective rate of phase variation at the receiver resulting in a change in received frequency due to the change in TEC along the path is the ionospheric Doppler.

We have seen in Eqn (7.17b) that the excess path due to TEC is $dl = 40.3 \, \text{TEC}/f^2$. Hence, the rate of change in this path is $d(dl)/dt$. The corresponding ionospheric Doppler will therefore be

$$df = 1/\lambda \left(40.3/f^2 \right) d(\text{TEC})/dt \tag{7.24}$$

There are techniques to estimate the ionospheric path delay utilizing this ionospheric Doppler after segregating it from the Doppler due to geometric path variation (Acharya, 2013).

7.4.1.1.4 Ionospheric scintillation
Scintillations are the rapid fluctuations in the phase and amplitude of the wave. These are caused by local rapid variations in the refractive index of the medium through which the wave is traversing. Scintillation is mainly caused by the local variation of the ionospheric electron density. In equatorial regions of the ionosphere after the local evening, there generates some abrupt depletion of the ionospheric density that moves from near the equator toward the higher latitudes. These depletions are called the ionospheric bubble. These are sites of ionospheric scintillation (Dasgupta et al. 1982). Details of these will also be discussed in Chapter 9. Both amplitude and phase scintillations are formed by destructive and constructive combinations of the signals occurring in a random manner.

It is very important to understand here that the ionospheric scintillations do not affect the code-based range measurements of the navigation signals like the ionospheric delays, since it does not affect the propagation time. However, heavy scintillation may lead to sudden lowering of the received power such that it results in loss of lock. This in turn may degrade the dilution of precision and thus affect the accuracy. The scintillation effects, like the delay, are more severe in the equatorial regions and during the high solar activity time (Bandyopadhayay et al. 1997). Typically, the lower L band frequencies are more affected than the higher frequencies.

7.4.1.2 Tropospheric effects

The troposphere is the part of the atmosphere that lies nearest to the earth's surface. It extends up to about 12 km from the earth's surface and comprises the gaseous components most important for living beings. It is mainly made up of nitrogen and oxygen, while other gases like CO_2, water vapor, and traces of a few other gases are also present. There is also liquid water and other precipitations or precipitable elements present in the troposphere at definite heights. These constituents remain mixed in almost fixed proportions. However, over time and space they vary by little amount. As navigation signals pass through these dry and wet components of the troposphere, they incur some related impairments that affect the ranging.

7.4.1.2.1 Tropospheric delay

As the navigation signal travels through the different medium of the atmosphere, it experiences different refractive indices, which are also different from that of the vacuum. Therefore, these electromagnetic waves passing through the troposphere experience a velocity different from the velocity in the vacuum, c. So, according to the Fermat principle, the ray bends to minimize the total optical path (Ghatak, 2005). Consequently, the satellite signals transmitted by the satellites experience an additional delay in reaching the receiver in the ground. The excess refractive index of the medium over the vacuum is $\Delta n = n - 1$. The additional delay accounted for by the signal as it passes through a length dl of this layer of the atmosphere is given by

$$dt = dl/v - dl/c$$
$$= dl(n/c - 1/c) \tag{7.25a}$$
$$= (dl/c)\,(n - 1)$$

So the effective excess path added by the troposphere during ranging is

$$\int dr = \int c\,dt = \int (n - 1)\,dl \tag{7.25b}$$

It is interesting to note that the tropospheric refractive index n exceeds that of the vacuum by a very trivial amount. So, $n - 1$ is a very small number. To make it convenient for use, this term is scaled up by 10^6 to generate the tropospheric refractivity N. This term N may now be directly used for calculating the delay with appropriate scaling.

The parameter N can be divided into dry and wet components, N_d and N_w, respectively. The dry component is contributed by the gases while the wet components are from the water vapors and liquid water present in this region. Although the absolute values of the delay are comparatively small, the relative contribution of the dry component is much more than the wet component. However, the variability of the wet component is many times more than the dry component. Both of these components result in delay in the signal and hence introduce error in navigation receivers during ranging. But, unlike the ionosphere, here the delays are not dispersive, i.e. they do not depend upon the frequency of the wave traveling through it. So, as $dn/d\omega = 0$, the same delay is experienced by all the frequencies.

This erroneous excess range needs to be eliminated from the measurements, and for that we need to know the delay exactly. There are models to find out the delays offered by the dry and wet components of the troposphere.

7.4.1.2.2 Attenuation

The tropospheric components like rain, fog, clouds, vapor, etc. are all responsible for the attenuation of the waves traveling through the troposphere. In this part of the atmosphere, the main causes of attenuation are absorption and scattering. These phenomena occur when the wave interacts with the tropospheric elements mentioned above.

Absorption involves dissipation of the energy of the wave as it passes through the tropospheric elements. In conductors or imperfect dielectrics having finite conductivity, this occurs in the form of the movement of the electrons, i.e. conduction current due to the incident field causing dissipation. Each water droplet present in the rain, cloud, fog, etc. may be treated as an imperfect dielectric and the traversing wave loses a part of its energy as it is incident on them during propagation.

Scattering involves diversion of the directed energy of the wave to different directions. Under the influence of the electric field of the propagating signal, the water molecules exhibit electronic polarization using a part of the incident energy. It occurs during half of the complete phase cycle of the incident field. Consequently, the energy of the wave is transferred to the molecule as potential energy. This energy is released and transferred back by the medium in the subsequent other half of the phase cycle of the wave when these molecules in the medium act as secondary sources of radiation. However, the released energy not only is radiated in the direction of the actual wave but gets scattered in all other directions. Hence, there is an effectual cause for the loss of the wave energy.

The more the wave interacts with the water particles, the more the attenuation, and hence the effect increases with heavier rain rates and effective paths through the medium. It is also to be noted that there are associated scintillations experienced by the wave on passing through the troposphere. However, the attenuation and the tropospheric scintillations are low at frequencies typically used for navigation and also do not affect the ranging process.

7.4.1.3 Multipath

Multipath is the phenomenon of fluctuations in signal strength formed by incoherent combination of signals coming from different directions through reflection or scattering with the direct signal. These reflected or scattered signals have different amplitudes and phases compared to the one received directly. This is due to the extra path traveled by the multipath signals and also due to the fact that during reflection or scattering of this signal there occurs an abrupt phase change of the wave and consequent reduction in the amplitude of the signal, depending upon the site from which the reflection takes place.

In Chapter 5, we found that the received direct signal enters the correlator, where it is correlated and the delay in the prompt version of the code required to achieve the peak is identified and utilized to find the range. You may recall that this identification of the peak was done by comparing the correlation values with early and late versions of the local code with the incoming signal. In the discriminator, the early and late correlation values exactly balance and reduce to zero, when the prompt version of the code at the receiver is exactly aligned with the incoming one.

In the case of multipath, the incoming signal is a compound signal containing the direct and a reduced portion of the reflected signal, which is also different in phase and amplitude.

Now, let us see how a delayed signal due to multipath affects the ranging. It does so by shifting the balanced condition due to the presence of the extra signal. Let N be the total numbers of bits in the sequence of interval T. Δ is the offset of the prompt code from the true incoming code and τ is the early and late offsets. We shall consider only small values of the offset Δ such that $|\tau| > |\Delta|$. With this offset, the correlation Rc of the early code with the incoming signal will be

$$Rc(\Delta + \tau) = 1/N \int c(t + \Delta + \tau)\, c(t)\, dt$$
$$= 1/N\, [1 - |+\tau + \Delta|/T] \qquad (7.26a)$$
$$= 1/N\, [1 - (\tau + \Delta)/T]$$

Similarly, the correlation value with the late code is

$$Rc(\Delta - \tau) = 1/N \int c(t + \Delta - \tau)\, c(t)\, dt$$
$$= 1/N\, [(1 - |-\tau + \Delta|/T)] \qquad (7.26b)$$
$$= 1/N\, [1 - (\tau - \Delta)/T]$$

So, from the above we find that the balanced conditions can only occur when $|\Delta| < |\tau|$ and the condition required to be satisfied for it may be given as:

$$Rc(\Delta + \tau) = Rc(\Delta - \tau)$$

or,
$$\Delta + \tau = \tau - \Delta$$

or,
$$\Delta = 0 \qquad (7.27)$$

Now, if in addition, another delayed version of the signal is received due to multipath, the total autocorrelation function becomes different. We shall estimate the error thus occurring for the simplest case when the delay δ due to the multipath is small such that $|\tau| > |\delta + \Delta|$. Since the time argument in the signal reduces by δ when the signal gets delayed by the time δ, we have the correlation with the delayed signal for the early code as

$$Rc(\Delta + \tau) = 1/N \int c(t + \Delta + \tau)\, c(t - \delta)\, dt$$
$$= 1/N\, [1 - |\tau + \Delta + \delta|/T] \qquad (7.28a)$$
$$= 1/N\, [1 - (\tau + \Delta + \delta)/T]$$

For the late section of the correlator, the correlation value becomes

$$Rc(\Delta - \tau) = 1/N \int c(t + \Delta - \tau)\, c(t - \delta)\, dt$$
$$= 1/N \,(1 - |-\tau + \Delta + \delta|/T) \qquad (7.28b)$$
$$= 1/N \,[1 - (\tau - \Delta - \delta)/T]$$

Now if we assume that both the direct and the multipath signals reach the receiver with the same strength, then the condition for the balance is

$$(\tau + \Delta) + (\tau + \Delta + \delta) = (\tau - \Delta) + (\tau - \Delta - \delta)$$

or, $$\tau + \Delta + \tau + \Delta + \delta = \tau - \Delta + \tau - \Delta + \delta$$

or, $$4\Delta = -2\delta \qquad\qquad (7.28c)$$

or, $$\Delta = -\delta/2$$

So, instead of $\Delta = 0$, the balance condition is achieved when the offset of the local code with the true signal is $\Delta = -\delta/2$.

Figure 7.5 illustrates the relative code phase conditions for both the scenarios in which the delay due to multipath is δ.

It follows from the above that in such a case, the discriminator does not go to zero exactly when the two codes are aligned but at some skewed value. This is because, even when the early and late components of the autocorrelation balance for the direct component, the multipath component adds a non-zero component to it and hence biases the zero crossing condition. So, this causes the range error to occur.

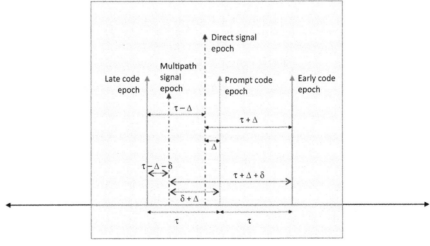

FIGURE 7.5

Relative code phase conditions for multipath scenario.

Here, we have implicitly considered that the received power of the direct and the multipath signals are exactly the same at the point of correlation.

For the most general case where the multipath is received with a power reduced by a factor α and where the θ_c is the composite phase of the direct plus multipath signal and θ_m is the phase of the only multipath component, then the discriminator output becomes (Braasch, 1996):

$$D(\tau) = [R(\Delta + \tau) - R(\Delta - \tau)] + \alpha \, [R(\Delta + \tau - \delta) - R(\Delta - \tau - \delta)] \cos(\theta m - \theta c)$$

$$(7.29)$$

7.4.2 **User segment error**

7.4.2.1 *Receiver noise*

Receiver hardware always has some noise associated with it. It also picks up noise from the external environment in the receiver band while receiving the signal. This noise adds up with the signal and corrupts the measurement process and hence the measurements deviate from their true values. The effect of the receiver thermal noise and the phase noise on the measurement process has already been discussed in Chapter 5. Recall that we estimated the error that occurred during the code tracking as a result of noise in the discriminator of the loop. Although the long-term average of this noise is zero, the individual discrete measurements have random errors, which cause the receiver to assume the peak of the correlation at a point slightly deviated from the true peak. This in turn deviates the estimation of the propagation delay, eventually resulting in ranging error leading to wrong estimation of the user position. The detailed descriptions of the different sources of receiver noise have already been discussed in Chapter 5 in connection with receiver performance.

7.4.2.2 *Receiver clock bias*

Receiver clocks, being typically inexpensive are relatively less precise and drift over time. Hence, at any instant the clock shows a relative bias and drift compared to the satellite or system time. This has already been discussed in Chapter 6 and its effect is included in nominal observation equations for estimation during position fixing.

7.4.2.3 *Receiver hardware bias*

As the signal passes through the hardware of the receiver, due to the finite hardware delay time, it experiences some lag between the instant of being received at its antenna phase center and being identified at the correlator. This adds up as an additional propagation delay and hence results in some very small but finite ranging error. These delays are generally dependent upon the signal frequency. So, in a receiver, especially with multiple frequencies, the effects may cause significant effect and should be taken care of very well.

7.4.3 Overall effect

Consider an observation equation in the form given in Eqn (7.6), rewritten here for convenience,

$$R = \sqrt{(x_s - x_u)^2 + (y_s - y_u)^2 + (z_s - z_u)^2} + c\delta t_u - c\delta t_s + \delta r_{eph}$$

We have already mentioned that in such an equation we always equate the measured value of the range R, i.e. the left-hand side of the equation, to the theoretical expression of the same, giving the geometric range on the right-hand side. Since the measured range gives the pragmatic value in the given situation adding up all the impairments which are acting concurrently on the signal, it is required to equate this with the truth model of the range where all the acting impairments have been accounted for. The receiver clock error, satellite clock error, and ephemeris error effects have already been added to the equation.

The propagation and the user segment introduce considerable errors, which now need to be compensated here by equivalent terms in the theoretical side for balancing the equation completely. In cases where any of these impairment parameters go unaccounted for or the necessary parameters are not used, it leads to error.

So, the pragmatic observation of the range R can be represented with an equation where the measured range is equated to the true geometric range plus the errors in ranging and reference positioning. So,

$$R = \sqrt{(x_s - x_u)^2 + (y_s - y_u)^2 + (z_s - z_u)^2} + c\delta t_u - c\delta t_s + \delta r_{eph}$$
$$+ \delta r_{ion} + \delta r_{trp} + \text{other errors} + \varepsilon \tag{7.30}$$

where δr_{ion} and δr_{trp} represent the effective range error due to the ionospheric and tropospheric delay of the signal. The other errors include the ranging errors due to multipath, etc. and ε represents the error due to noise.

Due to the total effect of all the errors, the true range is not known exactly. Thus, the true range from satellite to receiver is an estimated range with errors and is called pseudorange. The following Box 7.4 demonstrates the effect of random errors and bias type errors in ranges.

BOX 7.4 MATLAB

The MATLAB program estimation_error.m was run with 10 m (1σ) of random error in the ranges, which has normal distribution. The plot of the errors obtained for random bias for different ranges is shown below (Figure M7.4).

   Observe how the estimated position spreads about the true position. The spread similarly shows a normal distribution in its radius. Also note that the position estimation errors are larger in magnitude than the random error introduced.

   The same program was run with random but discretely different bias, with biases −5, 0, +5, and +10 for four ranges used, and the above plot was obtained. Observe that the errors are now collated into groups. However, they show abrupt change with time to form another collation. The changeover occurs when the satellites used for estimation are changed to different sets.

   Run the MATLAB program estimation_error.m with different values of 1σ error with random error option and observe how the estimated positions spread about zero values. Run the same with random bias option. See the errors. Compare them.

BOX 7.4 MATLAB—cont'd

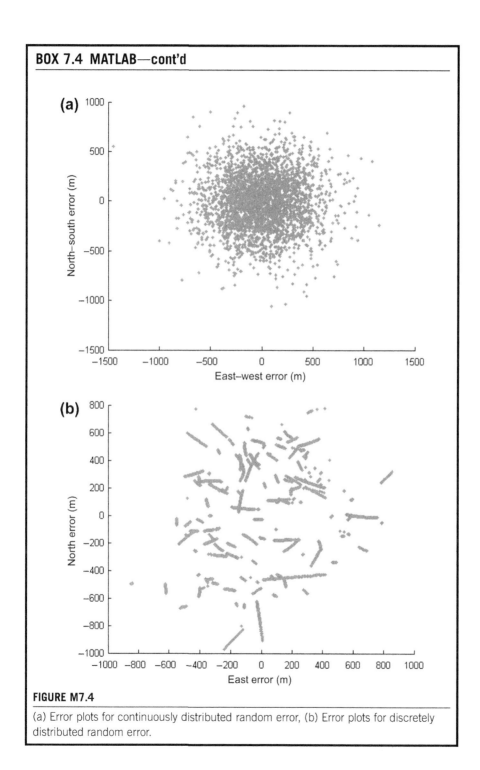

FIGURE M7.4

(a) Error plots for continuously distributed random error, (b) Error plots for discretely distributed random error.

7.5 **Techniques of error mitigation**

You must agree that the range values thus obtained through measurement cannot be directly used at the receiver for position fixing. This is because a wrong solution of user position will be obtained unless all the error terms are removed. This readily follows from Eqn (7.2). So, our next objective will be to remove the excess error terms from this pseudorange to get the true solution. The need for correcting errors in the pseudorange equation can also be understood from the above exercise in Box 7.2. So, the extra terms in addition to the geometric range and receiver clock bias are required to be eliminated from the equation. Correction of error terms is generally made in two ways as discussed in the following subsections.

7.5.1 **Reference-based correction**

In reference-based correction, there exists a reference station whose position is exactly known. The true position of this reference can be obtained by using some different independent means like ground-based surveying. The reference station can derive the values of the relevant correction parameters and updates this information as frequently as required and also can disseminate it to the users around it.

As we have seen, the error in the solution for position X is given by Eqn (7.2) as

$$\underline{X} + \underline{dX} = \underline{G}^{-1}\,\underline{R} + \underline{G}^{-1}\,\underline{dR}$$

or,
$$\underline{dX} = \underline{G}^{-1}\,\underline{dR}$$

Here, dX is the error in position solution, G is the observation matrix depending upon the satellite—receiver geometry, and dR is the ranging error.

The ranging errors dR of the users may be assumed to be equal to the dR of the reference station. Furthermore, if the users are adequately near to the reference, even the G may be assumed to be almost the same. In such a case, the dX will also be the same between the user and reference. Then, corrections are generally carried out in which the reference station informs the user only about the position errors dX. The reference station obtains this error by comparing his true position with that estimated using the satellites. This difference is transmitted to the users located near the reference. The users utilize the same errors in their estimated position to correct them. The users derive their position X, obtained by using their own measured range R and own value of G and using the correction dX transmitted by the reference station.

For most practical cases, G cannot be the same for the users and the reference receivers. In such cases, the individual range error for different satellites, dR, is derived by the reference station, which is transmitted to the users. These ranging errors are used by the users as appropriate in generating their own observation equations. So, they correct their measured range R using this and generate their own G to find their position estimate, X. However, here only the constant and the correlated errors can be corrected with certain accuracies. There are still uncorrelated errors, which differ drastically between positions of reference and those of the users, and these errors cannot be corrected in this way. All these and other methods of differential corrections will be discussed later in Chapter 8 in an elaborate manner.

We have already learned about the satellite clock shift in Section 7.3.1. What is important here is that these clocks are continuously tracked by the ground system and the corrections for the clock drifts are generated. The clock biases and drift parameters, estimated at the master control station (MCS) are then transmitted to the user through navigation parameters at appropriate intervals. This is also a kind of reference-based correction where the MCS acts as the reference station for all users, estimating and disseminating the corrections in real time. The MCS utilizes sophisticated methods and tools to generate the accurate deviations in the satellite clocks. Although these methods are very accurate, still there remains some error in estimating the clock drift and bias by the ground segment. So, some amount of residual error, however trivial, remains uncorrected while range measurements are done at the receiver. As these errors are constant errors and the same for all users, both the reference station and the users may correct these errors in their respective ranges. Therefore, only its residual errors along with the correlated errors from other sources are communicated between the reference station and the user.

7.5.2 **Direct estimation of errors**

Save the clock error, which is estimated and transmitted typically by the system itself, the reference-based correction of errors over a large geographical area demands for augmented facilities to cater the error information to the users. However, this causes the limitations of the usage, in terms of area and compatibility; puts additional compulsion on the receiver to receive reference station data; and obviously increases the cost of usage and operation as well.

Direct estimations are made using models and real-time measurements. The various error terms that are directly estimated are described in the following sections.

7.5.2.1 *Ionospheric error*

In the navigation services, generally two or more frequencies are used for the transmission of the data. However, depending upon the requirements, the receivers can be of single frequency type, or dual or multiple frequency type. The former can receive signals only in one of the available frequencies, while the latter can receive and handle signals in two or more frequencies. The ionospheric delay being dispersive, provides a great leverage to the dual- or multiple-frequency users. It enables them to estimate the ionospheric delay by itself. Here is how it is done.

We know that the ionospheric delay adds an equivalent extra range δr_{ion} to the true geometric range ρ during the range measurements, which is a function of frequency f and TEC, as given in Eqn (7.21) as

$$\delta r_{ion} = \left(40.3/f^2\right) \text{TEC}$$

So if the range measured by a receiver in one of the frequencies f_1 added with ionospheric delay is δr_{1ion}, then

$$R_1 = \rho + \delta r_{1ion}$$
$$= \rho + \left(40.3/f_1^2\right) \text{TEC} \tag{7.31a}$$

where ρ is the true geometric range. However, there is no way that this excess path δr_{1ion} can be derived from this single measurement, since the TEC values along the path through which the signal passes cannot be known anyway.

For dual-frequency receivers, where there is available a separate and independent measurement in the other frequency f_2, let the range measured there be R_2; then

$$R_2 = \rho + \delta r_{2ion}$$
$$= \rho + \left(40.3/f_2^2\right) \text{TEC} \tag{7.31b}$$

Now, if these two measured ranges R_1 and R_2 are differenced, the result is the difference in the excess path incurred in the two frequencies as the common true geometric range present in both the measurements gets canceled. The difference thus yields

$$R_1 - R_2 = 40.3 \,(\text{TEC})\left(1/f_1^2 - 1/f_2^2\right) \tag{7.31c}$$

In this expression, since all the other parameters save the TEC are known, we can derive the TEC values from it. The exact relation turns into

$$\text{TEC} = (R_1 - R_2)/\left\{40.3\left(1/f_1^2 - 1/f_2^2\right)\right\} \tag{7.32}$$

So, once the TEC value at any instant along any definite path is known, the true range R can be derived from any of the measured ranges as

$$\rho = R_1 - 40.3/f_1^2 \,(\text{TEC})$$
$$= R_1 - 40.3/f_1^2 \times (R_2 - R_1)/\left\{40.3\left(1/f_2^2 - 1/f_1^2\right)\right\} \tag{7.33a}$$
$$= R_1 - (R_2 - R_1)/\left(f_1^2/f_2^2 - 1\right)$$

Calling $(f_1^2/f_2^2) \, \mu$, we get

$$\rho = R_1 - (R_2 - R_1)/(\mu - 1) \tag{7.33b}$$

On the other hand, single-frequency receivers can correct ionospheric errors using the models of the ionosphere. In these receivers, the corrections are obtained by using parametric models, which provide either the ionospheric delay or the TEC for any location and time, using predetermined coefficients. The two important such models are the Klobuchar model and the NeQuick model. In the Klobuchar model (Klobuchar, 1987) the vertical ionospheric delay is obtained as a function of the local time and geographic location of the receiver. It is a semi-empirical model expressed as a half cosine function, given as

$$D(\lambda, \varphi, \tau) = a + b \cos\left\{(\tau - c)/d\right\} \tag{7.34}$$

where D is the vertical delay at time τ and the parameters a and c are constants. Parameters b and d, representing the amplitude and width of the cosine function, are dependent upon the geomagnetic latitude λ and longitude φ of the receiver location and can be derived in the form of their polynomials, where the coefficients are known as Klobuchar coefficients and given for a specific day. The parameter b becomes zero for a certain period of the day when $|(\tau - c)/d| \geqq \pi/2$ and then the delay remains constant as 'a'.

However, it is understandable and does not need any special mention that the dual-frequency receivers do better ionospheric corrections than the single-frequency receivers.

7.5.2.2 Tropospheric error
Tropospheric errors are corrected for both wet and dry tropospheric delays using models. Some well-known models are the Hopfield model (Hopfield, 1969) and the Saastamoinen model (Spilker, 1996).

This model presents the height profile of tropospheric refractivity N ($N = 10^6 \times (n - 1)$, where n is the index of refraction). This profile may be used to derive the excess path traversed by the navigation signal. The model is theoretically based on an atmosphere with constant lapse rate of temperature. It treats "dry" and "wet" components of N separately, giving their profiles with the height of the surface above sea level relative to the heights h_d and h_w, respectively, where the indices have been actually measured. The expression for the total zenith delay is given by

$$\delta r_{trp} = 10^{-6} \left[\int_0^h N_d(h_d)(1 - h/h_d)^4 dh + \int_0^h N_w(h_w)(1 - h/h_w)^4 dh \right] \quad (7.35a)$$

To estimate the zenith delay up to the satellite height, the wet component may be integrated to the tropospheric height of about 12 km while the dry component needs to be integrated up to the stratosphere due to the presence of refracting gases at these heights, although in a minor amount.

7.5.2.3 Receiver hardware delay
The receiver hardware delay changes slowly. These delay values may be estimated offline a priori using different available techniques and the estimation may be updated at large intervals.

7.5.2.4 Other errors
Other errors, such as multipath, can be neither modeled explicitly nor measured, and hence cannot be corrected in this manner. Removal of multipath can be done by using some intelligent design of the receiver, as in a rake receiver. Other methods include using a choke ring for large receivers, reducing the tracking loop filter bandwidth, or using a properly modeled Kalman filter.

Receiver noise can be reduced by using qualified components or by using a Kalman filter, or both.

Upon obtaining these errors from the reference stations or from the models or through direct measurements or models, the error values are corrected from the measured range to derive the true range. Then the position is calculated using these corrected ranges.

7.6 Effect of errors on positioning

In the previous chapter, we found that the position estimated is nothing but the common intersecting point of the four spheres defined with their centers at the locations of each reference satellite and having a radius equal to their respective ranges. This narrows it down to a single point when the ranges are exact. In cases where we have a certain ambiguity in the range, the derived positions are also ambiguous by a certain width. It follows readily that due to the range ambiguity, the intersection sprawls from a point to a finite volume and the required true position should lie somewhere in this common volume defined by the ambiguous range. The extent of this ambiguous volume will define the accuracy of estimation. So, the intention would be to reduce this uncertainty space to get better accuracy. This is depicted in Figure 7.6.

In the following section, we shall study, both qualitatively and quantitatively, how this volume, defining the extent of error in position fix, is dependent upon the relative orientation or geometry of the respective four satellites participating in the position fixing. So, the effect of errors on position measurements is defined in terms of dilution of precision (DOP), which relates the range errors to the position coordinate error. It is a function of geometry that the satellite bears with the receiver and that we shall derive here in mathematical terms.

Let us start with a very simple analogy. Suppose A and B are asked to get the position of a particular place on a map. It was found that A is fairly confident that the longitudinal location of the place is between φ and $\varphi + d\varphi$, but has large ambiguity over its latitude. On the other hand, B knows almost exactly that the latitude of the place they are looking for is between λ and $\lambda + d\lambda$ but has no knowledge of its longitude. So, combining these facts, the location of the place boils down to a

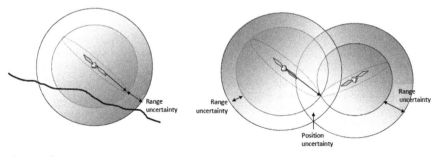

FIGURE 7.6

Orientation of error volume.

location between φ and φ + dφ and λ and λ + dλ. Thus the total uncertainty reduces to an area of dφdλ and hence the location can be very easily identified. Now, think what would happen if both of them were uncertain about the latitudinal position of the place. Then, even after combining the information that both of them have, there would have remained a large uncertainty about the latitude of the place. This indicates that better positioning can be done if the ambiguities of the information, by combining which we derive the position, are independent and differently directed, preferably orthogonal.

In the real scenario of position fixing in a 3D environment, after correcting the range error to the maximum possible extent, there still remain some residual range errors, which causes ambiguities while deriving the positions. This indicates that the true position should lie within an annular spherical cavity where the width of the annularity is the range ambiguity, σ_r. Now, notice that after correction the residual error can only be specified by the statistical parameters of standard deviation σ_r or variance, σ_r^2, assuming zero mean error over a large ensemble.

Now, the range ambiguity from another satellite will also produce similar annular ambiguity space. If the satellites are so placed that these ambiguity spaces have much overlapping, it will result in large effective uncertainly in position fixing. But, had the other satellite been positioned such that the ambiguity spaces intersect each other orthogonally or nearly orthogonally, it will consequently reduce the common uncertainty region, thus improving the precision of the position estimate. It is achievable when the satellites remain well separated from each other. So, with reference to the user point, if we consider a tetrahedron, formed by the user as its apex and four satellites as the four base corners, the above argument refers to the maximum volume of the tetrahedron for the highest precision. This is shown in Figure 7.7.

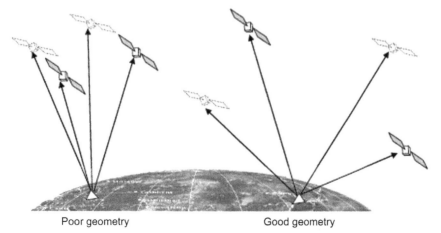

Poor geometry   Good geometry

FIGURE 7.7

Desirable positions of satellites for better accuracy.

7.6.1 Dilution of precision

Let us now deduce the fact mathematically that we have stated in the previous subsection. Assuming that the range errors are independent and identical, the solution for position may be obtained as discussed in Chapter 6. However, now, due to the presence of the residual error in the range, there will be some consequent errors in the position estimate. Here, we derive how much these errors will be present in the solution.

For this, let the measured pseudorange R be corrected for its errors and the residual error that remains after correction be dR. Note that this error is not definite but a stochastic value (otherwise it could have been corrected). So, the pseudorange error is given by

$$dR = [\partial R/\partial x \ \partial R/\partial y \ \partial R/\partial z \ \partial R/\partial b] \cdot [dx \ dy \ dz \ db]^T$$
$$= G \cdot dX \tag{7.36}$$

where G is the observation matrix depending upon the user–satellite geometry and dX is the geometric error vector including the clock bias error, db.

This equation relates the position estimation error dX that occurs as a result of the ranging error dR through G. Furthermore, this may be used to obtain the relationship between the covariance of these two sets of errors (Axelrad and Brown, 1996; Conley et al. 2006). Using the above equation,

$$dX = G^{-1} \ dR \tag{7.37}$$

Multiplying the above equation with its transpose,

$$dX \ dX^T = G^{-1} dR \ dR^T \ G^{-T}$$
$$= \left[G^{-1} G^{-T}\right] dR \ dR^T \tag{7.38}$$
$$= \left[G \ G^T\right]^{-1} dR \ dR^T$$

Now, we take the expectation value on both sides. Considering that the elements of G are definite and fixed, G matrices should remain out of the expectation operation. Hence,

$$E\left[dX \ dX^T\right] = \left[G \ G^T\right]^{-1} E\left[dR \ dR^T\right]$$
$$= \left[G \ G^T\right]^{-1} \sigma_R^2 \tag{7.39}$$

or,
$$\sigma_G^2 = H \ \sigma_R^2$$

where $H = (G_1^T G_1)^{-1}$ and $\sigma_R^2 = E\ [dR \ dR^T]$, and we have assumed that dR are the same for all satellites.

Notice that the diagonal elements of the expectation values of the matrix $E[dX \ dX^T]$ are nothing but the covariance of the individual errors along different axes. So, equating them to the diagonal elements of H matrix, we get

$$E[dx\ dx] = \sigma_x^2 = H_{11}\sigma_R^2$$

$$E[dy\ dy] = \sigma_y^2 = H_{22}\sigma_R^2$$

$$E[dz\ dz] = \sigma_z^2 = H_{33}\sigma_R^2 \tag{7.40}$$

$$E[db\ db] = \sigma_b^2 = H_{44}\sigma_R^2$$

These equations show that the variances in position estimates depend upon two factors:

i. The variance of user range error (σ_R^2).
ii. The elements of matrix H that depends on the user–satellite geometry.

From the above derivations, we can express the standard deviation of errors along any coordinate axes in terms of standard deviation of the error in range measurement. So,

$$\sigma_x = \sqrt{H_{11}}\sigma_r, \quad \sigma_y = \sqrt{H_{22}}\sigma_r, \quad \sigma_z = \sqrt{H_{33}}\sigma_r, \quad \sigma_b = \sqrt{H_{44}}\sigma_r \tag{7.41a}$$

The standard deviation in total position estimation is:

$$\sigma_p = \sqrt{\sigma_x^2 + \sigma_y^2 + \sigma_z^2}$$

$$= \sigma_R\sqrt{H_{11} + H_{22} + H_{33}} \tag{7.41b}$$

$$= PDOP \cdot \sigma_R$$

where the position dilution of precision, $PDOP = \sqrt{H_{11} + H_{22} + H_{33}}$
Similarly, the standard deviation in time estimation is:

$$\sigma_b = \sigma_R\sqrt{H_{44}}$$

$$= TDOP \cdot \sigma_R \tag{7.41c}$$

where the time dilution of precision, $TDOP = \sqrt{H_{44}}$
So, the standard deviation of total error is

$$\sigma_G = \sqrt{\sigma_x^2 + \sigma_y^2 + \sigma_z^2 + \sigma_b^2}$$

$$= \sigma_R\sqrt{H_{11} + H_{22} + H_{33} + H_{44}} \tag{7.42}$$

$$= GDOP \cdot \sigma_R$$

where the geometric dilution of precision, $GDOP = \sqrt{H_{11} + H_{22} + H_{33} + H_{44}}$

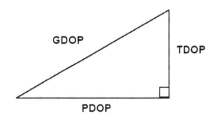

FIGURE 7.8

Relationship between the DOPs.

Thus, the various "dilutions of precision" parameters are defined as above on the basis of the stochastic error equations to provide a quantitative contribution of the user–satellite geometry on the error. Comparing, we can see that

$$PDOP^2 + TDOP^2 = GDOP^2 \qquad (7.43)$$

Recalling the Pythagoras theorem, the relationship between GDOP, PDOP, and TDOP may be looked upon as the three sides of a right-angled triangle, where PDOP and TDOP form the two perpendicular sides of the triangle and GDOP is the hypotenuse (Figure 7.8).

The dilution of precision depends upon the geometry of the participating satellites. We have seen intuitively that the wider the satellites are separated, the better the accuracy. Now, let us see the same from the mathematical expressions that we have developed. In terms of GDOP, for better accuracy of position, for any definite range error, the elements of the H matrix should be small. Now, $H = (GG^T)^{-1}$ and hence as H is generated by the inverse of the G matrix and its transpose. So, for the above condition to be satisfied, the G matrix should be well conditioned, i.e. G should not be singular or near singular, or in other words the value of the determinant of G should be much away from zero, as already mentioned in Section 7.1. This is possible when the two rows in the G matrix are not similar or close to each other and preferably orthogonal. Now, when can the two rows of the G matrix be similar? The G elements are constituted by the directional cosines of the satellites at the receiver point, which are actually the geometrical elements of the satellites with respect to the user. To satisfy the above condition, the directional cosines of two satellites, constituting two rows of the matrix, should not be equal or near to each other. Therefore, it implies that positions of the different satellites must be very different from each other, i.e. the satellites should have the widest separation, as seen from the receiver.

The standard deviation of the range error, incurred by the user referred to as "σ_r" above, is also called the σ_{UERE}, where UERE stands for user equivalent range error.

7.6.2 Horizontal and vertical dilution of precision

The geometric dilution of precision or its positional component, i.e. PDOP, can be resolved into components about the conventional axes of coordinates, viz. X, Y, and Z, as given in Eqn (7.41a).

$$\sigma_x^2 = H_{11} \ \sigma_r^2,$$
$$\sigma_y^2 = H_{22} \ \sigma_r^2,$$
$$\sigma_z^2 = H_{33} \ \sigma_r^2,$$
$$\sigma_b^2 = H_{44} \ \sigma_r^2$$

But these components, although easy to derive, are not in a convenient form of representation. Expected errors or their precisions about the earth-centered X, Y, or Z direction do not give a feel of the error magnitude and sense to one who is measuring the position at a local point on the ground. So, these DOP parameters are converted to a more convenient form in local coordinates and they are along the local horizontal and vertical directions, and so are accordingly referred to as horizontal dilution of precision (HDOP) and vertical dilution of precision (VDOP).

7.6.3 Weighted least squares solution

The position solution obtained in Chapter 6 through linearization was based upon a fundamental assumption that the ranges have no error in them. However, the same approach is valid when the range errors for all the satellites are statistically independent and identical. It is on this assumption that the DOP values have been developed in the last subsection. However, this is not true in pragmatic cases. Often, this error, also represented by the σ_{UERE}, is neither independent nor identical. So under such conditions, the least squares position estimate does not hold good as an optimal estimate.

If the pseudorange errors are Gaussian and the covariance of ranging errors ε for the visible satellites is given by the matrix M, then the optimal solution for user position relative to the approximated point Xa, as described in Section 6.4 is given by the weighted least squares estimate (Strang, 1988)

$$dX = \left(G^T M^{-1} G\right)^{-1} \left(G^T M^{-1}\right) dR \qquad (7.44)$$

Check that this expression converts to the conventional form when the errors are identical.

7.7 Error budget and performances

It is during the design phase that the maximum permissible range errors allowable in the system are fixed and apportioned. The upper hard limit is determined by the system capabilities and defines the system performance. This is called error budgeting. Any excess error that is introduced into the system during operation is correctable and may be eliminated through appropriate methods. The errors are removed through different processes, as we have discussed in this chapter Focus 7.2 explains an error budget for a typical satellite navigation system.

FOCUS 7.2 ERROR BUDGET

Assuming that the average error over the ensemble is zero, the typical values of the standard deviations of the residual error may be expressed as:

Ephemeris error	2 m
Clock error	2 m
Ionospheric error	5 m
Tropospheric error	1 m
Receiver noise	1 m
Multipath, etc.	1 m
Total	6 m

Notice that the errors being individually independent to each other, the root sum squared values of the errors are taken as the effectual error.

So, if we take the HDOP and VDOP as 3 and 4, respectively, for a system at a place, the horizontal and vertical precision of the position estimation at that place becomes 6 m × 3 = 18 m and 6 × 4 = 24 m, respectively.

Conceptual questions

1. Show that the GDOPs derived from the local DOPs are equal to the sum of the GDOPs derived in the earth-centered earth-fixed system.
2. Derive the expressions for DOP along north, east, and up directions in an east-north-up system in terms of elevation and azimuth.
3. Pragmatically, the multipath signal comes with a reduced power compared to the direct one. Find the condition of correlation balance when the multipath signal power is k times (k < 1) the direct signal and compare with Eqn (7.29).

References

Acharya, R., 2013. Doppler utilized kalman estimation (DUKE) of ionosphere. Advances in Space Research 51 (11), 2171−2180.

Axelrad, P., Brown, R.G., 1996. GPS navigation algorithms. In: Parkinson, B.W., Spilker Jr., J.J. (Eds.), Global Positioning Systems, Theory and Applications, vol. I. AIAA, Washington, DC, USA.

Bandyopadhayay, T., Guha, A., Dasgupta, A., Banerjee, P., Bose, A., 1997. Degradation of navigational accuracy with global positioning system during periods of scintillation at equatorial latitudes. Electronics Letters 33 (12), 1010−1011.

Braasch, M.S., 1996. Multipath effects. In: Parkinson, B.W., Spilker Jr., J.J. (Eds.), Global Positioning Systems, Theory and Applications, vol. II. AIAA, Washington, DC, USA.

Conley, R., Cosentino, R., Hegarty, C.J., Leva, J.L., de Haag, M.U., Van Dyke, K., 2006. Performance of Standalone GPS. In: Kaplan, E.D., Hegarty, C.J. (Eds.), Understanding GPS Principles and Applications, second ed. Artech House, Boston, MA, USA.

Dasgupta, A., Aarons, J., Klobuchar, J.A., Basu, S., Bushby, A., 1982. Ionospheric electron content depletions associated with amplitude scintillations in the equatorial region. Geophysical Research Letters 9 (2), 147–150.

Feynman, R.P., Leighton, R.B., Sands, M., 1992. Feynman Lectures on Physics. Narosa Publishing House, India.

Ghatak, A., 2005. Optics. Tata Mc Graw Hill Publishing Limited, New Delhi, India.

Hopfield, H.S., 1969. Two quartic tropospheric refractivity profile for correcting satellite data. Journal of Geophysical Research 74 (18), 4487–4499.

Klobuchar, J.A., 1987. Ionospheric time-delay algorithm for single-frequency GPS users. IEEE Transactions on Aerospace and Electronic Systems AES-23 (3), 325–331.

Parkinson, B.W., 1996. GPS error analysis. In: Parkinson, B.W., Spilker Jr., J.J. (Eds.), Global Positioning Systems, Theory and Applications, vol. II. AIAA, Washington, DC, USA.

Reitz, J., Milford, F.J., Christy, R.W., 1990. Foundations of Electromagnetic Theory, third ed. Narosa Publishing House, India.

Spilker Jr., J.J., 1996. Tropospheric effects on GPS. In: Parkinson, B.W., Spilker Jr., J.J. (Eds.), Global Positioning Systems, Theory and Applications, vol. I. AIAA, Washington, DC, USA.

Strang, G., 1988. Linear Algebra and Its Applications. Harcourt, Brace, Jovanovich, Publishers, San Diego, USA.

Differential Positioning

CHAPTER OUTLINE

In Chapter 6 we learned how to estimate absolute positions from range measurements while in Chapter 7 we described the aspects of having errors in the measurement from which these positions are derived. In this chapter we will find how to reduce the effect of the errors in positioning using *differential positioning* techniques. As the name itself suggests, in contrast to absolute positioning, here we require positioning is done with respect to a reference receiver. This reference receiver is located at a precisely known position. This reference receiver is located measurements which are done in a nominal user are available along with some of their derivatives. Here, we shall first explain the basic idea of differential positioning and briefly mention the options of the correcting ranges differentially. Then we will quickly provide to a brief review of the types of error. The theoretical aspects of the variation of these errors are essential for understanding the correction techniques and also form the basis of different possible classifications of differential positioning. The different techniques of differential positioning and their related intricacies will be discussed subsequently in detail, including both the code-based and carrier-based procedures. Finally, we provide some overview of the typical implementations of these methods for different applications and systems. It is worth mentioning that most of what we discuss here, including the techniques and implementations, has been conceived and developed with GPS. Here, we have generalized them under different theoretical perspectives, complying with our approach in this book.

8.1 Differential positioning

To estimate the position of a user, we need two main things, the position of the satellite and the range of the user from it. We have also seen that the range measurement at the receiver is impaired by different factors that add errors to these measurements. Although these errors are either estimated or modeled for correcting the ranges, the obvious uncertainty involved with the process results in retaining some residuals of these errors, which effectively deteriorates the position estimation.

The differential system is the method of alleviating, if not eliminating, these errors with the help of the measurements and their derivatives determined at a reference point whose position is known a priori and very accurately. So the differential positioning in satellite navigation may be defined as a technique to estimate precise positions using additional information derived from the measurements at the reference receiver situated at a known position.

In such reference-based correction, these reference stations should be able to derive the correct values of the relevant parameters needed for correction and can disseminate and update this information to the users. These parameters may either be the position error or the individual range errors. It may also be the measured ranges, as the case may be, depending upon the technique used for correction.

The correction parameters at the reference station may be required to be transmitted to the users in real time or they can be stored for posterior use through post processing. The schematic for a differential positioning system is shown in Figure 8.1.

8.1.1 Overview of differential corrections

In Chapter 7 we already learned about the different sources and types of error that affect the position estimation at the user receiver. We have also learned that if these errors can be corrected, the accuracy of the position estimation is enhanced, and one of the techniques of correction is by using a reference that knows its own position very precisely. Here, we shall detail the kind of corrections that can be obtained from these reference receivers to correct the parameters at the users.

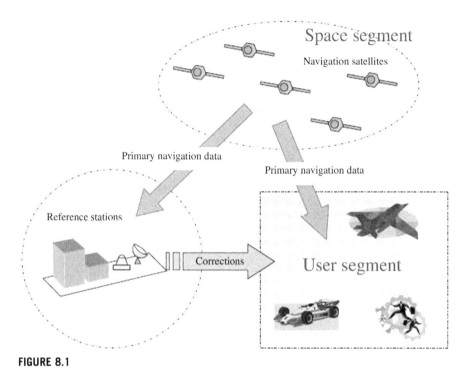

FIGURE 8.1

Schematic for differential system.

We have already mentioned that corrections can be done at different levels of the data. They can be done on the final position estimates as well as at the level of measured ranges. To understand the rationale of using the corrections at different levels, we need to recall how we estimated the position from the range measurements. We know that the observation equation is given by

$$R_m = h(X) \tag{8.1}$$

Here, R_m is the measured range, X is the position vector, and h is the quadratic expression for the range in terms of the satellite and user positions. Differentiating the above equation with respect to the true position, $X = X_t$, we get

$$\begin{aligned} dR_m &= (\partial h/\partial X)\big|_{X_t} dX \\ &= G\big|_{X_t}\, dX \end{aligned} \tag{8.2}$$

This equation states that any difference in measured range by δR will lead to the estimation error of δX, where δR and δX is related by

$$\delta R = G\big|_{X_t}\, \delta X \tag{8.3}$$

With the known position of the reference receiver, these differential values may be readily obtained. However, this is true for the errors incurred at a fixed reference location. The error values will vary for a different position of measurement. We will now look at how this is done.

Let the range measurement have a finite error of δR, which resulted in a position error of δX. So, using Eqn 8.3, this error of δR in range measurement is related to δX by the relation $\delta X = G^{-1}\delta R$. Now let us see if the error at the user position located near the reference may be obtained from the same at the reference station. To get that, let us find out how this ranging error δR varies spatially.

$$\frac{d}{dX}(\delta X)dX = \frac{d}{dX}(G^{-1}\delta R)dX$$

$$= \delta R \frac{d}{dX}(G^{-1})dX + G^{-1}\frac{d}{dX}(\delta R)dX$$

(8.4)

So, in order that this positional error remains invariant between two locations, $\frac{d}{dX}(\delta X)$ should be zero. Hence, the conditions to be satisfied are

$$\frac{d}{dX}(\delta R) = 0$$

(8.5a)

and

$$\frac{d}{dX}(G^{-1}) = 0$$

(8.5b)

Therefore, for the positional error at two places to be equal, it does not suffice only for the range error δR to be the same but also the matrix G, determined by the user satellite geometry to be equal at these two places.

So, only if the above criteria are fulfilled, the estimated positional error at the reference may be used to correct the estimated positional error at the user. However, this technique has certain major limitations.

First of all, the basic assumption that the user and the reference are at such close proximity that G remains identical is unrealistic. The equality can only be an approximation, and for most practical cases, they remain different.

Equality of G implies two things. First, the best set of satellites seen by the user should be the same as the best set of satellites among those visible by the reference. Under typical closeness, these two sets may not necessarily be the same. In such a case, the reference station has to derive its position and the corresponding position error values for all possible sets of four among the visible satellites so that the user can select one that corresponds to the set it is using. So the number of combinations for which the reference station has to derive the solution is $^nC_4$, where n is the numbers of visible satellites. Typically, "n" remains around 7, and hence the total number of solution set is 35 which is fairly high. Second, recall that the elements of G are constituted by the directional cosine terms α, β, and γ. These terms are functions of position. So, even if the satellite set remains the same, the geometry does not remain exactly the same, thus changing the values of the elements of the G matrix.

The second requirement is the equality of δR, which is nothing but the equality of the ranging errors. As we know from previous descriptions, many of the errors determined by the reference station vary with space and time in a correlated manner. Even the correlated errors retain their correlation within a certain range, and they worsen as the distance between the user and the reference increases. So, while the reference and the user have different ranging errors, the corrections derived in terms of position will also not be the same.

Besides space decorrelation, the time latency between the estimation instant and the instant of applying it for correction plays an important role in determining the accuracy of the technique. So, even if the G matrix remains equal between the two locations at a certain instant, it may not be the same after a time interval. A similar statement holds for δR. That is why, for all practical systems, the accuracy of differential corrections is required to be specified for a given maximum range of correction and stipulated latency period.

In addition, the estimation technique of the position solution must remain the same for the reference and the user for the position errors to be corrected directly. Even if the measurement-based errors remain the same, for different techniques used at the reference and user terminal, the derived solution will result in different estimation errors. Therefore, the correction values derived at the reference will not befit the error at the user's end and may even lead to worsening of the effective error under some cases. So, the method of correcting position errors has serious drawbacks.

If we restrict ourselves to Eqn (8.3), it follows from the equation that

$$G^{-1}\delta R = \delta X$$

This implies that the position estimation errors may be reduced by reducing δR. Again δR_u may be expressed in terms of δR_r, using Taylor's series, as

$$\delta R_u = \delta R_r + \frac{d}{dX}(\delta R_r)\big|_{X_r}dX + 1/2\frac{d^2}{dX^2}(\delta R_r)\big|_{X_r}dX^2 + \dots \qquad (8.6)$$

Thus, neglecting the higher order terms, we can derive the range errors at the user position using those at the reference receiver and its first order derivatives. This enables the user to correct the measured range, using only the range error information of the reference and the geometry-dependent derivatives evaluated there. Once δRu is obtained, the matrix G is generated with the own values derived by the user to find the correct values of X using Eqn (8.3).

Recall again that a differential position shift of the receiver by ΔX will result in a difference in measured range by ΔR, assuming G to be constant. Removing the errors from the measured range, the true range difference between the user and the reference, ΔR, may be obtained. Again this is related to the differential position ΔX by the relation

$$G^{-1}\Delta R = \Delta X \qquad (8.7)$$

So, instead of reckoning the ranging errors, knowing the true differential range ΔR between the reference and the user, we can find the relative position, ΔX. For this, we only need enough information to construct G, defined at the reference point. Thus, relative position may be accurately obtained by differencing the whole range measured by the reference station from the range measured by the user. Thus, it avoids finding the exact position of the reference station. The inaccuracy in reference position, however, affects the value of G.

On the whole, we found that the spatial variation of the ranging errors plays a pivotal role in all occasions. Only certain types of error can be truly assumed to be spatially invariant. We have assumed only the linear variations for the rest of the errors, while higher-order variations are also possible. The residual errors remaining from these sources lead to effective inaccuracy. In the following subsection we shall learn about the different classes of error, depending upon their characteristics of spatial variations, and the methods of eliminating them.

8.1.2 Error review

Before describing the theoretical aspects of corrections that we need to do in a differential system, it is first important to know the characteristics of the errors. More precisely, we need to know the spatial characteristics of the errors with respect to the reference and the user. These characteristics are important for correction of range at the user location. So, in this section, we shall briefly review the different types of errors that we learned in Chapter 7, specifically concentrating on their spatial and temporal correlation, which forms the basis of their differentiation (Parkinson and Enge, 1996).

8.1.2.1 Common errors (ε_0)

Common errors are those that are common to both reference and user. These are the errors that remain spatially unaltered irrespective of the user position. So, they are independent of the relative geometry of the satellite and the user and hence cannot be related to the positions of the satellite or the user or even the propagation path between them. However, they can vary with time. So, the common error $\varepsilon_0(x, t)$, experienced by a user at a location x at any instant t can be expressed as

$$\varepsilon_0(x,\ t) = \varepsilon_0(t) \tag{8.8}$$

For example, the satellite clock error is a common error. These errors, once estimated at any point, can be used at any other point for correction at the same instant. Such errors can be completely removed in differential mode.

8.1.2.2 Correlated errors (ε_c)

Correlated errors are those for which a definite correlation exists between reference and user. These errors are not constant over space but have definite functional form of variation within this range. Thus, the error at any distance from the reference point may be derived from the error at the reference point, provided this spatial function and the distance between them are known. Therefore, the correlated error ε_c at any point x at any time t may be expressed as:

$$\varepsilon_c(x,t) = \varepsilon_c(x_0,t) + \partial\varepsilon_c/\partial x\big|_{x0,t}dx + 1/2\partial^2\varepsilon_c/\partial x^2\big|_{x0,t}dx^2 + \ldots \tag{8.9}$$

Here, dx is the spatial difference between the reference point x_0 and the user point x. The first- and higher-order differentials are derived at x_0 and may vary with time. So once $\varepsilon_c(x_0,$ t) is known, $\varepsilon_c(x,$ t) may be estimated by knowing $\partial\varepsilon_c/\partial t$ etc. at the point (x_0, t). Thus these errors are partially removed after deriving the differentials at the reference point and using them to get the exact error value at the user location. The finite residual that still exists is due to the fact that the derivatives are typically limited to the first order. This effect of neglecting the higher-order derivatives gets worse with increasing distance from the reference point. Hence the correction degrades with distance.

Propagation errors are correlated errors such that the shorter the distance between the two receivers, the more similar the errors are. Ephemeris errors are also correlated errors as their values change with the position of the users. However, this change is so small over short distances that for small variations, the errors may be considered to be constant between the reference and user positions. Referring to the above equation, it means the first-order derivatives are negligibly small. We have already derived this quantitatively in Chapter 7, to demonstrate this further.

8.1.2.3 Uncorrelated errors (ε_{uc})

There are some errors that are not constant, nor are they correlated between the reference and the user receivers. These errors are independent between reference and user and spatially vary in a completely arbitrary manner. They cannot be related by any definite function and are called uncorrelated errors. Due to this nature of the errors, they can never be estimated for a point from the knowledge of the others. Hence these errors at the user location cannot be derived from those at the reference station and hence they cannot be removed at all. Multipath is this type of error.

All these divisions are based upon the nature of the errors and the manner in which they vary over space. In other words, the basis of this classification is whether the variations in these errors between the reference and the user can be expressed as the functions of their separation or not. Figure 8.2 shows the spatial variations of different types of errors.

In a similar fashion, the errors vary with time. Time correlation of the errors is also important as the latency between the time of estimation of the error at the reference station and the time of application at the user receiver will then determine the accuracy of the corrections. Some of the errors, like the clock errors vary quickly, and the corresponding corrections are required to be updated at frequent intervals and used immediately. The other errors like the ephemeris errors vary rather slowly, such that even with considerably large latency between the correction generation and its application, accuracy is not affected much. These errors may be updated at a slower rate.

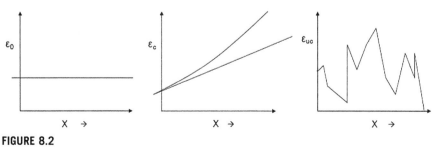

FIGURE 8.2

Variation of different types of errors.

8.1.3 **Classifications of differential positioning**

In the differential positioning, the errors of the individual users are eliminated or their impact is significantly reduced by utilizing additional information from the reference station, thus improving the position estimation accuracy. However, the differential positioning can be made using different techniques, which can form the basis of its classification.

In this section, we provide a brief overview of the different bases upon which the differential techniques of positioning can be categorized. Although these are not standard classifications, such systematic distinction helps in theoretical understanding of the system. We shall discuss these important classes in finer detail in our next section.

8.1.3.1 *Position/range-corrected differential positioning*

This category is based upon the parameter using which the differential corrections are made. The corrections to be applied at the user receiver can be done at different levels of data. The correction can be done on the final position estimates made in a user receiver, making it a "position differential correction." It may also be applied at the level of measured ranges, which may be referred to as the "range differential correction."

Position corrections are only possible for receivers with very close proximity to the reference. Such a correction reduces the computational load at the receiver and is very simple to apply. However, the range-based correction is a better option than doing corrective operations in the estimated positions, the reasons of which we have already discussed in Section 8.1.1. The range errors may be provided by the reference as a whole in a consolidated fashion or they may be resolved into their components due to various sources and these components may be supplied to the users for corrections.

The corrections parameters are derived by the reference station and are obtained from its own measured ranges. Sometimes, ancillary equipment to measure the atmospheric parameters is also collocated at the reference station from which the corresponding atmospheric delays may be derived. For the most effective applications of these correction parameters, the measurement methods need to be the same in the reference and at the user receiver. Therefore, for position corrections, the method of derivation must be identical, while for range corrections either the reference and the user must both use code-based ranging or both should employ the carrier-based ranging method.

8.1.3.2 *Absolute/relative differential positioning*

For the type of differential positioning in which the corrections are done on the measured range before deriving the positions, there are two different approaches for correction. In one of them, the correction is done at the user receiver, in which the absolute error parameters are disseminated by the reference. On correcting the measured ranges with these error values, the receiver obtains the true ranges, which are absolute in terms of the system reference frame - earth-centered, earth-fixed, for instance. Consequently, the positions thus derived from it are also absolute with respect to this frame.

Alternatively, the reference may convey its total measured range. So, on differencing this at the user from its corresponding measurement of range, the resultant becomes a differential range of the user relative to the reference station. This relative range being a function of the relative separation as per Eqn 8.7, all observations are transformed relative to the reference station, i.e. with respect to a new frame centered at the reference position. Consequently, the positions thus derived are also relative, with respect to this reference station only.

Accordingly, the aforementioned types of differential positioning is referred to as "absolute differential positioning," or "relative differential positioning," respectively. For relative positioning the vectorial difference between the user receiver and the reference position is determined. It is called the baseline. Baseline determination in relative positioning is possible with both code-based and carrier-based measurements. It is also possible either when the user receiver is static and does not change its position with time or when it is dynamic and roving around the reference.

8.1.3.3 Code-based/carrier phase-based differential positioning

Depending upon the types of measurements at the receiver and the references that are used for the corrections, the differential positioning may be classified into code-based and carrier phase-based differential positioning.

In code-based differential positioning, the code-based measured ranges are exchanged for corrections, which are absolute but noisy measurements. On the other hand, the carrier phase-based differential positions employ the measured phase delay corresponding to the range at each receiver for the purpose. Since the measured carrier phase has ambiguity, as we have learned in Chapter 5, the differential ranging also needs resolving for this relative integer ambiguity. However, the carrier phase-based positioning is more precise than the code-based method.

8.1.3.4 Real-time/post processed differential positioning

Depending upon the application for which the differential system is being used, the corrections may be applied in real time or may be done through post processing. Real-time differential positioning requires a radio communication link to communicate and the results are obtained almost instantaneously. It is therefore important in applications like real-time surveying, although the final accuracy available in such a process is less than that obtained through post processing.

Post processing is generally done in cases of carrier phase-based ranging, in which both the reference and the user receiver collect the data, store them, and later share the same for generating the precise corrective values. These corrections are then applied for getting accurate positions of the user a posteriori. Postprocessing results are more accurate (El-Rabbani, 2006) and need no link between the reference and user, while giving more options for data correction and editing.

8.1.3.5 Single-/multiple-reference differential positioning

Reference receivers are responsible to provide the corrective data to the individual receivers attached to it. Differential positioning may be distinguished by the numbers of reference receivers it uses. Some may have a single reference to serve a local area, while others may be served by multiple reference stations. It is obvious that the latter option is used to overcome the constraint of the limited spatial boundary that the system can serve. Multiple references are required for larger areas of service, in which additional reference stations are effective to combat the spatial and temporal decorrelations of the error. Consequently, they become equivalent to local area and wide area differential positioning, respectively.

Although, in most of the cases, the reference stations are used as generating and disseminating correction parameters only, there are systems that use this reference station as an additional source of ranging data improving the geometric dilution of precision and hence the accuracy of the estimation, as well as its availability. It also aids in estimation purposes for carrier phase-based ranging.

8.2 Differential correction techniques

In this section, we shall explore the different methods employed for differential corrections. We shall start by defining the term "baseline," which is the vector joining the user receiver and the reference station. Throughout the discussion, we shall consider that this baseline, indicating the position difference between the reference and the user, is small. This is referred to as a short baseline. This warrants that the range difference between the two receivers may be expressed as a linear function of their positional difference, i.e. the baseline length. Further, for short baselines, the radio waves incident on the two receivers from the satellite may be assumed to be parallel and so the angles that the vectors pointing toward the satellite from the reference and the user receivers make with the baseline are the same. It also implies that the ephemeris error is almost constant between the two receivers and keeps the other correlated errors like the ionospheric and tropospheric errors within the first order of variation. In this discussion, we neglect the effects of the uncorrelated errors such as the multipath.

8.2.1 Code-based methods

8.2.1.1 Absolute differential methods

While estimating positions with code-based range measurement, we have already seen that the results are corrupted with ranging errors. Further, in Chapter 5 we have learned that these types of range measurements are noisy, with the data largely scattered, revealing poor precision. But, unlike carrier-based measurements, these range values are unambiguous. The noise terms include mainly the receiver self noise. In addition, the ranges are biased by propagation errors like the ionospheric and tropospheric errors and clock bias. So the measured range R_u at the user may be represented as:

$$R_u = \rho_u + c\delta t_u - c\delta t_s + \delta r_{ion,u} + \delta r_{trp,u} + n_u \qquad (8.10a)$$

where R_u is the measured range at the user compared to ρ_u, which is the true geometric range. $c\delta t_s$ and $c\delta t_u$ are the range equivalent of the satellite and receiver clock bias values, respectively, with respect to the system time. Similarly, $\delta r_{ion,u}$ and $\delta r_{trp,u}$ are the excess path equivalents for ionospheric and tropospheric delays, respectively, at the user receiver. n_u is the receiver noise at the reference.

Similarly, in a reference receiver, the measured range is obtained as

$$R_r = \rho_r + c\delta t_r - c\delta t_s + \delta r_{ion,r} + \delta r_{trp,r} + n_r \qquad (8.10b)$$

But, the reference station has its own position estimated. So it can find out the true geometric range ρ_r of the reference from the satellites. However, if the corresponding satellite positions are simply derived from the transmitted ephemeris, then the estimations will contain the ephemeris error. Then Eqn (8.10a) becomes

$$R_u = \rho'_u + c\delta t_u - c\delta t_s + \delta r_{eph,u} + \delta r_{ion,u} + \delta r_{trp,u} + n_u \qquad (8.10c)$$

However, the reference can also use the transmitted ephemeris only, and hence the corresponding range Eqn (8.10b) becomes

$$R_r = \rho'_r + c\delta t_r - c\delta t_s + \delta r_{eph,r} + \delta r_{ion,r} + \delta r_{trp,r} + n_r \qquad (8.10d)$$

Here, ρ'_u and ρ'_r are the geometric estimation of ranges using the transmitted values of the ephemeris.

So, differencing the estimated range, obtained using the provided ephemeris from the measured ranges at the reference station, we get

$$\begin{aligned} \varepsilon &= R_r - \rho'_r \\ &= c\delta_u - c\delta t_s + \delta r_{eph,r} + \delta r_{ion,r} + \delta r_{trp,r} + n_r \end{aligned} \qquad (8.11)$$

Now, the collective range errors ε can be used by the users to correct their own ranging errors. On applying this correction, the corrected pseudorange of the user R_{cu} becomes

$$\begin{aligned} R_{cu} &= \rho'_u + c(\delta t_r - \delta t_u) + (\delta r_{eph,u} - \delta r_{eph,r}) + (\delta r_{ion,u} - \delta r_{ion,r}) \\ &\quad + (\delta r_{trp,u} - \delta r_{trp,r}) + (n_u - n_r) \\ &= \rho'_u + c\delta t_{ur} + \delta r_{eph,ur} + \delta r_{ion,ur} + \delta r_{trp,ur} + n_{ur} \end{aligned} \qquad (8.12)$$

Since the reference and the user differ in their error amounts due to their location difference, there remain the error residuals, however small. This is observed in the above Eqn (8.12). Here, δt_{ur} is the differential clock bias between the reference and the user. This Eqn (8.12) is obtained by differencing the collective error term ε, as a result of which the satellite clock bias term δt_s, being common to both, has gotten completely eliminated and the user clock bias and other biases have become relative to the clock bias of the reference station.

Typically, the reference station clock is atomic and it is corrected and disciplined with the satellite clock. Hence, this part contains the error due to the receiver clock error only. The ephemeris error changes very slowly with separation, and hence the relative ephemeris error may be assumed to be zero for very short baselines. The ionosphere has large correlation over space and time. Hence, over smaller distances, the relative ionospheric delay $\delta r_{ion,ur}$, which is nothing but the difference in ionospheric errors at the two stations, is small. For most of the practical scenarios, $\delta r_{ion,ur}$ thus can also be taken as zero. The same argument holds for $\delta r_{trp,ur}$ also for short baselines. In such cases, the equation turns into

$$R_{cu} = \rho'_u + c\delta t_{ur} + n_{ur} \tag{8.13}$$

But, this assumption of canceling ionospheric and tropospheric delays may not be true for certain geographic regions where the ionosphere gradient is large and for partic- ular periods of time. The ionospheric delay at these locations and times may substan- tially change within a limited spatial or temporal extent. We shall learn about the ionospheric variations and their characteristics for different global regions in Chapter 9. For regions of rapid spatiotemporal variations, the delay at the correction site, i.e. the user position, is not the same as that of the reference site, and hence the errors do not cancel out. However, the same errors at the user location may be derived in terms of those at the reference station using Taylor's series to represent the spatial variation as

$$\delta r_{ion,u} = \delta r_{ion,r} + \partial\left(\delta r_{ion,r}\right)/\partial X\big|_{X_r}dX + 1/2\,\partial^2\left(\delta r_{ion,r}\right)/\partial X^2\big|_{X_r}dX^2 + \ldots \tag{8.14a}$$

This makes the difference $\delta r_{ion,ur}$ be equal to

$$\delta r_{ion,u} - \delta r_{ion,r} = \partial\left(\delta r_{ion,r}\right)/\partial X\big|_{X_r}dX + 1/2\,\partial^2\left(\delta r_{ion,r}\right)/\partial X^2\big|_{X_r}dX^2 + \ldots \tag{8.14b}$$

For short baselines, the higher-order differentials generally do not contribute significantly and hence can be neglected. So, the delay difference may be expressed in terms of the first-order differentials with respect to baseline length only. This de- rivative is thus required to be derived along with the errors, and transmitted to the users for appropriate error cancellations. Similarly, for the temporal variations, the difference in ionospheric delay between the time of estimation t_1 and the time of application t_2 may be obtained as

$$\delta r_{ion}(t_2 - t_1) = \partial r/\partial t\big|_{t\,=\,t_1}\Delta t \tag{8.15}$$

where, $\Delta t = (t_2 - t_1)$. This temporal derivative may be derived at the reference time. The same argument holds true for the tropospheric errors. These terms help to remove the unbalanced errors between the reference and the user receivers from Eqn. 8.12. The changes can be seen in Eqn. 8.13.

In cases where these distances are large or the derivatives are large enough, the higher-order terms are so conspicuous that they cannot be neglected. We shall see in Chapter 9 that this characteristic is manifested by the ionosphere in equatorial regions. In such cases, the best result is achieved when the user receiver itself is able to estimate its own ionospheric delay. Alternatively, many reference stations may be used such that wherever the user receiver is, it remains within the region of first-order variation of the ionospheric delay for any one of the receivers. This forms the basis of wide-area differ- ential systems that we are going to learn about in the next section.

It is important to note here that the corrections at the user receiver leading to def- inite derivative terms are only restricted to the same specific satellites for which the errors have been derived at the reference. This is inconvenient to use especially when the area to serve is large. Nevertheless, there is an option to segregate this confluence of the errors with the specific satellites. The option is to separate out the components present in the composite range error at the reference station. The ionospheric delays may be obtained by using the dual-frequency receivers. The tropospheric delays may similarly be obtained from the empirical models using parameters obtained from the meteorological instruments installed at the reference station. Further, systems with extensive architectural set-up can also track the locus of individual satellites and estimate their true current position and thus derive the satellite's positional deviation.

In case the individual errors are obtained along the ray path for each satellite, the ionospheric and tropospheric delay errors calculated along the individual slant rays can now be converted to vertical delays before dissemination to the users for correction. The corresponding slant factors may be used for the purpose. For an ionospheric slant delay of δr_{ion}, observed over a path along an elevation angle of θ, the corresponding vertical delay is obtained as

$$\delta r_{ion,vertical} = f_{ion}(\theta) \, \delta r_{ion,slant} \tag{8.16a}$$

Here, $f_{ion}(\theta)$ is the ionospheric slant factor at the user location for the elevation angle θ. One of such slant factors for the ionosphere is given by RTCA (1999) as

$$f_{ion} = \sqrt{\left\{1 - R_e^2 \Big/ (R_e + h_e)^2 \, \cos Z\right\}} \tag{8.16b}$$

Here, R_e is the mean radius of the earth, h_e the ionospheric effective height above the earth's surface, and Z the elevation angle. The vertical total electron content is obtained at the ionospheric pierce point (IPP). IPP is the point where the line joining the user to the satellite intersects the horizontal plane at the effective ionospheric height h_e. Its position is obtained using the following equation (RTCA, 1999):

IPP latitude:

$$\Phi_{pp} = \sin^{-1}\left\{\sin(\Phi_u)\cos\left(\Psi_{pp}\right) + \cos(\Phi_u)\sin\left(\Psi_{pp}\right)\cos(Z)\right\} \tag{8.17a}$$

IPP longitude:

$$\lambda_{pp} = \lambda_u + \sin^{-1}\left\{\sin\left(\Psi_{pp}\right)\sin(Z)\Big/\cos\left(\Phi_{pp}\right)\right\} \tag{8.17b}$$

where $\Psi_{pp} = \pi/2 - Z - \sin^{-1}\{R_e/(R_e + h_e) \cos Z\}$, Φ_u and λ_u are the receiver latitude and longitude, and Z is the azimuth angle. It assumes horizontally stratified ionospheric layers.

Similarly, the tropospheric delays may also be transmitted in terms of vertical delays. Likewise, effective ephemeris range errors may be replaced by total vectorial deviation of the satellite position. This enables the users to convert the vertical delay into the slant delay information for any visible satellite at his position. The slant factor $f_{ion}(\theta)$ may be used for converting the vertical delays to delays along the elevation angle θ at the user points. Therefore, the corrected range at the user location now becomes

$$\begin{aligned} R_{cu} = {} & \rho_u + c(\delta t_r - \delta t_u) + \left(\delta r_{eph} - g\delta r_{sat}\right) + \left(\delta r_{ion,u} - f_{ion}(\theta_u)\delta r_{ion,v}\right) \\ & + \left(\delta r_{trp,u} - f_{trp}(\theta_u)\delta r_{trp,v}\right) + (n_u - n_r) \end{aligned} \tag{8.18}$$

Here, δr_{sat} is the total satellite deviation and g is the geometrical factor that converts it to the radial range error at the user position. $\delta r_{ion,v}$ is the vertical ionospheric delay derived from the reference measurements at the user location. $f_{ion}(\theta_u)$ is the ionospheric slant factor at the user location for the elevation angle θ_u. Similarly, $\delta d_{trp,v}$ and $f_{trp}(\theta_u)$ are, respectively, the vertical delay and the slant factor for the tropospheric delay for the user elevation angles θ_u toward a satellite.

For spatially varying propagation errors, the derivations done at the reference station are vertical delays and can be obtained from any set of satellites visible there. The vertical delays can be conveniently converted for any look angle at the user location for any other satellite. So, this disjoins the errors with the satellites and removes the criterion of common satellite visibility at the two places. However, there are two issues associated with it. First, the vertical delay at any instant is a continuous function of space. So, the vertical delay at which point is suitable to be disseminated to the user is a question. On the other hand, when such vertical delays at finite numbers of spatial points are given, then how to use these vertical delays so that the user can get the one at the IPP created by the satellite the user is looking at is another question to be answered.

A convenient approach toward the same, that is used in the augmentation system is to divide the whole service area into finite grids and to provide the vertical delays at these definite grip points. This may be obtained by using the IPP points obtained from the observations of the reference station. Further, from these grid points, the user may interpolate the given delays to the user IPPs and convert them to slant delays.

This way, a liberty is obtained to convert vertical errors to any direction of visible satellite, thus making the information independent of any satellite direction. But, it is at the cost of accepting the errors incurred in converting the errors from slant to vertical at the reference and from vertical to slant at the user positions.

Similarly as in Eqn (8.13), considering the residues are negligible, the equation turns into

$$R_{cu} = \rho'_u + c\delta t_{ur} + n_{ur} \tag{8.19}$$

So, now we have an equation with three unknowns in ρ'_u and an unknown δt_{ur}. These parameters can be easily estimated using the technique of linearization described before. Compared to the standalone case, this technique only uses information derived by the reference station for error correction.

The n_{ur} represents the difference in the self noise between the receiver and the reference. These noises do not have any correlation and hence do not cancel out. So, upon differencing the reference correction, the effective noise thus produced is the difference between the individual noises at the user receiver and the reference station. Since the two noises are independent, the effective noise σ after differencing at the user receiver will be

$$\sigma_{eff} = \sqrt{\sigma_u^2 + \sigma_r^2} \tag{8.20a}$$

where σ_u and σ_r are the standard deviations of the individual measurements at the user and of that derived for correction at the reference, respectively. If the two σ values are the same, $\sigma_u = \sigma_r = \sigma$, then this becomes

$$\sigma_{\text{eff}} = \sqrt{2}\sigma \tag{8.20b}$$

FOCUS 8.1 ADDED NOISE

To understand the effect of differencing the error values at the user, let us consider that the noise at the reference and the noise at the receiver are both Gaussian with equal standard deviation, σ. In this composite noise, the effectual noise of value n will be produced when one of the noise components will assume value n_1 and the other will assume a value $(n - n_1)$ So, the total probability of occurrence of noise value n is equal to the sum of all possibilities of one acquiring value n_1 and the other value $n - n_1$. So, the effectual probability is

$$p(n) = \int p(n_1 = n_1)\, p(n_2 = (n - n_1))\, dn_1$$
$$= n_1 * n_2$$

The integration runs from $-\infty$ to $+\infty$. Thus, the effective probability distribution is the convolution of the individual distribution.

So, writing the probabilities in terms of Gaussian distribution, we get

$$p\left(n\right) = \int\left[A\exp(-x^2/2\sigma^2) \times A\exp\left\{-(n-x)^2/2\sigma^2\right\}\right]dx$$

$$= A^2\int \exp - \left(n^2 - 2nx + 2x^2\right)/2\sigma^2\, dx$$

$$= A^2\int \exp - \left(n^2/2 - 2nx + 2x^2 + n^2/2\right)/2\sigma^2\, dx$$

$$= A^2\left[\int \exp - \left(n^2/2 - 2nx + 2x^2\right)/2\sigma^2\, dx\right]\exp\left(-n^2/4\sigma^2\right)$$

$$= A^2\left[\int \exp - (x - n/2)^2/\left\{2\left(\frac{\sigma}{\sqrt{2}}\right)^2\right\}dx\right]\exp\left(-n^2/4\sigma^2\right)$$

The integral in the above equation is definite. It is similar to the integral of a Gaussian distribution of a variable x about the mean n/2 and standard deviation $\sigma/\sqrt{2}$. We know that the integral of a Gaussian function is independent of the mean value and hence this definite integral becomes independent of n and equal to $\left(\frac{\sigma}{\sqrt{2}}\right)\sqrt{2\pi} = \sqrt{\pi\sigma^2}$ (Papoulis, 1991). Calling this constant value K, this probability p(n) becomes

$$p(n) = K\exp(-n^2/4\sigma^2)$$
$$= K\exp\left\{-n^2/2\left(\sqrt{2}\sigma\right)^2\right\}$$
$$= K\exp\left\{-n^2/2\sigma_{\text{eff}}^2\right\}$$

where K is a constant. This is nothing but a scaled Gaussian distribution with effective σ given by

$$\sigma_{\text{eff}} = \sqrt{2}\sigma$$

This effect is illustrated in Box 8.1.

BOX 8.1 MATLAB EXERCISE

The MATLAB program noise_addition.m was run to observe the effect of adding two signals carrying noise of unity standard deviation value, i.e. $\sigma = 1$. The effective noise thus produced gets enlarged compared to the original noise. This is shown in Figure M8.1(a) and (b) below. Figure (a) shows how the time variation gets effectively amplified while figure (b) shows the increment in the effective standard deviation value, σ.

Understand the program and identify the different parameters used. Run the program with different original σ and for different sample numbers and observe the effect.

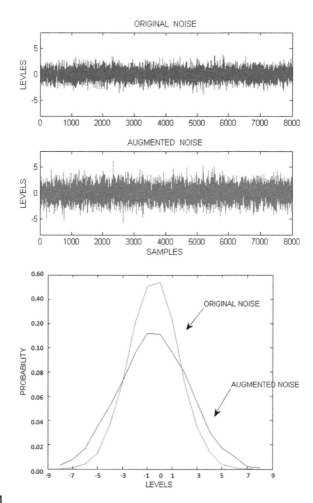

FIGURE M8.1

(a) Time variation of original and augmented noise, (b) Probability distribution of original and augmented noise.

8.2.1.2 Relative differential methods

Next consider the case of relative differential positioning. Here, instead of disseminating the range errors only, the reference station provides the complete measured ranges to the users. So, on differencing, the common errors are eliminated and the differential range is obtained, which is the function of the relative distance between the reference and the user, i.e. the function of the baseline. However, the differential terms for correlated errors are still contained in it. So, this method is used for cases where the user is interested to know his or her location with respect to the position of the reference station, in which we define and estimate the differential positions in terms of the baseline vector between the user and the reference. However, if the reference station position is also known to the user, one can always obtain the user's position with respect to absolute references from this.

This relative differential positioning can be done with code-based ranging. Rewriting the uncorrected measured range again for both the reference and the user location, we get

$$R_r = \rho_r' + c\delta t_s - c\delta t_r + \delta r_{eph} + \delta r_{ion,r} + \delta r_{trp,r} + n_r \qquad (8.21a)$$

$$R_u = \rho_u' + c\delta t_s - c\delta t_u + \delta r_{eph} + \delta r_{ion,u} + \delta r_{trp,u} + n_u \qquad (8.21b)$$

Unlike the previous case, where only the range error was transmitted by the reference, here the whole range R_r measured at the reference is made available to the user. The user utilizes this total range measured by the reference to estimate the relative position through different differencing techniques that we discuss below.

The code-based relative differential method is generally used in conjunction with the carrier phase-based differential methods as an aid to resolve the integer ambiguity, a related issue that we shall come across while learning the carrier phase-based methods. However, it can also be used as an independent method for relative positioning, but with lower accuracy compared to the phase-based methods.

8.2.1.2.1 Single differencing technique

A single difference equation is formed when the range equation obtained at the user is differenced once from the range equation obtained at the reference for the same satellite and at the same instant. Now, differencing Eqn (8.21a) from Eqn (8.21b), we get

$$\begin{aligned}\Delta R_{ur} &= R_u - R_r \\ &= \Delta\rho_{ur}' + c\Delta\delta t_{ur} + \Delta\delta r_{eph,ur} + \Delta\delta r_{ion,ur} + \Delta\delta r_{trp,ur} + \Delta\varepsilon \qquad (8.22) \\ &= \Delta\rho_{ur}' + c\Delta\delta t_{ur} + \Delta\delta r_{eph,ur} + \Delta\delta r_{ion,ur} + \Delta\delta r_{trp,ur} + n_{ur}\end{aligned}$$

Concentrating on the errors due to the ionospheric delay and the tropospheric delay, these errors are supposed to be different at the user and reference locations, leading to the above differential terms. The general approach that may be used to mitigate the effect here is also to provide the delay rate information, as we have discussed above. The second- and higher-order terms can be neglected.

Considering the case that the differential ephemeris error is negligible and the ionospheric and tropospheric delays cancel completely due to their correlation, the difference becomes

$$\Delta R_{ur} = \Delta\rho'_{ur} + c\Delta\delta t_{ur} + n_{ur} \tag{8.23}$$

Now, let us see how we can replace the range difference $\Delta\rho$ as a function of the baseline. The variation in range may be expressed in terms of the positional difference ΔX as

$$\rho'_u = \rho'_r + \partial\rho'/\partial X|_{Xr}\Delta X + \tfrac{1}{2}\,\partial\rho'/\partial X^2|_{Xr}\Delta X^2 + \text{higher order terms} \tag{8.24a}$$

Thus the range difference may be written as

$$\Delta\rho' = \partial\rho'/\partial X|_{Xr}\Delta X + \tfrac{1}{2}\,\partial\rho'/\partial X^2|_{Xr}\Delta X^2 + \text{higher order terms} \tag{8.24b}$$

For short baselines, only the first-order derivative is significant while the higher-order derivatives are vanishingly small. So, the difference for short baselines becomes

$$\Delta\rho' = \partial\rho'/\partial X|_{Xr}\Delta X \tag{8.24c}$$

This can be expanded in terms of individual coordinates as

$$\Delta\rho' = \partial\rho'/\partial x|_{Xr}\Delta x + \partial\rho'/\partial y|_{Xr}\Delta y + \partial\rho'/\partial z|_{Xr}\Delta z \tag{8.24d}$$

But these differentials of the range ρ with respect to the coordinates at the reference position Pr are nothing but the direction cosines of the range vector with respect to the coordinate axes. This becomes obvious on writing the derivatives in terms of the functional expression. So, writing the geometric range ρ in terms of the coordinates and then differentiating at the reference point, we get

$$\rho' = \sqrt{(x_s - x_r)^2 + (y_s - y_r)^2 + (z_s - z_r)^2}$$
$$\partial\rho'/\partial x_r|_{xr} = -(x_s - x_r)/\rho = -\alpha$$
$$\partial\rho'/\partial y_r|_{xr} = -(y_s - y_r)/\rho = -\beta \tag{8.25}$$
$$\partial\rho'/\partial z_r|_{xr} = -(z_s - z_r)/\rho = -\gamma$$

where x_s, y_s and z_s are the satellite positions derived from the ephemeris values and α, β, and γ are the cosines of the angles that the range vector makes with the coordinate axes. The latter parameters are also the components of the unit vector along the satellite at the reference position X_r, along the three coordinate axes. So, if the unit vector and e_x, e_y, and e_z are its components along the x, y, and z axes, respectively, then $e_x = \alpha$, $e_y = \beta$, and $e_z = \gamma$. So, the differential range becomes

$$\Delta\rho' = -e_x\Delta x - e_y\Delta y - e_z\Delta z \tag{8.26}$$

Recalling that the baseline b is nothing but the vectorial difference between the two stations, it can be expressed as a variation of their position coordinates, P_r and P_u respectively. Thus,

$$b = P_r - P_u$$
$$= -(\Delta x\,\hat{x} + \Delta y\,\hat{y} + \Delta z\,\hat{z}) \tag{8.27}$$

i.e. Δx, Δy, Δz are the components of the baseline b along the reference axes. So,

$$\Delta\rho' = -\bar{e}\cdot\bar{b}$$
$$= b\cos\theta$$
(8.28a)

This is evident from Figure 8.3, where $\Delta\rho$ can be expressed as $b\cos\theta$, where b is the baseline length and θ is the angle between the direction of the satellite from the receivers and the baseline. For short baselines, the incident rays from the satellite are parallel and this angle θ is equal at both the receivers. Then

$$\Delta\rho' = b\cos\theta$$
(8.28b)

So, using Eqn 8.28b in Eqn 8.23 we get

$$\Delta R_{ur} = b\cos\theta + c\,\Delta\delta t_{ru} + \varepsilon$$
$$= \begin{bmatrix} -e_x & -e_y & -e_z c \end{bmatrix}\begin{bmatrix} \Delta x & \Delta y & \Delta z & \delta t_{ur} \end{bmatrix} + \varepsilon$$
(8.29)
$$= G_1\begin{bmatrix} \Delta x & \Delta y & \Delta z & \Delta\delta t_{ur} \end{bmatrix} + \varepsilon$$

where $G_1 = \begin{bmatrix} -e_x & -e_y & -e_z & c \end{bmatrix}$ is defined at the reference position. Since the reference position is known, G_1 can be generated from its coordinates. Hence G_1 becomes a known parameter.

The solution for the differential coordinates Δx, Δy, Δz, and $\Delta\delta t_{ur}$ thus may be obtained using standard least squares method. This method has the advantage that transmission of the measured range satisfies the need of the algorithm and the individual errors are not required to be estimated at the reference. This eliminates much of the computational load at the reference and reduces latency. For users explicitly requiring relative distance, the reference location need not be surveyed a priori and

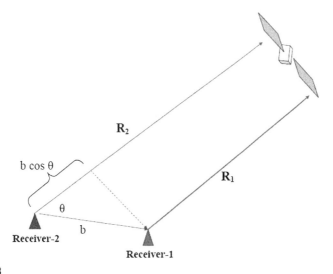

FIGURE 8.3

Geometric relation of range difference with baseline.

may also be relocating with time. However, this method has more use in carrier-based techniques, which we shall learn about in subsection 8.2.2.

8.2.1.2.2 Double differencing technique

A double difference equation is formed when the single differenced range equations obtained for two different sets of satellites, but for the same pair of receivers, are differenced.

For m numbers of visible satellites which are common to the reference and the user, m such single difference equations can be formed for a definite pair of reference and user. If θ_1 and θ_2 are the corresponding angles that the baseline makes with the direction of the two satellites S1 and S2, respectively, at the reference receiver location, then using Eqn (8.29), the equation can be written as

$$\Delta R_{ur}^1 - \Delta R_{ur}^2 = b\cos\theta_1 - b\cos\theta_2 + \varepsilon_{ur}^1 - \varepsilon_{ur}^2$$

$$\text{or,} \quad \nabla\Delta R_{ur}^{12} = b(\cos\theta_{\overline{1}} - \cos\theta_2) + \varepsilon_{ur}^{12}$$

(8.30a)

where R_{ur}^1 and ΔR_{ur}^2 are the single differenced measured ranges for satellite S1 and S2, respectively, between the user and the reference. Notice that the common term of relative receiver clock bias has been eliminated. The geometry is evident from Figure 8.4.

Similarly, as in the case of single difference, this equation may be written in the matrix form as

$$\nabla\Delta R_{ur}^{mn} = G_2[\Delta x\ \Delta y\ \Delta z] + \varepsilon_{ur}^{m\ n}$$

(8.30b)

where $G_2 = [e_{x2} - e_{x1}\quad e_{y2} - e_{y1}\quad e_{z2} - e_{z1}]$. Notice that the relative receiver clock bias that was present in the single difference equation has been eliminated of further differencing. G_2 in this case becomes the difference of the corresponding elements in G_1 for single differencing for satellite S1 and S2. The aspects and approach for single and double differencing is illustrated in focus 8.2.

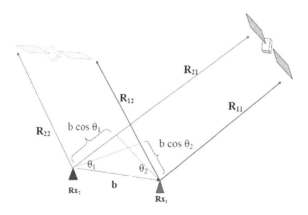

FIGURE 8.4

Double difference in range and relation to baseline.

FOCUS 8.2 ASPECTS OF DIFFERENCING

The range of a definite satellite S1 measured using code phase in two different frequencies of 1575.42 MHz (L1) and 1227.6 MHz (L2) for station A situated at a known location are, respectively

$$22646708.32 \text{ m}$$
$$\text{and} \quad 22646719.32 \text{ m}$$

The same parameters when measured at another station B located at an unknown location are

$$22658409.71 \text{ m}$$
$$\text{and} \quad 22658419.71 \text{ m}$$

From the above, the ionosphere free measured range from station A and station B becomes, respectively

$$22646690.99 \text{ m}$$
$$\text{and} \quad 22658393.54 \text{ m}$$

The ionospheric delay causes more than 15 m of excess range in L1 and more than 25 m in L2. So differencing the ionosphere free ranges at the two stations, we get

$$dR_1 = 22658393.54 - 22646690.99 \text{ m}$$
$$= 11702.55 \text{ m}$$
$$= 11702.5 \text{m}$$

Instead, differencing the total measured range would have given us

$$dR_1' = 22658409.71 - 22646708.32$$
$$= 11701.39 \text{m}$$

The difference between dR_1 and dR_1 is small indeed. So, even if we directly difference the ranges, the ionospheric delay for small baselines being almost the same, irrespective of its magnitude (15 and 25 here), gets almost eliminated.

So, if b is the baseline, then

$$b \cos \theta 1 + \delta t_{ur} = 11,702$$
$$\text{or,} \quad -(e_{x1} \cdot b_x + e_{y1} \cdot b_y + e_{z1} \cdot b_z) + \delta t_{ur} = 11,702$$

The ranges measured at the same time toward a different satellite S2 for station A and B, respectively, in frequency L1 are 22,713,461.72 and 22,711,149.54 m.

The corresponding difference in range dR2 becomes

$$dR_2 = 22713461.72 - 22711149.54 \text{ m}$$
$$= 2312.18 \text{ m}$$

So,

$$b \cos \theta 2 + \delta t_{ur} = 2312$$
$$-(e_{x2} \cdot b_x + e_{y2} \cdot b_y + e_{z2} \cdot b_z) + \delta t_{ur} = 2312$$

Differencing the two single differences, we get

$$b(\cos \theta_1 - \cos \theta_2)$$
$$= [(e_{x2} - e_{x1}) \cdot b_x + (e_{y2} - e_{y1}) \cdot b_y + (e_{z2} - e_{z1}) \cdot b_z]$$
$$= 11,702 - 2312$$
$$= 9390$$

Now, the surveyed position of the reference station is

$$x_r = \quad 2067.9947 \text{ m}$$
$$y_r = 5921962.0445 \text{ m}$$
$$z_r = \quad 377.8407 \text{ m}$$

Then the corresponding calculated ranges with the two satellites from the reference station become

$$R_1 = 21182426.32 \text{ m}$$
$$R_2 = 22712324.35 \text{ m}$$

Note that the measured ranges are different from these calculated ranges. This is due to the presence of errors in the measurements and the effect of ephemeris errors in the calculations. However, here the errors are exaggerated compared to the pragmatic values for convenience and highlighting.

Then, using x_r, y_r, z_r and the ephemeris derived satellite position, we get

$$e_{1x} = 0.03036, \; e_{1y} = 0.86960, \; e_{1z} = 0.49281$$
$$\text{and} \quad e_{2x} = -0.45908, \; e_{2y} = 0.57847, \; e_{2z} = -0.67424$$

So,

$$e_{2x} - e_{1x} = -0.48945$$
$$e_{2y} - e_{1y} = -0.29112$$
$$e_{2z} - e_{1z} = -1.16706$$

However, these three equations are not linearly independent. Hence solutions for three unknowns cannot be obtained from these equations only. More similar measurements are required for solving.

8.2.2 Carrier phase-based methods

Carrier phase-based differential methods are those in which both the reference station and the user receiver use their carrier phase measurements for estimation of the range of the receivers at their respective positions.

We have already read in Chapter 5 that the phase of the received signal compared to a local reference gives ranges that are many times more precise than those obtained using the code phases. However, the cost one has to pay for it is the added computational complexity. We can recall that the carrier phase-based ranging is precise but has the limitation of integer ambiguity. It is the integral number N of the wavelengths present in the range at the instant when the receiver starts measuring the phase changes of the incoming signal. Different techniques are used to undertake this problem of ambiguity resolution. Here we shall see how the same techniques are also used for differential estimations.

The carrier phase-based differential estimation of position is significantly useful when precise positioning is done with respect to the reference station, i.e. for relative differential positioning. Here too, like the code-based technique, single differencing and double differencing and even triple differencing are done. The basic idea is to reduce the numbers of unknowns to be solved for utilizing the commonalities between the user and the reference.

The measured parameter in the carrier phase technique is the phase of the carrier. If φ is the measured fractional phase at the receivers, obtained by comparing the received phase relative to that of the local carrier. The receiver's local clock being synchronized with the satellite, this becomes the difference between the current received and the current transmitted phase. ρ is the true range, and N the integer ambiguity, then the measurements done by the reference and user receivers, respectively, from the same satellite s may be written as

$$(\lambda/2\pi)\varphi_r = \rho_r + N_r\lambda + c\delta t_r - c\delta t_s - \delta r_{ion,r} + \delta r_{trp,r} + n_r \quad (8.31a)$$

$$(\lambda/2\pi)\,\varphi_u = \rho_u + N_u\lambda + c\delta t_u - c\delta t_s - \delta r_{ion,u} + \delta r_{trp,u} + n_u \quad (8.31b)$$

The suffixes u and r represent the above parameters for the user and the reference, respectively. The rest of the notations in the above equations signify usual parameters. Notice here that the ionospheric error has the opposite sign compared to that obtained for the range measurements. It is because, unlike the code, which gets delayed as it passes through the ionosphere, the carrier phase experiences equal phase advancement. Here we have deliberately removed the time variable or any indication of the satellite from which the data is being received for brevity of expression, but shall introduce them at an appropriate time.

8.2.2.1 Single differencing technique

We have already seen in connection to the code-based relative methods how the relative range equations are formed using the respective individual measured ranges. Similar difference equations may be formed by using the respective phase measurements, too.

The measured phase at the reference receiver with signals from satellite s may be expressed as in Eqn (8.31a) and the measured phase at the user receiver is expressed as in Eqn (8.31b). On differencing these two, we get

$$(\lambda/2\pi)(\varphi_r - \varphi_u) = \rho_r - \rho_u + N_r\lambda - N_u\lambda + c\Delta\delta t_{ru} - \Delta\delta r_{ion,ru} + \Delta\delta r_{trp,ru} + n_{ru}$$
$$(8.32)$$

Thus, the common satellite clock bias gets removed. Now, considering the short baseline and keeping the first-order derivatives only, the geometric range may be expressed in the form

$$\rho_u = \rho_r + \partial\rho/\partial x\, dx + \partial\rho/\partial y\, dy + \partial\rho/\partial z\, dz\big|_{X_r}$$
$$\text{or,} \quad \rho_u - \rho_r = \partial\rho/\partial x\, dx + \partial\rho/\partial y\, dy + \partial\rho/\partial z\, dz\big|_{X_r} \quad (8.33)$$

As before, the above single difference equation turns into

$$(\lambda/2\pi)\Delta\varphi_{ru} = (\lambda/2\pi)(\varphi_r - \varphi_u)$$
$$= \big(\partial\rho/\partial x\, dx + \partial\rho/\partial y\, dy + \partial\rho/\partial z\, dz + N_{ru}\lambda \quad (8.34)$$
$$+ c\Delta\delta t_{ru} - \Delta\delta r_{ion,ru} + \Delta\delta r_{trp,ru} + n_{ru}\big)$$

Similarly, as in Eqn (8.25) and Eqn (8.26), $\partial\rho/\partial x_r = -e_x$, $\partial\rho/\partial y_r = -e_y$ and $\partial\rho/\partial z_r = -e_z$, where the differentials are derived at X_r while dx, dy, and dz are the differential elements of the baseline b. N_{ru} is the difference of the integer ambiguity for the two stations. So, assuming the ionospheric and tropospheric errors nearly vanish on differencing, the equation can be written as

$$(\lambda/2\pi)\Delta\varphi_{\,ru} = (\overline{e}_r \cdot b) + N_{ru}\lambda + c\Delta\delta t_{ru} + \varepsilon_{ur} \tag{8.35}$$

where $\overline{e}_r$ is the unit vectors from the reference to the satellite s. ε_{ur} is the error considering the receiver noise and all other differential residual errors between the two receivers.

For n number of visible satellites, which are common to the reference and the user, m such single difference equations can be formed for a definite pair of reference and user. The matrix form of the equation becomes

$$(\lambda/2\pi)
\begin{bmatrix}
\Delta\varphi_{u\,r}^{S_1} \\
\Delta\varphi_{u\,r}^{S_2} \\
\vdots \\
\Delta\varphi_{u\,r}^{S_n}
\end{bmatrix}
=
\begin{bmatrix}
-e_x^1 & -e_y^1 & -e_z^1 & c \\
-e_x^2 & -e_y^2 & -e_z^1 & c \\
\vdots & \vdots & \vdots & \vdots \\
-e_x^n & -e_y^3 & -e_z^1 & c
\end{bmatrix}
\begin{bmatrix}
dx \\
dy \\
dz \\
\delta t_{ur}
\end{bmatrix}
+
\begin{bmatrix}
N_{u\,r}^1 \\
N_{u\,r}^2 \\
\vdots \\
N_{u\,r}^n
\end{bmatrix}
+
\begin{bmatrix}
\varepsilon_{u\,r}^1 \\
\varepsilon_{u\,r}^2 \\
\vdots \\
\varepsilon_{u\,r}^n
\end{bmatrix}$$

or

$$(\lambda/2\pi)\Delta\phi_{ur}^s = G_1 \cdot dX + N + \varepsilon \tag{8.36a}$$

These single difference equations have three relative positional elements and a relative time offset element in dX and n integer ambiguity values in N as unknowns. ε contains the vector noise elements whose individual values are not known but the variance of which may be used for solving for dX. This cannot be readily solved using least squares method due to the presence of N. However, techniques like the carrier smoothed range measurements may be used for estimating N, following which dX may be solved.

Alternatively, one can solve for dX by tactfully eliminating the N. If we take similar measurements at two different instants, in the interval of which the measurement of the phase remains continued, then the initial integer ambiguity, which remains common for both, vanishes on differencing. Therefore, writing Eqn (8.35a) for two time instants k_1 and k_2, we get

$$(\lambda/2\pi)\Delta\Phi_{ur}^{si}(k_1) = G_1(k_1) \cdot dX + N + \varepsilon(k_1) \tag{8.36b}$$

$$(\lambda/2\pi)\Delta\Phi_{ur}^{si}(k_2) = G_1(k_2) \cdot dX + N + \varepsilon(k_2) \tag{8.36c}$$

Subsequently, differencing the two, we get

$$(\lambda/2\pi)\delta\Delta\Phi_{ur}^{si}(\Delta k) = (\lambda/2\pi)\left[\Delta\Phi_{ur}^{si}(k_1) - \Delta\Phi_{ur}^{si}(k_2)\right]$$
$$= [G_1(k_1) - G_1(k_2)] \cdot dX + \delta\varepsilon(k) \tag{8.37}$$
$$= G_{1\Delta k} \cdot dX + \delta\varepsilon(k)$$

where $G_{1\Delta k} = [G_1(k_1) - G_1(k_2)]$ The integer ambiguity N, which remains the same over the whole session, thus gets eliminated from the equation on differencing.

Now the general least squares solution for dX may be derived out of it as

$$dX = (\lambda/2\pi)\left[G_{1\Delta k}^T G_{1\Delta k}\right]^{-1}(G_{1\Delta k})^T \delta\Delta\Phi(k) \tag{8.38a}$$

Further, if the variance of the error $\delta\varepsilon$ is known for each satellite and if they form a matrix M, the weighted least squares solution may be obtained as

$$dX = (\lambda/2\pi)\left[G_{1\Delta k}^T M^{-1} G_{1\Delta k}\right] - 1\left(G_{1\Delta k}^T M^{-1}\right)\delta\Delta\Phi(k) \tag{8.38b}$$

Here, $G_i(k_2)$ and $G_i(k_1)$ need to be widely separated to provide a well-conditioned value of $G_{1\Delta k}$ that determined the precision of the estimation of dX. This means that the time interval needed between two sets of measurements should be appropriately large. So, this technique can be applied only for non-real-time applications where the solution may be obtained a posteriori after a long interval.

It is important to appreciate and remember once again that the convenience of the linearity is obtained by considering short baseline. Due to this assumption, it becomes possible to express $\rho_u - \rho_r$ as a linear function of dx, dy, and dz to form G_1, which eases the problem to a great extent.

8.2.2.2 Double differencing technique

The differential method has the intrinsic advantage of cancellation of the common errors. We have seen this in the code range method as well as in the single difference phase method. However, the relative clock bias parameter still remains in the equation as a menace for solving for dX. Many of the applications that use this relative positioning technique are interested in obtaining precise position. Hence, the relative clock offset parameter $\Delta\delta t_{ru}$ is not mandatory for being in the equation and may be removed during the estimation process. This may be done by generating the double difference (DD) phase equations. As for the case of range equations, the DD is obtained by generating another single difference equation with a separate satellite and then differencing the two single difference equations.

Using Eqn (8.34), the two single difference equations between the user and the reference receiver for satellites m and n may be respectively expressed as

$$\begin{aligned}(\lambda/2\pi)\Delta\varphi_{ur}^m &= \left(-e_r^m \cdot b\right) + N_{ur}^m \lambda + c\Delta\delta t_{ru} + \varepsilon_{ur}^m \\ (\lambda/2\pi)\Delta\varphi_{ur}^n &= \left(-e_r^n \cdot b\right) + N_{ur}^n \lambda + c\Delta\delta t_{ru} + \varepsilon_{ur}^n\end{aligned} \tag{8.39}$$

Differencing it once, we get the DD equation as

$$(\lambda/2\pi)\nabla\Delta\varphi_{ur}^{mn} = \left(-e_r^{mn} \cdot b\right) + N_{ur}^{mn}\lambda + \varepsilon_{ur}^{mn} \tag{8.40}$$

This forms the conventional double difference equation, where we have assumed that the ephemeris, ionospheric, and tropospheric error residuals have already been removed on single differencing. Any residual thereof is insignificant and random in nature, to be treated as a noise and included in ε_{ur}^{mn}. Notice that here the relative clock bias of the $\Delta\delta t_{ur}$, being the same between the same user receiver and reference, gets completely canceled out.

Considering the discussion we had in reference to the linearity of the geometric range variation for short baseline, and the derivation of the range difference in Eqn 8.28, this equation can also be expressed in terms of the angle θ that the baseline b makes with the unit vector e. Thus the expression becomes

$$\left(-e_r^{mn} \cdot b\right) = b(\cos \theta_m - \cos \theta_n) \tag{8.41}$$

Here, θ_m and θ_n are the respective angles with satellites m and n, respectively. So, putting this relation from Eqn (8.40) in the above Eqn (8.39), we get

$$(\lambda/2\pi)\nabla\Delta\varphi_{ur}^{mn} = b(\cos \theta_m - \cos \theta_n) + N_{ur}^{mn}\lambda + \varepsilon_{ur}^{mn} \tag{8.42}$$

Eqn (8.39) can also be written as

$$(\lambda/2\pi)\nabla\Delta\varphi_{ur}^{mn} = b\left(G_1^m - G_1^n\right) \cdot b + N_{ur}^{mn}\lambda + \varepsilon_{ur}^{mn}$$
$$= G_2^{mn} \cdot b + N_{ur}^{mn}\lambda + \varepsilon_{ur}^{mn} \tag{8.43}$$

where $G_{2r}^{mn} = (G_{1r}^m - G_{1r}^n)$, the geometric observation matrix for double difference. But, apart from eliminating the relative receiver clock bias, we have not gained much in this differencing until now, since the problem term N still remains in the equation.

This linear equation may be solved in the same manner as we have done for the single difference case by using equations separated by time. This strategically avoids the problem of finding the integer number N. Differencing two double difference equations separated by time gives a triple difference equation.

8.2.2.3 Triple difference method
For the same pair of satellites, DD can be estimated at a large time difference. Then these two obtained double differences may be again differenced to get the triple difference of the range observation equation. Thus, for triple difference it becomes

$$\Delta^3\varphi_{ur}^{mn}(\Delta k) = \nabla\Delta\varphi_{ur}^{mn}(k_2) - \nabla\Delta\varphi_{ur}^{mn}(k_1)$$
$$= \left[G_{2r}^{mn}(k_2) - G_{2r}^{mn}(k_1)\right] \cdot b + \left[\varepsilon_{ur}^{mn}(k_2) - \varepsilon_{ur}^{mn}(k_1)\right] \tag{8.44}$$
$$= G_{3r}^{mn}(\Delta k) \cdot b + \varepsilon_{ur}^{mn}(\Delta k)$$

So, for each pair of satellites, similar differences may be made at considerably large intervals. By this time, both the satellites, their positions having changed, will give a new set of independent equations. The scenario is described in the schematic in Figure 8.5.

So, the phase equation can also be written as

$$\Delta^3\varphi_{ur}^{mn}(\Delta k) = b[(\cos \theta_{mk1} - \cos \theta_{nk1}) - (\cos \theta_{nk1} - \cos \theta_{nk2})] + \varepsilon_{ur}^{mn}(\Delta k) \tag{8.45}$$

where θ_{jk} is the angle made by the baseline with the unit range vector toward satellite j and at time k. So to solve for the three unknowns, three such triple differences are required to be established and hence three independent pairs of visible satellites need to be measured for range over a large interval of time. As mentioned for the case of single differencing, this method is not possible where solutions are required in near

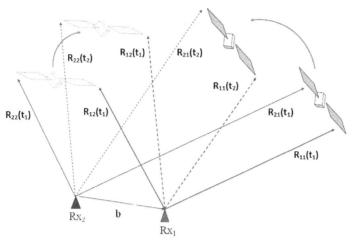

FIGURE 8.5

Schematic of the Scenario to Generate Triple Difference Equations.

real time. In such cases, it is required to resort to the DD method where some ambiguity-resolving technique is required to be employed to estimate N. Then the value of dX may be solved for.

Resolving ambiguity is nothing but solving for integer values for N. It is one of the key issues in relative positioning, mainly due to the constraint that N has to be an integer. Two approaches are mainly followed for the purpose (Consentino et al. 2006). In one approach, it is done by finding the optimum value of N, disregarding the integer constraint for N. This involves solving for the float value of N. The code-based measurements are used in this process. Further, the code-based double differences are smoothed using the carrier-based measurements. The Kalman filter, discussed in Chapter 9, may be used in the process. Finally, this float solution is tuned to the nearby integer, which offers minimum error to the equations formed to get the true value of N. Otherwise, it is done by first limiting the estimate region for N to a set of possible integer values of N and then searching for the best value.

Once the estimate of integers is given, it is straightforward to solve for dX. Table 8.1 lists the different popular methods of differential positioning. However, this list is not exhaustive.

8.3 Implementation of differential systems

Until now, we were discussing the different techniques of applying the correction to the errors at the user receivers. In this chapter, we have learned the basic principles of using both the code-based ranges and the carrier phase-based measurements for differential corrections. Here, we shall talk more about the implementation part of it and we shall see that depending upon the applications, areas of coverage, etc.

Table 8.1 Some Popular Methods for Differential Positioning

Methods	At Reference Station (RS)	At User End
Absolute Methods		
Method 1	1. Calculate position X, Y, Z for all possible sets of 4 satellites visible 2. Compute position error (ΔX, ΔY, ΔZ) with respect to known location for each set separately 3. Broadcast position correction data to respective users for each set	1. Select optimum set of 4 satellites and measure pseudoranges 2. Compute uncorrected position 3. Receive and apply RS position correction (ΔX, ΔY, ΔZ) for the corresponding set of the satellite actually used by it 4. Obtain corrected position solution
Method 2	1. Measure pseudoranges (PR) for all visible satellites 2. Calculate PR correction with respect to known location 3. Broadcast range corrections for all satellites to all users	1. Select optimum set of 4 satellites and measure pseudoranges 2. Receive pseudorange corrections from RS 3. Apply appropriate pseudorange correction to pseudorange data 4. Obtain corrected position solution
Relative Methods		
Method 3	1. Measure all satellite pseudoranges 2. Broadcast pseudorange to all users	1. Select optimum set of 4 satellites and measure pseudoranges 2. Receive pseudoranges from RS 3. Difference the respective pseudoranges to get corrected relative range data 4. Solve for corrected relative position solution
Method 4	1. Carry out carrier phase measurements of all visible satellites 2. Disseminate the carrier phase measurements to users along with PR measurements	1. Carry out carrier phase measurements for an optimum set of 4 satellites 2. Use double difference techniques 3. Resolve ambiguity using PR data 4. Estimate relative position from reference station

for the differential corrections, the types as well as the architecture needed for implementation become different.

Implementation of the differential positioning methods mainly depends upon a few aspects, viz. the accuracy required, real-time or post processed solution required, static or kinematic user, and area to be covered. Typical navigation receivers measure the pseudoranges using the code-based techniques. It is because code-based measurements of pseudoranges are easier to implement in a receiver than measuring carrier phase. So, it is also more convenient to provide the differential corrections using range-based techniques. The major requirement here is the measurement of range at a well-defined reference location while the subsequent processing part is rather simple. However, the accuracy of the solution is not as good as the phase-based techniques due to the presence of the intrinsic noise in the measurements.

In contrast, solving for positions with phase-based measurements is much more complex and computationally intensive. Even the most convenient method of using data with temporal separation, which avoids the estimation of the integer ambiguity, bars it from being used for near-real-time applications. In such cases, ambiguity resolution is unavoidable. It adds intricacies to the process of working out the position.

This makes the range-based differential positioning popular for most of the applications. Provisions of quick solutions, frequent updates, and simple computational algorithms with precision enough for nominal movements make this range-based method very suitable for vehicle navigations using differential systems.

Thus, the applications restricted to near real time with moderate precision are mainly supported by range-based differential techniques for navigation. These applications may be limited to local area as well as extended area of usage.

On the other hand, applications like surveying, attitude determination etc. demanding high accuracy precision of obtained position (Leick, 1995), must go for carrier-phase methods. Even for the carrier-phase techniques, the static and post processing methods may provide better results while the performance degrades comparatively for kinematic user and real-time processing. Applications including static survey, geodesy etc., which do not require position solutions in real time can hence avail the advantage of post processing. They can also use data integrated over large intervals or separated by large temporal differences. This gives improved accuracy and reduces operational complexity as well since they do not require real-time communication of data. However, applications employing phase-based corrections like real-time kinematic (RTK) cannot afford to avoid it and may also have multiple reference stations for the same reason of extending the area of application.

A single reference station is sufficient for the purpose where the total area required to be served is small. The limit of the spatial extent that a single reference station can serve is based upon the factor of the distance the error correlation exists. Similarly, the temporal correlation of these errors determines the latency of using the corrections of these errors and hence answers how often the correction must be provided. Thus, the number of references required is based upon the extent of the area of service as well as the accuracy needed.

We have seen that many of the errors are spatially and temporally correlated and remain almost constant for small distances between the reference and the users. At the most, the spatial variation may be of first order. Thus using the same range errors as the reference or with additional first-order derivatives, as necessary for appropriate correction, only a small residual error is left to the corrected parameters. Such cases constitute the local area differential systems, where the user receives real-time correction from reference when it remains within the line of sight of the reference station. This is easiest to establish and corrections are valid only within certain limited distance, defined by the correlation of the propagation conditions.

For larger extents of service area, the above methods start showing large errors (Kee, 1996). The estimates get worse with increasing distance between the reference station and the user. The variation of the correlated errors becomes large and there remains considerable nonlinearity in its spatial variation. So, the area of service is of such an extent that the errors cannot be corrected by a single reference station.

Additionally, in the worsening of the errors, there is also another constraint pertaining to large areas of service. The requirement of the user to remain within the transmission range of a single reference station may not be satisfied every time. Moreover, due to the limited visibility of the single reference station, it may see different sets of satellites than a user at large distance from it.

To alleviate the problems faced in the local area differential systems, the simplest method is to establish a number of such discrete local area differential systems with a network of reference stations distributed over the total area under service. Then, the user receiver obtains the corrections from its nearest reference transmitters and corrects its errors. The underlying assumption is that at least one of the reference stations remains located so close to the user, irrespective of its location, that the spatial separation is within the first-order differential zone to at least one of them. But fulfillment of this condition demands the establishment of a large number of reference stations to cover a large area, like that of a whole country.

So the concept of coalition of many local area differential systems to serve a large area is required to be graduated to a modified approach of a wide area differential system. The problems of establishing a large numbers of reference stations may be alleviated with the approach that, instead of transmitting the total consolidated range error, the components of the errors are separated and individual errors are transmitted to the user. This idea has already been mentioned in connection with the code-based reference positioning. We repeat it here with elaborations for the sake of continuity and completeness.

With this approach, the geographically correlated errors become independent of any particular satellite or look angle. Instead, here the total service area is divided into numbers of grids with orthogonal grid lines intersecting perpendicularly at the grid points. The total vertical error at these predefined points is transmitted. The vertical delays are estimated from the slant delays experienced by the signals at the reference using some proper algorithm. The separation is determined from the assumption that the variations of these errors between these points are small enough that linear extrapolation holds good. So, wherever the user receiver is, it has some estimation of vertical delay values around it that can be used as reference. This enables the user to calculate his or her own absolute errors appropriate for the user's own location from the received individual component errors. So, the effects of errors like the ephemeris error or ionospheric and tropospheric errors are directly estimated from the user position relative to the measured points by interpolating them from these points to the points relevant to the user's measurements, e.g. to the user's IPPs (Enge and Van Dierendonck, 1996; RTCA, 1999). Thus, the effect of higher-order nonlinearities of the spatial variation during their estimation is eliminated. This also liberates the error correction at the user from adhering to the same satellite from which the error has been estimated. However, this is at the cost of additional computational and modeling activities at the reference side.

The satellite clock and ephemeris errors for individual satellites and propagation errors for different wide regions are separately determined here. This warrants the need of many reference stations located at precisely determined locations across the entire service area at suitable locations on the ground. This network can estimate precisely the locations of each of the satellites individually, with their timing errors. Thus, the absolute satellite position errors and the satellite clock errors can be easily determined separately and may be disseminated to the users.

Thus, after correcting the errors, the corrected range at the user location becomes as Eqn 8.18, which is given below:

$$R_{cu} = \rho_u + c(\delta t_r - \delta t_u) + (\delta r_{eph} - g\delta r_{sat}) + (\delta r_{ion,u} - f_{ion}(\theta_u)\delta r_{ion,v})$$
$$+ (\delta r_{trp,u} - f_{trp}(\theta_u)\delta r_{trp,v}) + (n_u - n_r)$$

Here, $\delta r_{ion,v}$ is the vertical ionospheric delay derived from the reference measurements at the user location. $f_{ion}(\theta_u)$ is the ionospheric slant factor at the user location for the elevation angle θ_u. Similarly, $\delta d_{trp,v}$ and $f_{trp}(\theta_u)$ are, respectively, the vertical delay and the slant factor for the tropospheric delay for the user elevation angles θ_u toward a satellite.

The same set of reference stations can participate in ephemeris and clock error measurement of many satellites. Moreover, each reference station generates a number of propagation error data at any instant from differential range measurements with different satellites. So, with few strategically located reference stations, the whole purpose may be served and thus it requires a lesser number of stations than it would have required to serve the same area just by adding up numbers of discrete local area differential systems.

For larger areas of service, this dissemination takes place simultaneously over the entire area, preferably by a satellite that has suitably large coverage. Figure 8.6 illustrates an arrangement for providing correction data over a geographically large region.

This is the basis of the augmentation system, in which various system resources are set up to augment the primary service, which efficiently provides the differential corrections to the primary users over a wide area, consequently improving the accuracy and precision of positioning.

Conceptual questions

1. Is the carrier phase-based ranging only useful for differential positioning and not for the standalone case?
2. Is it necessary to derive the IPP points while converting given vertical ionospheric delays to slant delays along the satellite path? If yes, why?
3. What are the factors that determine the accuracy in an RTK system where both the user and the reference are moving with time?
4. What similarities do you find in the ground resources required for establishing a differential positioning system over a wide area with code-based ranging with the general control segment architecture that we read about in Chapter 2?

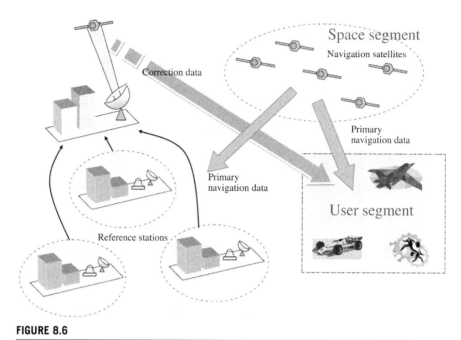

FIGURE 8.6

Dissemination method for geographically large area.

References

Cosentino, R.J., Diggle, D.W., de Haag, M.U., Hegarty, C.J., Milbert, D., Nagle, J., 2006. Differential GPS. In: Kaplan, E.D., Hegarty, C.J. (Eds.), Understanding GPS Principles and Applications, second ed. Artech House, Boston, MA, USA.

El-Rabbani, A., 2006. Introduction to GPS, second ed. Artech House, Boston, MA, USA.

Enge, P.K., Van Dierendonck, A.J., 1996. Wide area augmentation system. In: Parkinson, B.W., Spilker Jr, J.J. (Eds.), Global Positioning Systems, Theory and Applications, vol. II. AIAA, Washington DC, USA.

Kee, C., 1996. Wide area differential GPS. In: Parkinson, B.W., Spilker Jr, J.J. (Eds.), Global Positioning Systems, Theory and Applications, vol. II. AIAA, Washington DC, USA.

Leick, A., 1995. GPS Satellite Surveying, second ed. John Wiley and Sons, New York, USA.

Papoulis, A., 1991. Probability, Random Variables and Stochastic Processes, third ed. McGraw Hill, Inc, New York, USA.

Parkinson, B.W., Enge, P.K., 1996. Differential GPS. In: Parkinson, B.W., Spilker Jr, J.J. (Eds.), Global Positioning Systems, Theory and Applications, vol. II. AIAA, Washington DC, USA.

RTCA, 1999. Minimum Operational Performance Standards for GPS/WAAS Airborne Equipments. DO-229B, A-34. USA.

Special Topics

CHAPTER OUTLINE

This chapter consists of special topics. Experienced readers know that it is customary to keep a separate chapter for those items that do not fit into any other chapter in a book. The topics to be discussed in this chapter are somewhat like that. But, that they do not fit elsewhere does not mean that they are irrelevant. Rather, their relevance can be seen in many topics throughout this book, and the chapter title is really not just for the sake of giving a name to it. These topics are of special interest as far as satellite navigation is concerned, and that is what we wish to bring out here. Our discussion in this chapter will be restricted only to two topics: the Kalman filter and the ionosphere. But what is so special about them? Though

Understanding Satellite Navigation. http://dx.doi.org/10.1016/B978-0-12-799949-4.00009-9

they are categorically antagonistic in nature in terms of their influence, they have a large bearing upon satellite navigation in their own ways. So it is their immense significance that makes them special here. In the opening section of this chapter, we will discuss the theory and working of the Kalman filter in a nutshell. Then we will briefly talk about some of its applications relevant to satellite navigation that improve system performance. In contrast, in the next section we will cover a subject that deteriorates navigation performances - the ionosphere. The main focus of this topic will be its equatorial aspects, where it has a major influence. However, here we will deliberately avoid mathematical expressions and complex technical jargon, since our main aim is to understand the potential each of them has to improve or deteriorate navigation system performance. It is important to repeat here that readers may happily skip this chapter without any loss of continuity of the topics laid out in this book, but of course at the price of losing many interesting portions of it.

9.1 Kalman filter

9.1.1 Introduction to the Kalman filter

The Kalman filter is a mathematical tool for estimating the dynamic states of a system from the noisy measurements that have known relations to these states and aided by some prior knowledge about the progression of the states over time. It also provides the confidence value associated with every estimate.

It is a state-space approach in the sense that it assumes that the dynamic behavior of a system can be described by some finite states that completely define the condition of the system. The filter can estimate the values of these states at any instant and also their temporal variation, provided it has some essential information about how these system states evolve over time and some related measurements are available along with it.

It is necessary at this point to define the terms mentioned so that we can conveniently use them later while describing the working of the filter. The *state* of a system is defined as the variables describing the condition of the system., and can be represented quantitatively by distinct values that may change with time under the effect of factors that influence these states. These states may either be independent of each other, so that the change in one does not affect the change in the other; or they can even be dependent, with the variation of one state related to the change in the other. The dependent state variables may be connected through a linear or nonlinear relation.

For example, let us take a solid metallic disc of iron. The physical condition of the disc may be described by a few properties like its physical state (solid), color (red), mass (1000 g), and radius (15 cm). These features describe the state of the disc and the entries in the square brackets represent the values of the variables representing the state that the disc is currently in. Notice that the first three variables are independent of each other while the last two are related nonlinearly through the constant density terms.

Now, we leave the disc to roll on its edge down a slope without slipping. In this condition, the dynamic state of a body is defined by the motion it executes under the action of a constant gravitational force, as shown in Figure 9.1. Its state dynamics are our interest now, and hence accordingly a different set of variables now forms the system states.

The condition of the motion of this body may be represented by three variables: position s, velocity v, and acceleration α along the slope. Among these terms, the acceleration term is constant and independent of the other two variables.

Velocity is represented as the time integral of the constant acceleration, and hence it is linearly dependent upon acceleration, α. When there is no slippage, the angular velocity ω and the angular acceleration ω' are related to linear velocity v and acceleration α through the constant radius term, "r". So they are alternative representations of the same dynamic states and hence are not distinctly separate state variables. Similarly, the position s is the linear time integral of instantaneous velocity and acceleration α, and hence has a linear dependency with both of these variables. So, the temporal evolution of the states from any arbitrary discrete instant k to k + 1 separated by time Δt may be represented using Newton's equation of motion as:

$$\alpha_{k+1} = \alpha_k \tag{9.1a}$$

$$v_{k+1} = v_k + \alpha_k \Delta t \tag{9.1b}$$

$$s_{k+1} = s_k + v_k \Delta t + \tfrac{1}{2} \alpha_k \Delta t^2 \tag{9.1c}$$

where s_j, v_j, and α_j are the position, velocity, and acceleration, respectively, of the disc at instant j.

The state variables may or may not be measurable themselves. However, they must manifest themselves through some measurable variables of the system. In other words, these measured variables of the system must be a function of the identified state variables from which the estimates of the state may be obtained. The functions relating the measurements to the states may also be linear or nonlinear.

In the above example, we can directly have the measurements of position, velocity, and acceleration of the particle as measurements. Alternatively, we can also measure the kinetic energy, K, and potential energy, P, of the disc (by some means). The former

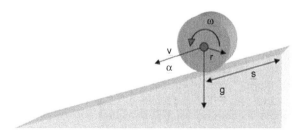

FIGURE 9.1

Description of states.

is a nonlinear function of its velocity while the latter is a linear function of its position. So, this can be expressed as $K = f_1(v)$ and $P = f_2(s)$. The above-mentioned variables of the dynamic states may be derived from these measurements using a Kalman filter.

With this understanding of the states and measurements, we can start to describe the Kalman filter. Further explanations of other terms will be made as we come across each of them.

9.1.2 Kalman filter basics

9.1.2.1 Initial concepts

The Kalman filter, as we have defined it, is basically a tool that estimates the state variables of a system in an optimal manner and the confidence in the estimation using the dynamic behavior of these variables and some relevant measurements as input, even in the presence of noise.

Let the system S be defined by a set of state variables X. Thus, X is a collection of variables defined by $X = (x_1, x_2, x_3, ...x_n)$. X defines completely the state of the system that we are interested in and vary distinctly over the time of our interest, such that at any instant k, it takes values X_k. The set Z consists of the measurable variables $Z = (z_1, z_2, z_3, ..., z_m)$, available from the system that are related to the state X. The measured values of the set of variables at time instant k are denoted by Z_k.

The values of the variables in X change with time, and the Kalman filter, to estimate the values, needs to know how. For this purpose, it requires an equation that approximately models how the values of the state variables evolve from one instant to the next. Accordingly, this equation should be able to define the state variables X_k of any instant as a function of the state variables of just its previous instant, X_{k-1}. In our example of states in Section 9.1.1, Eqn (9.1a–9.1c) thus are the equations that serve this purpose.

Pragmatically, these relationships are approximate. There are always some unmodeled elements that play a role in varying the exact state values from those determined by the relations while propagating from one instant to the next. Consider, in our example, that there are random frictional forces due to the uneven nature of the path that act upon the disc oppositely to its movement, whose effects are not included in the effective acceleration, α. Moreover, there are also some untraceable forces that are not quite consistent in a practical sense. In all pragmatic cases, these random forces remain acting upon the body yet unaccounted for, causing random variation of the dynamic states by a minute amount. As these variations cannot be represented by any definite model, in a Kalman filter these are taken as noise and are called the "system process noise."

Therefore, considering the known relation for the system state transition and allowing for the process noise, the true transition may be expressed as the combination of these two. This complete relation is called the system dynamic process model, and is given by:

$$X_k = \Phi(X_{k-1}) + w_k \qquad (9.2a)$$

Here Φ is the definite function called the "state transition function," which relates the current states to the preceding ones. It can be linear and nonlinear as well. "w" is the process noise. For a linear case, the new state becomes:

$$X_k = \varphi_k X_{k-1} + w_{k-1} \tag{9.2b}$$

Here, φ_k is a discrete matrix called the state transition matrix, and w_k denotes the process noise at instant k. The latter includes all of the errors that may have occurred in using this definite model, which is a mathematical approximation of the true variation. Whatever the nature of the effect of w is, its individual values cannot be known separately. But we need to know the statistical nature of w. For a Kalman filter, w is assumed to be a zero mean white Gaussian noise with Q as the variance. We will henceforth assume this in all our subsequent derivations. Thus, the variables in X propagate across time from one instant to the next following Eqn 9.2(a) or 9.2(b) for the linear case and as shown in Figure 9.2.

The system S has some measurable variables. Among these, some are functions of the system states of our interest. Measurements of such variables are necessary for our purpose, as it is from these measurements that the states are to be derived by the filter. However, such measurements are added with noise which cannot be segregated by any common means. So, our measurement Z_k at any instant k can be represented through a measurement equation in terms of X_k as:

$$Z_k = h(X_k) + r_k \tag{9.3a}$$

This is the generalized measurement equation, where r is the measurement noise and h is the function that relates the current state with the measurements. "h" is called the observation function. For a linear system, it can be represented as:

$$Z_k = HX_k + r_k \tag{9.3b}$$

where H is called the "observation matrix" or the "measurement sensitivity matrix" and defines the linear relationship.

The system noise at instant k that includes all the errors present in the measurements is r_k. However, we cannot separately identify what the values of r are at any instant. Had this been possible, we could have corrected the errors. For the Kalman filter to operate, we need to know at least the statistical nature of this noise. So, we assume r is a zero mean Gaussian noise with R as the measurement error variance.

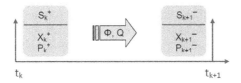

FIGURE 9.2

Time propagation of states.

For multivariate state X, φ and H are matrices. X becomes a [n × 1] vector, where the number of state variables is n, and φ is a [n × n] matrix relating the values of n current states with those of the same n state variables of the next instant. The element φ_{ij} of the matrix relates the ith component of X_k with the jth component of X_{k-1}. This takes zero value if these two elements of the states are mutually independent. H is a [m × n] matrix where m is the numbers of measurements. Similarly, H_{ij} represent how the ith measurement is related to the jth state variable in X. H becomes identity when the measurements are the direct gauge of the variable themselves and zero if they are not related. Accordingly, w becomes [n × 1] and r becomes [m × 1] matrix representing the noise in each individual element in the dynamic equation and each measurement in the measurement equation, respectively. The corresponding covariance matrices Q and R thus become [n × n] and [m × m] matrices, respectively.

Fortunately, the filter can estimate the states and can update them over time even in the presence of noise when it knows the form of φ and H explicitly and simultaneously knows the values of only the statistical variances Q and R of the noise terms. Although a conventional Kalman filter demands these noises to be white Gaussian, there are techniques by which even non-Gaussian noises may be incorporated into the filter through noise shaping (Grewal and Andrews, 2001; Maybeck, 1982).

9.1.2.2 Estimation process

The Kalman filter starts with initialization of the state variables in X and with an appropriate error covariance associated with the same. Subsequent to this, the state values are calculated and propagated over time by updating the variables successively through two recurring processes, viz. the measurement update and the temporal update. Consequently, the error covariances also get modified and converge to a limiting value.

9.1.2.2.1 Temporal update

Temporal updating is the process in which the state derived or assumed by the filter at any instant are propagated through time according to the system process model as given in Eqn (9.2a). The state transition matrix is used to obtain the best possible values of the new state variables at the immediate next instant from the current ones. These time propagated state variables now form the a priori estimate of the state for the next instant.

9.1.2.2.2 Measurement update

Measurements done at any instant are assimilated with the a priori estimate of the state variables already available to the filter from the last temporal update. This further updates the state, and the result thus formed constitutes the posterior estimate of the states. Addition of independent information thus taking place during this measurement update leads to further improvement of the estimation accuracy compared to the a priori estimate. The measurement equation is utilized during this update to relate the measurement with the state variables.

So, if we call the temporal update a prediction process, the measurement update may be referred to as a correction process. It is important to mention here that along with the state estimates, the Kalman filter also updates the confidence bounds of each of the estimates it generates in both the processes. The above-mentioned updates continue iteratively while the Kalman filter offers an optimal solution for the state variables of the dynamic system. The complete iterative process consisting of the temporal update and the measurement update is shown in Figure 9.3 and is described in Section 9.1.3.

For the purpose of convenience, we demarcate every instant by the point at which the measurements are done. So the times marked by τ_1, τ_2, τ_3, etc. are those specific instants at which the measurements are done. The instant just before any particular measurement instant τ_k is denoted by τ_k^-, and that momentarily after the measurement instant is denoted by τ_k^+. So, the a priori estimate X_k^- of the instant τ_k is available at τ_k^-, while the posterior estimate X_k^+ is available at τ_k^+.

9.1.3 Derivation of the filter equations

With this basic idea of the working of the Kalman filter, let us find out the quantitative expression for the different updates of the state of a system. We assume a linear system and start from an initial guess, X_0^+ of the state variables. The error variance of the estimates is taken as P_0^+ when we are starting the filter process afresh at the instant τ_0. From this initial state X_0 or from any intermediate time τ_k^+, at which the state variables X_k^+ are already determined, the immediate next task is to hop to the next instant and to get the time updated states. The values of the state variables to be arrived at will be X_{k+1}, where k can be zero or any other positive value, as the case may be. Thus, we propagate the state over time using our knowledge of the system dynamics given in the process equation, Eqn 9.2b.

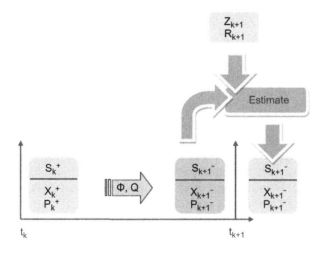

FIGURE 9.3

Complete iterative process for Kalman filtering.

But how do we proceed when we do not have the values of w? With φ as the state transition matrix that carries the system state variables X_k^+ from τ_k^+ to τ_{k+1}^-, the Kalman filter simply updates the states as

$$\widehat{X}_{k+1}^- = \varphi X_k^+ \tag{9.4}$$

$\widehat{X}$ is the ensemble average of X_{k+1} and is the best possible estimate that the filter could do with the available information. The noise term w_k, being zero mean vanishes in the process of finding the expectation value. However, this exclusion of the non-deterministic component in the update must result in some deterioration in the estimate precision. So the error covariance is updated by the filter in such a way that it accommodates and represents this effect. So the effective error covariance may be expressed as the sum of the covariance due to the errors that were already in the state estimate which propagated during this temporal update and the effect due to this probabilistic mean taken in Eqn (9.4).

From the process equation, it is clear that as X_k^+ is carried from τ_k^+ to τ_{k+1}^-, so are its associated errors. So, if X_k is the true value at instant τ_k^+, let the estimate be $\widehat{X}_k^+ = X_k + \varepsilon_k$, where ε is the associated estimation error. This is, however, the best estimated value for the instant. The estimate after the temporal update is $\phi(\widehat{X}_k^+) = \phi(X_k + \varepsilon_k^+)$. For a linear process, the propagated state may be represented as $\varphi X_k + \phi \varepsilon_k^+$. So the error due to the already existing error ε_k^+ is now $\varepsilon_{k+1}^- = \varphi \varepsilon_k^+$ after the update at τ_{k+1}^-. Therefore, its covariance is given by

$$\begin{aligned} P(\varepsilon_{k+1}^-) &= E[\varphi \varepsilon_k^+] \cdot [\varphi \varepsilon_k^+]^T \\ &= E[\varphi \varepsilon_k^+ \varepsilon_k^{+T} \varphi^T] \\ &= \varphi P_k^+ \varphi^T \end{aligned} \tag{9.5a}$$

Here, $E[\varepsilon_k^+ \varepsilon_k^{+T}] = P_k^+$, which is the error covariance of the last state X_k^+. Now the effect of vanishing of "w" will incorporate the value of the covariance of w into this in addition. As Q_k is the variance of w_k, and w_k is uncorrelated to the existing estimate errors ε_k^+, the total error covariance will be just the sum of the two individual covariances. So the effective error covariance after the temporal update is

$$P_{k+1}^+ = \varphi P_k^+ \varphi^T + Q_k \tag{9.5b}$$

Since P_k^+ has all positive diagonal elements and $\phi P_k^+ \phi^T$ is quadratic in nature, the above expression reveals that the estimates obtained during the temporal updating process does not improve the confidence level of the solution but rather deteriorates it. So at the instant τ_{k+1} and before the next measurement is done, the a priori estimate of the state at $\tau = \tau_{k+1}^-$ is given by

$$S_{k+1}^- = [X_{k+1}^-, P_{k+1}^-] \tag{9.6}$$

This completes the temporal update part of the process.

The measurements are taken at τ_{k+1}. Let Z_{k+1} be the measured value and let r be the error in measurement with variance R. From the measurement equation, we know that:

$$Z_{k+1} = HX_{k+1} + r_{k+1} \tag{9.7}$$

Now, we already have an estimate of the value of X_{k+1}^-, which was obtained by time propagating the state variables of the previous instant. Additionally, we have its corresponding error variance, P_{k+1}^-, which has also obtained by time propagating P_k^+. This forms the a priori estimate of the state.

So, at this point, we have two sets of data at τ_{k+1}. One of them is from the a priori knowledge of the system state, X_{k+1}^- and its covariance P_{k+1}^-, while the other is the direct measurement, consisting of Z_k and R.

These two pieces of data are intelligently combined by the filter such that it gets a new estimate of the updated posterior state at τ_{k+1}^+ that is better in precision than each individual constituent. In the Kalman filter, this is done by a weighted combination of the two states in accordance to the relative confidence to get the refined state X_{k+1} at $\tau = \tau_{k+1}^+$.

Now leaving the particular case of Kalman filtering for the moment, if we concentrate on any estimation, in general, with a weighted combination of available estimates where the weight values are w_1 and w_2 for the two estimates X_1 and X_2, respectively, the updated estimate will be

$$\hat{X} = w_1 X_1 + w_2 X_2 \tag{9.8a}$$

However, the question is what value of w_1 and w_2 will give us the optimal estimate? A simplified way to derive the weights is to assume that the two states are available with corresponding error variances as σ_1 and σ_2. Since the two states are derived from independent sources, the errors are uncorrelated. Moreover, with w_1 and w_2, as the weights assigned to the two estimates for combination, the normalized weight factor warrants $w_1 + w_2 = 1$. So the weights can be represented as $w_1 = w$ and $w_2 = (1 - w)$.

Since the estimate is a linear combination of the individual estimates, so the resultant error will also be the same linear combination of the errors. Then, if ε_1 and ε_2 are the errors in the individual state estimates, the resultant error will be

$$\varepsilon_{\text{eff}} = w_1 \varepsilon_1 + w_2 \varepsilon_2 \tag{9.8b}$$

So the variance of the resultant error will be

$$\begin{aligned}
\sigma_{\text{eff}}^2 &= E\left[\varepsilon_{\text{eff}} \cdot \varepsilon_{\text{eff}}'\right] \\
&= E[(w_1 \varepsilon_1 + w_2 \varepsilon_2) \times (w_1 \varepsilon_1 + w_2 \varepsilon_2)] \\
&= \left[w_1^2 \varepsilon_1^2 + w_2^2 \varepsilon_2^2\right]
\end{aligned} \tag{9.9a}$$

As w_1 and w_2 are definite numbers and the errors in the two estimates are uncorrelated, expectation values of their mutual product are zero. So it gives the value of σ_{eff} as

$$\sigma_{eff}^2 = w_1^2\sigma_1^2 + w_2^2\sigma_2^2$$
$$= w^2\sigma_1^2 + (1-w)^2\sigma_2^2 \quad (9.9b)$$

The combination should yield maximum confidence and hence the σ_{eff} should be the minimum. Now, to minimize this quantity, we differentiate the above expression in Eqn (9.9b) with respect to w and equate to zero. So we get

$$2w\sigma_1^2 - 2(1-w)\sigma_2^2 = 0$$

or,

$$w\sigma_1^2 = (1-w)\sigma_2^2$$

or,

$$w = \sigma_2^2/(\sigma_1^2 + \sigma_2^2)$$

so,

$$(1-w) = 1 - \sigma_2^2/(\sigma_1^2 + \sigma_2^2) = \sigma_1^2/(\sigma_1^2 + \sigma_2^2)$$

Thus, we get the two weights as

$$w_1 = \sigma_2^2/(\sigma_1^2 + \sigma_2^2) \quad (9.10a)$$
$$w_2 = \sigma_1^2/(\sigma_1^2 + \sigma_2^2) \quad (9.10b)$$

Using these values of the weights, we get the estimation as

$$\widehat{X} = w_1X_1 + w_2X_2$$
$$= X_1\sigma_2^2/(\sigma_1^2 + \sigma_2^2) + X_2\sigma_1^2/(\sigma_1^2 + \sigma_2^2) \quad (9.11)$$

Further, the minimized value of the variance is obtained as

$$\sigma_e^2 = \sigma_2^4\sigma_1^2/(\sigma_1^2 + \sigma_2^2)^2 + \sigma_2^2\sigma_1^4/(\sigma_1^2 + \sigma_2^2)^2$$
$$= \left\{\sigma_2^2\sigma_1^2/(\sigma_1^2 + \sigma_2^2)^2\right\} \times (\sigma_2^2 + \sigma_1^2) \quad (9.12)$$
$$= \sigma_2^2\sigma_1^2/(\sigma_1^2 + \sigma_2^2)$$

Now, coming back to our specific problem, in the case of KF, we have the a-priori estimate $\widehat{X}_k^-$ with proper covariance value P_k^-. But the other information available is in the form of the measurement Z and not as a state variable, X. Then how can we combine the two and get the corresponding variance? That can be done in the way we will now describe.

From the measurement equation, we know $Z = h(X) + r$. For linear measurements, this becomes

$$Z = HX + r$$
$$X = H^{-1}Z + H^{-1}r \quad (9.13)$$

The error in the state estimate due to the measurement error "r" is $H^{-1}r$. Hence, the corresponding error variance is $E[H^{-1}r][H^{-1}r]^T$. This makes the error covariance in the estimates derived from only the measurements at instants τ_{k+1} to be $P^m_{k+1} = H^{-1}RH^{-T}$.

Therefore, if X^-_{k+1} and X^m_k are the state values at instant k obtained from a-priori estimate and from measurement, respectively; and if P^-_{k+1} and P^m_{k+1} represent their respective error covariance values; then we can obtain the updated estimate X^+_{k+1} derived from the two individual estimates using Eqn (9.11) as

$$
\begin{aligned}
X^+_{k+1} &= \left(X^m_{k+1}P^-_{k+1} + X^-_{k+1}P^m_{k+1}\right)/\left(P^-_{k+1} + P^m_{k+1}\right)\\
&= \left[X^m_{k+1}P^-_{k+1} + X^-_{k+1}H^{-1}RH^{-T}\right]/\left[P^-_{k+1} + H^{-1}RH^{-T}\right]\\
&= \left[X^-_{k+1}P^-_{k+1} + X^-_{k+1}H^{-1}RH^{-T} - X^-_{k+1}P_{k+1-} + X^m_{k+1}P^-_{k+1}\right]/\left[P^-_{k+1} + H^{-1}RH^{-T}\right]\\
&= \left[X^-_{k+1}\left(P^-_{k+1} + H^{-1}RH^{-T}\right) + \left(X^m_{k+1} - X^-_{k+1}\right)P^-_{k+1}\right]/\left[P^-_{k+1} + H^{-1}RH^{-T}\right]\\
&= X^-_{k+1} + \left\lfloor\left(X^m_{k+1} - X^-_{k+1}\right)P^-_{k+1}\right\rfloor/\left\lfloor P^-_{k+1} + H^{-1}RH^{-T}\right\rfloor
\end{aligned}
$$

After this, we do a few manipulations, first by right-multiplying both the numerator and denominator of the second term in the above expression by H^T and then left-multiplying both by H, and this becomes

$$
\begin{aligned}
X^+_{k+1} &= X^-_{k+1} + \left[H\left(X^m_{k+1} - X^-_{k+1}\right)P^-_{k+1}H^T\right]/\left[HP^-_{k+1}H^T + R\right]\\
&= X^-_{k+1} + \left[\left(HX^m_{k+1} - HX^-_{k+1}\right)P^-_{k+1}H^T\right]/\left[HP^-_{k+1}H^T + R\right]
\end{aligned}
$$

Again using the measurement equation $Z = HX + r$ and making use of the definite part as the expected value of it, we get

$$
X^+_{k+1} = X^-_{k+1} + \left(Z_{k+1} - HX^-_{k+1}\right)P^-_{k+1}H^T/\left[HP^-_{k+1}H^T + R\right] \tag{9.14}
$$

$(Z - HX_{k+1})$ is the amount of new information available to the filter for the measurement Z. It is called the innovation. The expression $P^-_{k+1}H^T/[HP^-_{k+1}H^T + R]$ is called the Kalman gain, K. It represents the portion of the innovation that should be added to the a-priori estimate to yield the optimal posterior result. The corresponding error variance after update can be written using Eqn 9.12 as

$$
P^+_{k+1} = \left[P^-_{k+1}H^{-1}RH^{-T}\right]/\left[P^-_{k+1} + H^{-1}RH^{-T}\right]
$$

Doing similar manipulations as above, we get

$$
\begin{aligned}
P^+_{k+1} &= \left[HP^-_{k+1}H^{-1}R\right]/\left[HP^-_{k+1}H^T + R\right]\\
&= P^-_{k+1}R/\left[HP^-_{k+1}H^T + R\right]\\
&= P^-_{k+1}\left(HP^-_{k+1}H^T + R - HP^-_{k+1}H^T\right)/\left(HP^-_{k+1}H^T + R\right)\\
&= P^-_{k+1}(1 - HK)
\end{aligned} \tag{9.15}
$$

The same expression for X^+_{k+1} can be obtained with a different approach. The measured and the estimated states can be written as equations in matrix form, with X^+_{k+1} as unknown

$$
\begin{aligned}
Z_k &= HX^+_{k+1} + r\\
X^-_{k+1} &= IX^+_{k+1} + p
\end{aligned} \tag{9.16a}
$$

where p is the error in the a-priori estimation with covariance, P_{k+1}^-. For convenience, we use the simple notation P in place of P_{k+1}^- in the derivations below

$$\begin{pmatrix} Z_{k+1} \\ X_{k+1}^- \end{pmatrix} = \begin{pmatrix} H \\ I \end{pmatrix} X_{k+1}^+ + \begin{pmatrix} r \\ p \end{pmatrix}$$

(9.16b)

or, $\quad b = A X_{k+1}^+ + m$

where $A = [H^T \ I]^T$, $b = [Z_k^T \ X_{k+1}^{-T}]^T$ is the matrix of the available parameters $\widehat{X}$ and Z at the beginning of the temporal update, and $m = [r \ p]^T$ represents the error portion of the equation. So, considering the covariance of m as $M = E[m \cdot m^T]$, the optimal solution for X (Strang, 1988) is given by

$$X_{k+1}^+ = (A^T M^{-1} A)^{-1} A^T M^{-1} b$$

(9.17)

Replacing the values of m in M and considering r and p are uncorrelated, we get

$$M = E\{[rp]^T [rp]\}$$

$$= E \begin{bmatrix} r^2 & rp \\ pr & p^2 \end{bmatrix}$$

$$= \begin{bmatrix} R & 0 \\ 0 & P \end{bmatrix}$$

(9.18)

$$\text{or,} \quad M^{-1} = \begin{bmatrix} R^{-1} & 0 \\ 0 & P^{-1} \end{bmatrix}$$

Replacing Eqn (9.18) in Eqn (9.17)

$$X_{k+1}^+ = \left([H^T \ I] \begin{bmatrix} R^{-1} & 0 \\ 0 & P^{-1} \end{bmatrix} [H^T \ I]^T \right)^{-1} \left([H^T \ I] \begin{bmatrix} R^{-1} & 0 \\ 0 & P^{-1} \end{bmatrix} \right) [Z_k^T \ X_{k+1}^{-T}]^T$$

$$= \left([H^T R^{-1} \ P^{-1}] [H^T \ I]^T \right)^{-1} \left([H^T R^{-1} \ P^{-1}] [Z_k^T \ X_{k+1}^{-T}]^T \right)$$

$$= (H^T R^{-1} H + P^{-1})^{-1} (H^T R^{-1} Z_k + P^{-1} X_{k+1}^-)$$

$$= [H + RH^{-T} P^{-1}]^{-1} [Z_k + RH^{-T} P^{-1} X_{k+1}^-]$$

$$= [HPH^T + R]^{-1} [Z_k PH^T + RX_{k+1}^-]$$

$$= (Z_k - HX_{k+1}^-) PH^T / [HPH^T + R] + (HX_{k+1}^- PH^T + RX_{k+1}^-) / [HPH^T + R]$$

$$= [X_{k+1}^- (HPH^T + R) + (Z_k - HX_{k+1}^-) PH] / [HPH^T + R]$$

$$= X_{k+1}^- + (Z_k - HX_{k+1}^-) / [HPH^T + R]$$

(9.19)

Comparing this equation to Eqn (9.14), we can find that it leads to the same expression for X. This matrix solution is nothing but the projection of the true values of X onto a plane spanned by the Z vector. So another way to visualize the same is to assume the n-state system as an n-dimensional vector in a vector space and to interpret the same result geometrically.

This ends the measurement update process. At the end of one such iteration of a temporal update and subsequent measurement update, we get the updated state values X_{k+1}^{+} at time instant τ_{k+1}^{+} and its corresponding error covariance P_{k+1}^{+}.

In this way, the KF function starts from the estimate of the initial values of the state variables and then recurs through the cycles of measurement and temporal updates, as shown in Figure 9.4, to find the updated estimates of the state in a continuous manner. The process, improves the estimates on successive cycles and ultimately converges to the optimal values of the state variables. The time intervals between successive iterations may be the same or different from cycle to cycle. Accordingly the state transition matrix φ has to be established and the Q values may have to be changed, if necessary.

One important thing to notice here is the fact that the error covariance improves in a measurement update as it is evident from Eqn 3.15. We have already seen that the variance deteriorates during the temporal update. So for the solution to converge, the improvement in the measurement update must overcompensate the amount of deterioration during the temporal update.

We have only described here the working of the conventional linear Kalman filter. However, the filter may be modified to work even for nonlinear cases and also for conditions when the noises are not Gaussian.

Box 9.1 shows the use of a MATLAB program for Kalman filtering.

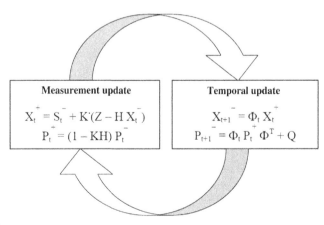

FIGURE 9.4

Iteration cycles of a Kalman filter.

BOX 9.1 KALMAN FILTERING

The MATLAB program Kalman.m was run to obtain the true value of a source voltage when its measurements were corrupted by noise of Gaussian random nature with $\sigma = 1$. The figure below represents the true voltage at 5 V. The measured samples are shown as (X) and the Kalman estimation of the true voltage is shown as diamonds ($\diamond$). Note the convergence of the estimates (Figure M9.1).

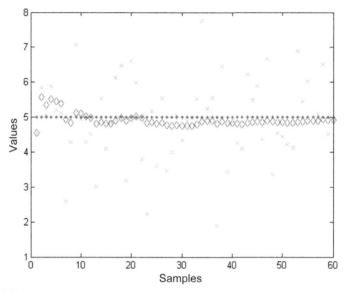

FIGURE M9.1

Estimations in a Kalman filter.

   Run the program with different values of σ and for different sample lengths and observe the variation. Put in some finite value of Q and also notice the difference.

9.1.4 **Use of the Kalman filter**

9.1.4.1 *Precision estimation in a Sat-Nav receiver*

The estimation of the position in a navigational receiver may be improved by using a Kalman filter. This is a proven technique and needs only proper implementation for the purpose of precise position estimation in the navigation receiver.

   A satellite navigation receiver will provide the PVT information to the users. So, the standard errors expected in such receivers are also present here. The objective of the Kalman filter will be to use the variables available in the receiver and by means of the filter operations reduce the associated errors to provide more accurate and precise values of the final PVT output along with a level of confidence on the estimation.

   A systematic approach may generate different alternative filter designs for such applications, each based on a particular set of system model. It is also recommendable to design simplified system models that retain the salient features of the states and its variations yet providing adequate estimation accuracy. In case of a navigation receiver, the system may be described through the observation equation that relates the pseudorange to the position coordinates and time offset. As the measured pseudorange is related nonlinearly to these state variables, an extended Kalman filter (EKF) model is appropriate here. The elements in the measurement matrix in such an EKF are obtained by the partial differentiations of the nonlinear measurement functions. So for an observation $Z = h(X) + r$, the measurement matrix H is given by

$$H_k = \partial h / \partial X | X = X_k \qquad (9.20a)$$

Partial differentiations, with respect to each of the states for a given measurement, will form the elements of a row in this Measurement Sensitivity Matrix, H. Similar row will be generated for different observations at the same instant of measurement update.

9.1.4.1.1 Choice of state variables

The output of the receiver will be the navigation parameters, PVT. So position S_k, velocity V_k, and clock bias B_k are obvious state variables of the filter. To obtain these values, the model also includes variables of one-order-higher derivatives. So the acceleration A_k and the clock drift D_k are also state variables to be considered here. However, this assumes that the derivatives of position from third order onwards and time beyond first order are either zero or negligibly small to impact the values of the parameters over the updating time. But the state can be extended to variables of higher orders, if necessary, for receivers with a higher order of dynamics. So the system state considered consists of are the following variables

$$
\begin{aligned}
&\text{1. Current Position} && S_k = [x \ y \ z]_k \\
&\text{2. Current Velocity} && V_k = [v_x v_y v_z]_k \\
&\text{3. Current Accelaration} && A_k = [\alpha_x \alpha_y \alpha_z]_k && (9.21) \\
&\text{4. Clock bias} && B_k = [b]_k \\
&\text{5. Clock drift} && D_k = [b']_k
\end{aligned}
$$

Thus, the complete state vector becomes $X_k = [S_k \ \ V_k \ \ A_k \ \ B_k \ \ D_k]$. It will constitute an $[11 \times 1]$ array to represent the dynamic state in a three-dimensional space.

9.1.4.1.2 Measurement equation

Here, the measurements are the corrected pseudoranges, ρ corrected for the ionospheric, tropospheric, and other available corrections. So

$$Z = \rho \qquad (9.22a)$$

Since at any instant the number of measurements depends upon the number of visible satellites, the size of Z is variable. All visible satellites can be used for the purpose with required cutoff in elevation angle.

The measurement done by the receiver, i.e. the pseudo range, which has a quadratic relationship with the state parameters. The measurement model equation is

$$Z = \rho$$
$$= \sqrt{(x_s - x_r)^2 + (y_s - y_r)^2 + (z_s - z_r)^2} + b_k + r \qquad (9.22b)$$
$$= h(S) + r \qquad (9.22c)$$

where h(.) is the nonlinear function for measurement model and r is the residual range error assumed to be normally distributed as N(0, R). Other notations carry their usual meanings.

From this equation, using Eqn (9.20), the observation matrix H may be obtained as the partial differentiation of h with respect to the individual state variable

$$H = dh/dX \Big|_{X^-(k)} \qquad (9.23a)$$

Or:

$$H = \left[\frac{\partial h}{\partial x} \; \frac{\partial h}{\partial y} \; \frac{\partial h}{\partial z} \; \frac{\partial h}{\partial v_x} \; \frac{\partial h}{\partial v_y} \; \frac{\partial h}{\partial v_z} \; \frac{\partial h}{\partial a_x} \; \frac{\partial h}{\partial a_y} \; \frac{\partial h}{\partial a_z} \; \frac{\partial h}{\partial b} \; \frac{\partial h}{\partial b'}\right]$$

Since the pseudo-range is only a function of position and clock bias and is independent of velocity, acceleration, or clock drift, we get

$$H = [\; -(x_s - x_r)/l \; - (y_s - y_r)/l \; - (z_s - z_r)/l \; \; 0 \; \; 0 \; \; 0 \; \; 0 \; \; 0 \; \; 0 \; \; 1 \; \; 0]$$
$$(9.23b)$$

where $l = \sqrt{\{(x_s - x_r)^2 + (y_s - y_r)^2 + (z_s - z_r)^2\}}\Big|t_k^-$. So we find that for the esti-

mation of H, both the satellite coordinates and the receiver positions must be known for the instant. The satellite coordinates are known to the receiver from the ephemeris. But, what should we use for receiver position coordinates when we are only trying to derive the same? We shall use here the a-priori knowledge of the receiver position that it has at the moment of measurement update τ_k^- before the position got refined by the process.

9.1.4.1.3 Process equation
Although the measurement equation is nonlinear, the process equation may be expressed as a linear relation between the current variables and the same at the next instant. So using Eqn (9.2b), we get

$$X_{k+1} = \varphi_k \cdot X_k + w_k \qquad (9.24a)$$

Where, φ_k is the linear relation for state dynamics and is taken as

$$\varphi = \begin{pmatrix} I(3) & dt.I(3) & 0.5.dt^2.I(3) & 0 & 0 \\ 0 & I(3) & dt.I(3) & 0 & 0 \\ 0 & 0 & I(3) & 0 & 0 \\ 0 & 0 & 0 & 1 & dt \end{pmatrix}$$

where I(3) is an identity matrix of dimension 3 and w is the dynamic model error normally distributed as N(0, Q).

9.1.4.1.4 Kalman filter process

The Kalman filter process starts from the point of initiation with an initial guess of the system state X_0 and its covariance, P_0. Then the standard algorithm is followed to get the evolution of the states through measurement and temporal updates. Following Eqns (9.4) and (9.5b), the temporal updates become

$$\text{State propagation} \quad X_k^- = \varphi_k X_{k-1}^+$$

$$\text{Error covariance propagation} \quad P_k^- = \varphi_k P_k^+ \varphi_k + Q_k \tag{9.25a}$$

Then, after the measurements are done, the measurement update is carried out according to the equation of the states and corresponding covariance

$$\text{State update} \quad X_k^+ = X_k^- + K_k[Z_k - h_k(X_k)]$$

$$\text{Error covariance update} \quad P_k^+ = [I - K_k H_k] P_k^- \tag{9.25b}$$

Here the Kalman Gain is given by $K_k = P_k^- H_k^T [H_k P_k^- H_k^T + R_k]^{-1}$. Thereafter, this iterative progress continues updating the values of X and P over time and measurements to obtain continually improving states of the system.

An important thing to observe is that with the extended Kalman filter, the Kalman gain K, being a function of the observation matrix H, is dependent upon the current state values. This is because H is defined by the derivatives of h derived at the current state values. However, for a linear case it was a constant independent of the state values. So for nonlinear cases, the covariance update, involving K during the measurement update process, cannot be done without the knowledge of the current state variable values, which is possible in the case of a linear system.

Box 9.2 shows the use of a MATLAB program to improve point estimation using the Kalman filter.

9.1.4.2 Integration of INS and Sat-Nav

In this section, we will learn a few techniques that intelligently combine the satellite navigation solution with an Inertial Navigation System (INS) using a Kalman filter. As a result, the individual systems work in a synergistic way in the combined form to yield better performance in terms of accuracy, precision and availability.

The endeavor to integrate different types of navigation systems started with the Global Positioning System (GPS), to harness the advantages of different systems to effectually obtain a navigation solution with enhanced performance. However, theoretically it is also possible to be done with other systems. So here we take a general view of such integration that need no special prerequisites of any definite system.

BOX 9.2 PRECISE POSITION ESTIMATION

The MATLAB program Kalman_Pos.m was run to understand the working and the improvement obtained by using the Kalman filter. Data were simulated from the visible satellites, using the "true" GPS navigation file from a receiving station named IISC at Bangalure in India. for an arbitrarily selected date. The variation of the satellite positions with time was derived from the files. The true position of the receiver is known a priori. Therefore, the true geometric ranges are calculated from them and added with zero-mean Gaussian measurement noise to create the pragmatic conditions. The standard deviation of the noise added to the ranges is 5 m (1σ).

A single-point positioning (SPP) estimator was also developed to compare the position estimations. Standard SPP-based estimation was done along with Kalman filter-based estimations, and the results were compared. Figure M9.2 shows the comparison of position estimations using the Kalman filter. The figure clearly indicates the improvement in terms of geographical position when the Kalman filter has been used.

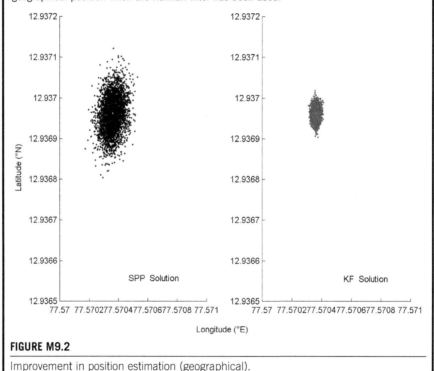

FIGURE M9.2

Improvement in position estimation (geographical).

Here, we will only mention how and where a Kalman filter can be used for the purpose. So in the following section, first we will point out the complementary nature of the two systems. Then an overview on the different integration strategies and configurations will be mentioned, with emphasis on the loose and tight architectures.

9.1.4.2.1 Integration goals

The basic motivation behind the integration of any two navigation systems remains the improvement of their combined performance as the two systems can perform in a synergistic manner. The integration of an INS and Sat-Nav has no different intentions as far as the objectives of integration are concerned. This enables the combined system to offer increased accuracy, precision, availability, and continuity (Grewal et al. 2001). In other words, it maintains a specified level of navigation performance, even during the outage of one of the systems.

We have come to understand that Sat-Nav has some characteristic drawbacks. Position estimates are typically made using the range measured with a code-based ranging technique in a commercial receiver. The code-based ranging is noisy, and hence the positions derived from them have very poor precision, although the accuracy of the estimates may be good enough. Further, it has a stringent requirement of visibility of four good satellites to obtain an accurate solution.

On the other hand, an INS, working on the principles of dead reckoning using the output data from the accelerometer and gyroscope, gives the values of instantaneous position and orientation, i.e. condition in six degrees of freedom. However, these results are very much prone to the bias of the instruments. Commercial instruments show relatively high bias after a considerable time in operation. Hence the positions derived from an INS have poor accuracy, although their precision may be much better than the Sat-Nav derived values. So for the INS, the short-term errors are relatively small, but they degrade rapidly and are unbounded over time; external aid is necessary to maintain their performance (Farrel, 2008). In an integrated system, the accuracy of a Sat-Nav system may be combined with the precision of an INS to offer improved performance.

We have already learned that it requires at least four good satellites to obtain a navigation solution with considerable accuracy in Sat-Nav. However, there are certain scenarios in which this criterion is not fulfilled. For example, in an urban canyon with tall skyscrapers surrounding the users, it is difficult to get four satellites, well dispersed in the sky to give a good Geometric Dilution of Precision (GDOP). Again, consider the case of banked turns of aircraft. During such situations, the navigation antenna, which is typically mounted on the back of the aircraft, remains directed toward one corner of the sky, and then it becomes difficult to satisfy this requirement. Further, in the equatorial region, there is severe scintillation in the signal due to the rapid variations in the ionospheric density in a local scale. This causes rapid fluctuations in the received signals and very frequent loss of lock of the receivers with very low-powered signals.

INS equipment, being passive and self-contained, is not sensitive to such outages. Therefore, under such conditions, the INS may take over and the integrated system may continue to offer solutions without much degrading its estimation accuracy and also offering enhanced continuity of position solution even in the presence of severe vehicle dynamics.

The typical estimation rate of a Sat-Nav receiver is about 1 s. In addition to that, the initial time to first fix (TTFF) for a typical receiver is quite high. For certain applications, this rate is not sufficient. For such applications, an integrated system may use the combined solution at the estimation rate of the Sat-Nav while continuing to offer solutions using the dead-reckoning feature of the INS for instants in between. Thus, it provides effective position outputs at a rate higher than conventional Sat-

Nav. The short-term data of the INS are fairly accurate and available at a very high rate. It effectively extrapolates the position locus derived from satellite navigation between the two successive updates of the latter, with the Sat-Nav offering the long-term bias correction aids to the INS at the instant of its updates. The solution accuracy will degrade for these intermediate instants, but the amount will be trivial compared to the large leverage that will be achieved in terms of performance of service.

Thus, Sat-Nav and INS, having complementary performance characteristics in terms of position estimation, can be integrated to improve the overall navigation performance.

9.1.4.2.2 Integration architecture

The constituent elements of the combined system are the basic navigation receiver, the inertial measurement units like accelerometer and gyroscope providing linear and angular accelerations, respectively, and a processor that performs the integration work. This processor typically contains a Kalman filter, which combines the parameters available from the two units using a preassigned algorithm.

The navigation receiver may work as a complete unit providing the position solutions for integration, or the integration can occur in its modular level, too. The "depth" of the integration depends upon the integration algorithm. For the INS, inertial sensors with moderate accuracies can be used as part of integrated navigation systems to perform even in difficult environments.

The extended Kalman filter is used to merge the satellite and inertial information. Based on the integration strategy adopted, the individual system units are either independent or dependently coupled. Independent coupling occurs when the integration is done by just combining the information separately derived, while dependent coupling adopts the strategy of sharing the information between the two participating systems. The integration architecture can be described as being made up of different interrelated units.

There are different standard modes of integration. The more common strategies (Petovello, 2003; Greenspan, 1996) are

- Uncoupled integration
- Loosely coupled integration
- Tightly coupled integration

But instead of starting with the different integration modes, we will first review the different elements of integration, including the information generated at different sections and resources available at different units that can be utilized for integration.

In a Sat-Nav receiver, the section where the position information is generated and used is the navigation processor (NP). The resident Kalman filter of the NP uses position, velocity, and acceleration as the state variables along with clock bias and drift. These state variables outputs are available for utilization for integration. This filter typically uses range as its input. However, it can accept individual state variables measurements from other independent sources as additional input for further bettering its estimates.

The tracking loop of the receiver is also a resource element where the information regarding the receiver dynamic can be utilized. The phase-locked loop may be aided with the velocity estimation from other external sources to identify the Doppler shift and hence can track the signals with better precision as well as accuracy.

In addition, the measurement unit of an INS system is generally equipped with the facility of strapping down the estimates, where the INS estimates can be forcefully modified to any definite value by application of correcting bias derived from external independent information. This helps the bias-prone estimates of the INS measurement units to reduce error.

Thus, the elements of integration in an INS and the possible information to exchange are primarily the measurement unit of the INS, consisting of the accelerometer and gyroscope and their output; and also its strap-down facility for feedback. For a Sat-Nav receiver, the correlator, estimating the Doppler, and the navigation processor, which houses the basic navigation Kalman filter, participate in the integration. With this basic knowledge of the candidate components, we are now in a position to describe the different modes of integration.

1. Uncoupled mode. In this mode of integration, Sat-Nav and INS function independently as stand-alone units. Here the integration processor only combines the individually derived positional data according to the predefined algorithm. The integration is typically a Kalman-based assimilation of the two data sources. The estimation confidence provided by the KF at the NP, and the specified measurement error at the INS, constitute the required error covariance for the filter. As we have seen when discussing the theory of the Kalman filter, the resultant error variance of the posterior estimate is better than the individual ones and it reduces until the convergence. It is the simplest form of all possible integration modes. Moreover, it is to some extent failure tolerant, in the sense that here each unit functions independently of and transparent to the other. So even if one unit fails, the integration process can continue with only the other. However, in the case of the failure of one of the systems, the integrated system performs equivalent to the surviving system along with its respective individual limitations.
2. Loosely coupled mode. In this mode of integration, the Sat-Nav and the INS unit still estimate position independently. But here the data are integrated in an integration processor, and the information produced therein is fed back to the individual units for corrective actions.

We have already learned that the inertial sensors have high propensity for bias after working for a certain period. Therefore the units are provided with strapping-down facilities, through which the position and velocity values derived by the dead-reckoning system can be calibrated with respect to any reference value. Using this facility, the solution of the integrated system is used to correct the bias of the INS system. This facilitates the system by providing subsequent solutions of higher accuracy, thus improving the overall performance.

Similar feedback data paths exist to different levels of the Sat-Nav receiver. Recall that the Sat-Nav receiver estimates the Doppler in the signal and removes it in the receiver loops for proper locking. This activity may be aided by the velocity and position information obtained from the integration processor as feedback. Moreover, the Kalman filter within the navigation filter of the Sat-Nav receiver has acceleration as a state, along with position and velocity. The measurement is done only on the range from which the states are derived through a nonlinear observation equation. In an integrated system, the position velocity, and acceleration derived by the integrated processor can be used as a separate measurement to do the measurement updating at the Kalman filter through a unitary observation matrix. The navigation solution derived by the Kalman filter in the navigation processor, thus aided by the feedback information from the integrated solution, produces better results. This in turn improves the state estimation in the final integrated system.

It is important to mention once more that here the feedback information flows from the integrated processor to the individual systems and no direct interface between the participating systems exists. Like the uncoupled mode, in loosely coupled mode, the units work independently, and the integration is also a robust approach. Therefore, the INS and Sat-Nav systems operating separately can continue to provide a navigation solution even if one system fails (Godha, 2006).

3. Tightly coupled mode. In tightly coupled mode, the functional elements of the two units, instead of working as separate individual systems, rather act as the components of an integrated estimation system. The measured parameters generated at each of such elements are assimilated in the system integration processor to derive a single navigation solution.

Here the feedback path exists from the INS acceleration sensors to the Sat-Nav receiver's tracking loop. With the short-term performance of the INS, it provides the latter with more precise velocities than the satellite navigation output for better Doppler correction. Further, the Sat-Nav estimations are also used to remove the INS bias. Other dependences accommodated in the integration processor result in better accuracy and reduced random noise in solutions even at higher update rates.

This mode of integration offers high tolerance to dynamics of the receiver. As the INS velocity solution is fed back to the receiver tracking loop, the variation is well compensated and hence it reduces the requirement of tracking loop bandwidth. This in turn reduces the noise. However, unlike the other two modes, here the two systems working as one unit fail completely when either one of the units fails to perform.

It is relevant to mention here that in both the coupled modes, the tracking loop of the navigation receiver is generally implemented with a Kalman filter. This makes the system work better even in the presence of large tracking noise and thus improves the subsequent range estimates. The feedback acts as an additional input during the measurement updates, expediting convergence and improving performance. Thus, it creates a synergy between the systems.

Figure 9.5 shows the different configurations of integration between the INS and the satellite navigation system in terms of the individual elements and the feedback paths.

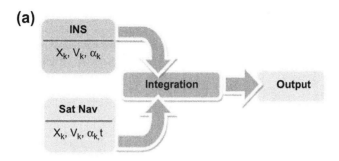

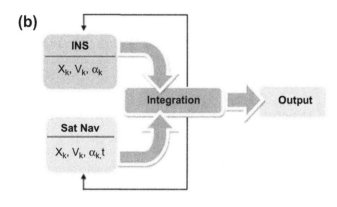

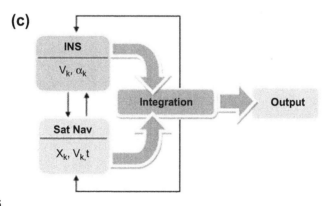

FIGURE 9.5

Configurations of integration in (a) uncoupled, (b) loosely coupled, and (c) tightly coupled modes.

Finally, to conclude this section on Kalman filtering, we can say that due to the generic nature of the filter and adaptive options of its design, the Kalman filter is very popular in satellite navigation systems. Although we have discussed it in relation to the receivers only, it finds its applications in other segments of the system as well.

9.2 The ionosphere

The ionosphere is one of the major issues for satellite navigation systems. This is due to the fact that the measurement principle of the satellite ranges in such systems, from which the positions are derived, is based upon the propagation time of the signals, and the ionosphere affects that in a conspicuous manner. We have already seen in Chapter 7 how the ionosphere adds delay to propagating radio signals used for navigation. Here, we will read in more detail about the structure of the ionosphere and its prominent characteristics. As the ionosphere poses a large threat to navigation applications, especially those meant for critical uses, it is important to understand its nature in a comprehensive fashion.

9.2.1 Basic structure of the ionosphere

The neutral molecular density in the atmosphere at heights more than 50 km from the earth surface is very rare, and it further decreases with height. Due to this fact, the sunlight coming from above remains almost unattenuated at these levels, with energy enough to dissociate these molecules into charged ions and free electrons. Moreover, due to the triviality of the existing particle density, the ions thus created spend a considerable lifetime before they recombine to vanish again. Thus, a region is created at these heights between 50 and 1000 km above the earth's surface where the atmosphere contains natural ions and is known as the ionosphere. The ionization, thus resulting from the photodecomposition of the thin upper atmospheric gases by the radiation from the sun produces electron—ion pairs This ionization process remain in a dynamic equilibrium with the recombination process to offer a definite electron or ion density, creating the ionosphere. The density of neutral atoms decreases exponentially with altitude, whereas the incoming solar radiation intensity decreases exponentially in the opposite direction as it is absorbed by the neutral particles. This gives the ionosphere a finite range of extent, with a definite peak as shown in Figure 9.6. The lower limit is defined by the penetration capacity of the high-energy solar radiation and the upper limit restricted by the availability of the neutral atoms for dissociation.

The ionization production function "q" is given by the well-known Chapman function (Chapman, 1931), expressed in terms of the reduced height "z" as

$$q(z) = q_0 \exp[1 - z - \exp(-z)\sec\chi] \qquad (9.26)$$

Here, $q(z)$ is the production rate of the ions at the reduced height z. The reduced height z is the representation of the true height h, relative to the peak height h_p under the vertical sun and normalized by the scale height H, i.e. $z = (h - h_p)/H$. The value q_0 is the production rate at the peak height, while χ is the solar incidence angle at the point in question.

The electron density is dependent upon the equilibrium due to this production and the loss due to recombination. For lower heights, where the neutral particles are found in abundance, the loss factor is proportional to the square of the electron density, N. Hence, if α is the constant of proportionality at equilibrium when the production and the losses are equal

$$q(z) = \alpha N^2$$
$$\text{or,} \quad N(z) = \sqrt{q(z)/\alpha} \tag{9.27a}$$

For higher altitudes, the loss due to recombination is linear with the electron density. Hence, with β as the proportionality constant for equilibrium,

$$q(z) = \beta N$$
$$\text{or,} \quad N(z) = q(z)/\beta \tag{9.27b}$$

At all heights in between, the loss varies intermediately between the linear and quadratic function of N, and the equilibrium density is hence accordingly determined (Rishbeth and Garriott, 1969). The variation of the ionospheric profile with height is shown in Figure 9.6. The peak production value is found to lie at an altitude of about 350 km. This height and the peak ionization value varies with solar

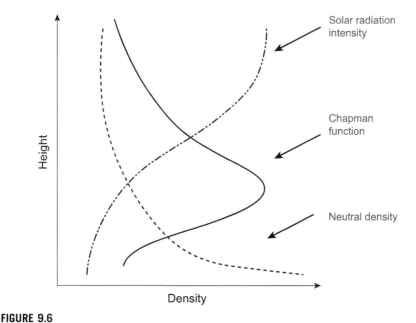

FIGURE 9.6

Vertical profile of the ionosphere.

zenith angle, χ. However, as the radiation is not constant, the earth is rotating and also there are physical transport of the ions, the electron density is variable both spatially and temporally. The distribution and dynamics of this ionospheric plasma are both dependent upon a complex combination of the neutral atmosphere, solar heating, photoionization, electrical conductivity, and neutral winds, all interacting with the magnetic field of the earth.

The structure of the ionosphere is horizontally stratified into layers of various ion composition and structure. Lower layers are ionized by highly penetrating components of the solar radiation, like hard X-rays; while in the upper layers, the soft X-rays or extreme UV component of the sun's rays do the ionization (Klobuchar, 1996). These ionospheric layers have been given the designations of D, E, F1, and F2 regions, from the earliest ground-based probing using radio-wave reflections. The peak production of plasma occurs in the F region, from 400 to 600 km. With a typical electron density of $\sim 10^{12}/m^3$ near the peak, this varies with time of day, season, solar activity periods, and for various other reasons. At altitudes above the peak, the ionospheric density decays monotonically with increasing altitude. Below this peak, the density at the lower portion of the ionosphere varies greatly from day to night, when the D and E regions of the ionosphere essentially disappear when production stops at sunset whereas the F1 and F2 layer combine to form a common F layer.

Characteristically, the ionosphere varies drastically with time and location. The total global distribution of the ionosphere may be divided into three distinct regions:

1. Equatorial region, extending up to $\pm 25°$ on both sides of the magnetic equator
2. Mid-latitude region, extending from latitudes $\pm 25°$ up to $\pm 65°$
3. High-latitude region, extending from above $65°$ up to the poles

Each of these regions has its own characteristics and has effects on propagating radio signals. However, considering few very important and interesting occurrences pertaining to navigation occurring in the equatorial region, and bearing in mind the brief scope we have to describe the ionosphere, we restrict our current discussion to the equatorial ionosphere and related phenomena.

9.2.2 Equatorial ionosphere

The equatorial region extends from the magnetic equator to about $\pm 25°$ geomagnetic latitude and is very dynamic in nature. This region has the highest ion production rate, as most of the solar flux is directly incident over this region. Moreover, it is very sensitive to variations in the geomagnetic conditions that control the movements of these charged particles. In addition to the severe irregularity, the equatorial region is characterized by the occurrence of a prominent phenomenon of equatorial anomaly. It has associated consequences of offering large and varying delay. In addition, the equatorial Spread-F phenomenon leads to severe scintillation to the propagating radio signals. Both of these phenomena affect the navigational signal propagating through this region, which will be discussed next.

9.2.2.1 Equatorial ionospheric anomaly

The ionized particles, i.e. electrons and ions, coexist in the ionosphere in the form of plasma. The ionospheric vertical profile of the electron density, being created by the solar influx, is expected to form higher density at locations with vertical solar flux, i.e. with zero solar zenith angles. But for the equatorial ionosphere, prominently during the equinoctial period when the sun remains vertically above this region, an anomalous behavior has been observed. It shows a trough of electron density at the equator with a dynamic enhancement in the density at a certain low latitude location. This occurs because the plasma is moved from the region of the magnetic equator to either side in a range of latitudes up to around 20° north and south of it. This is called the Equatorial Ionization Anomaly (EIA). This was first observed by Appleton, who showed the anomalous nature of the equatorial ionospheric electron density density distribution at noon with an F region depletion called the bite-out at the equator and peaks at around ±15° dip latitude (Appleton, 1946). Mitra (1946) was the pioneer to propose an explanation of the phenomenon, attributing the anomaly to the diffusion of the plasma from the equator toward the poles along magnetic lines of forces. Later, Martyn (1947) introduced the idea of equatorial vertical drift, and explained the anomaly in the light of the combined action of the vertical drift and diffusion. The general process by which this plasma movement occurs is now known and proved to be happening by the vertical drift of plasma followed by the diffusive movement toward the higher latitudes. Although much has been understood about this variation, definite functionalization of its day-to-day behavior is still lacking.

The development of the EIA depends on a complex interaction of a number of atmospheric processes and shows considerable variability with local time, longitude, and season. It sets off at the equator due to the transport of plasma by the vertical drift initiated by a natural phenomenon of Lorentz force acting upon the particles. The force is generated due to the movement of these charged particles aided by the neutral wind in the presence of the northward horizontal magnetic field. Both ions and electrons move in the upward direction, resulting in no net current. This drift thus lifts the charged plasma from the site of its abundance and puts it in a greater altitude. A second mechanism moves the raised plasma pole-ward from the equatorial region. The lifted plasma, now controlled by gravity-aided diffusion, moves along the magnetic field, sliding towards higher latitudes. The combination of vertical lift due to the drift and subsequent diffusion gives the plasma a fountain-like pattern and hence is called "fountain effect." It redistributes the electron density by transporting them from the equatorial region, resulting in a trough, and moving them toward higher latitude, where they are dumped, forming the crest around 15° to 20° latitude. This results in what is known as the equatorial anomaly (Figure 9.7).

The EIA exhibits day-to-day variability, and there are variations with seasons and solar activities as well. The daily variability affects both the extent of the anomaly, i.e. the latitudinal extent at which the anomaly crest is formed; and the strength of the anomaly, i.e. the peak density at crest with respect to the trough. The magnitude of this variability depends upon the strength of the controlling geomagnetic and solar parameters and the extent by which they affect the transport process of the plasma.

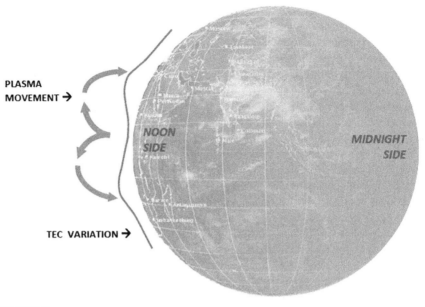

FIGURE 9.7

Plasma fountain effect and equatorial anomaly.

9.2.2.2 Equatorial spread F

Equatorial spread F (ESF) is a geonatural phenomenon in which the vertical profiles of the equatorial ionosphere are reshaped after sunset. A large vertical plasma density gradient developed in this region initiates the plasma instabilities responsible for equatorial spread F. It occurs on time scales ranging from seconds to hours and across a length ranging from centimeters to tens of kilometers.

With the sunset, the driving mechanism that maintains the vertical equilibrium condition of the ionosphere, holding denser layers over the relatively rarer ones, ceases to continue. The electron density hence now experiences instability and tends to get redistributed, with the plasma drooping downward due to the action of the gravity in a spatially random fashion, with different orders of magnitude and extent. This phenomenon is called the ESF, and causes random and abrupt penetration of high F-region density into lower-density regions adjacently below it as shown in Figure 9.8. The term "spread-F" was used because the F-region characteristics appeared to be spread over a finite height when probed through ground-based observations.

One of the distinguishing features of this phenomenon is the occurrence of local and small-scale plasma depletions known as ionospheric bubbles. They are small, confined regions with abrupt low electron density surrounded by relatively higher densities.

The plasma irregularities that occur in the F region also influence the performance and reliability of space-borne systems including navigation, causing

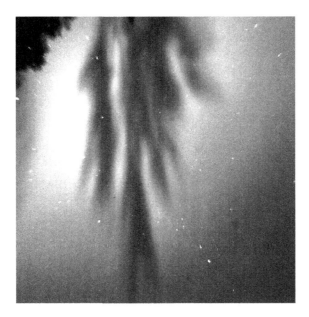

FIGURE 9.8

Equatorial spread F.

disruption to the satellite signals. It causes the waves propagating through the medium to experience a random local but rapid spatial variation of the electron density. It results in high scintillation, i.e. a rapid fluctuation in its amplitude and phase.

Thus, scintillations occur generally at night and normally prevail from after sunset up to midnight. The highest levels of scintillation are observed near the locations of the anomaly around the equatorial anomaly peak, where the vertical density gradient is the highest. Although it is not certain to occur every day even in equinoctial months, it is most probable during this time (Acharya et al. 2007).

The redistribution of electron density with occurrences of peaks and crests in an unpredictable manner in terms of locations and time not only causes variations in the delay of the navigation signals but also causes any definite model to fail. So models used for ionospheric delay compensations in satellite navigation perform badly in this region. Moreover, strong scintillation taking place at these locations causes loss of lock of the signals in the navigation receivers, which even in benign conditions experience very low received power.

9.2.3 **Models of the ionosphere**

We need to reiterate at this point that our main purpose here is to understand the effect of the ionosphere on navigation. So to keep the focus, we will only concentrate on those features of the ionosphere that have direct bearing on navigation systems.

Therefore, here we will only touch upon those aspects of models that have direct consequences to navigation.

In general, the ionospheric models can be divided into two main categories, the theoretical models and the empirical or semi-empirical models. Theoretical models utilize the physics behind the origin and distribution of the electron density to obtain the model output. They are most commonly developed considering physical processes including photochemical reactions, drifts, diffusion, collisions from neutral particles, winds, etc.

The empirical models, on the other hand, avoid the uncertainties of the theoretical understanding. They depend upon pragmatic data, partially or totally, to obtain the variational form of the electron density of the ionosphere. Due to their empirical nature, they are as good and representative as the data used for developing the models.

In addition, there are certain algorithms for near real-time estimation procedures of the ionospheric TEC, which are loosely termed models. However, they are actually algorithms in a strict sense.

Depending upon the portability of the models, they are either global or local in nature. Local ionospheric models are mainly developed to accommodate localized phenomena, and hence are mostly restricted to a definite zone.

Although all types of models have their respective importance with their merits and demerits, all are not relevant for navigational purposes. In addition to the requirement at the control segment, the navigational needs of the models are mainly at the receivers, where the computational resources and capacities are limited. For navigation receivers, we need some models that are described by very few inputs, computationally not intensive but still offering adequate accuracy of estimates. So after only mentioning the theoretical models, here we discuss some of the important global models of navigational interest and their salient features. Very popular theoretical models include the following (AIAA, 1998):

- Time-dependent ionospheric model (TDIM)
- Coupled thermosphere-ionosphere-plasmasphere model (CTIP)
- Global theoretical ionospheric model (GTIM)
- Sheffield University plasmasphere-ionosphere model (SUPIM)

These models give spatial electron density distribution as a function of location and time. They typically use inputs like neutral density, temperature, and wind. Some models also include drift pattern, electron energy model, neutral wind, and ion temperature as input in addition to other factors like solar activity, solar production rates, etc. Equations are thus formed for continuity, momentum, and energy, considering drift and diffusion to obtain the solution. Some models couple the ionosphere with the plasmasphere to derive the combined density behavior along geomagnetic flux-tubes, solving equations thereof.

9.2.3.1 Parameterized ionospheric model

Among the empirical models, the Parameterized Ionospheric model (PIM) is a popular global model for ionospheric density. In this model, the output from many theoretical ionospheric models, including TDIM and GTIM, are parameterized in terms of solar activity, geomagnetic activity, and season. The user inputs required by the model are location, time, and solar and geophysical conditions like F10.7 and K_p, etc. The output represents the ion density profiles, which are represented as linear combinations of empirical orthonormal functions. However, as it uses model output as data for modeling, its performance limitations include its inability to precisely represent the specific pragmatic situations like the ionospheric disturbances during geomagnetic storms.

9.2.3.2 International reference ionosphere

The International Reference Ionosphere (IRI) is an empirical ionospheric model providing the distribution of the electron density and other parameters in time and space. This model is recommended by the Committee on Space Research (COSPAR) and the International Union of Radio Science (URSI), and is based on a compilation of rocket and satellite data. Incoherent scatter data have been used for improved low-latitude performance. The model defines some specific profiles relevant to the real ionosphere and describes the electron density in terms of these functions. It uses global maps as inputs for certain parameters. These parameters can alternatively be provided by the user.

9.2.3.3 NeQuick

NeQuick is another empirical ionospheric model that uses solar parameters for estimation of the ionospheric electron density and therefore estimating delay or the total electron content. It is a four-dimensional ionospheric electron density model, which provides electron density in the ionosphere as a function of the position and time. It allows computation of ionospheric delays (TEC or STEC) as the integrated electron density along any ray path.

The input parameters of the model are the position (longitude, latitude, and height); the epoch, in month, date, and UT; and the solar activity, either expressed as F10.7 or R12. Alternatively, the NeQuick model may also be driven by the Effective Ionization Level, Az, as a surrogate that replaces the solar flux parameter and is also an empirical function of the location.

9.2.3.4 Klobuchar model

The Klobuchar model (Klobuchar, 1987) is the global parametric model for the TEC and is being used in navigational applications. Here, the total TEC is represented as a summation of half cosine terms with some prespecified quasistatic parameters. It is a parametric model, giving the vertical ionospheric delay δ in seconds of the day at any local time t given in seconds of day as:

$$\delta(t) = a + b \cos\left(\frac{t-c}{d}\right) \quad \text{when} \quad \left|\frac{t-c}{d}\right| < \pi/2$$

$$= a \quad \text{when} \quad \left|\frac{t-c}{d}\right| \geq \pi/2$$

The coefficients $a = 5 \times 10^{-9}$ and $c = 50{,}400$ are constants representing the constant bias and the time of peak, respectively. The parameters "b" and "d" are location-dependent and are derived using the geomagnetic longitude of the location and some specific coefficients that vary from day to day. The function indicates a constant delay during certain portions of local nighttimes. The variation is shown in Figure 9.9.

Box 9.3 shows the use of a MATLAB program for Klobuchar estimation.

9.2.4 Other methods of estimating the ionosphere

A dual-frequency receiver can directly estimate the delay using relative propagation time of the two frequencies from which the ionospheric TEC can be obtained. The method has been discussed in detail in Chapter 7 in connection to the correction of navigation ranging errors. The other methods of estimating the ionospheric TEC are by using Faraday's rotation of the signal polarization and by using Doppler.

TEC can also be derived from the effect that the free electrons have on the signal polarization in the presence of a magnetic field. Radio waves of the satellite signal propagating through the ionosphere experience a rotation in the direction of polarization as the effect of the interaction between the earth's magnetic field and the electric field of the signal. The free electrons of the plasma contribute to this rotation, which is proportional to the electron density, Ne, the magnetic field B, and

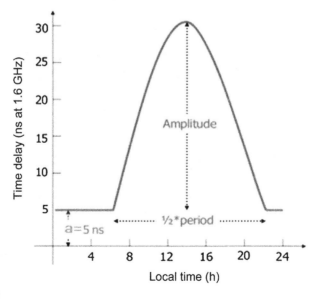

FIGURE 9.9

Variation of ionospheric delay in the Klobuchar model.

BOX 9.3 KLOBUCHAR ESTIMATION

The MATLAB program klobuchar.m was run to generate the following figure representing the Klobuchar estimation of ionospheric delay for a specific location. Notice the variation of the vertical delay amplitude with time and also the width of the half-period of the variation. The amplitude and the width vary from place to place (Figure M9.3).

Run the same program with different locations in terms of latitude and longitude and observe the changes.

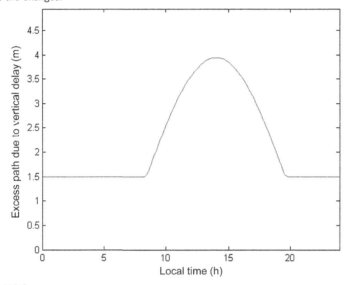

FIGURE M9.3

Vertical ionospheric delay from Klobuchar model.

the total path traversed through it. Thus, the integrated total rotation is proportional to the path integral of the electron density, i.e. TEC. The effect is more pronounced in comparatively lower bands than in those generally used for navigation. The total rotation of a signal of frequency f over a path dl with electron density Ne is thus given by:

$$d\beta = K/f^2 Ne \, dl \, B_{||} \qquad (9.29a)$$

where $B_{||}$ is the component of the earth's magnetic field parallel to the direction of the signal's propagation. As both $B_{||}$ and Ne vary along the path, the total rotation is given by:

$$\int d\beta = K/f^2 \int Ne \, B_{||} \, dl \qquad (9.29b)$$

The measurement of the total rotation is done by comparing the received polarization with respect to that supposed to be transmitted by the satellite. If we restrict our measurement over a path where $B_{||}$ may be considered to be constant, then:

$$\beta = \left(KB_{||}/f^2\right)\int N_e\,dl \tag{9.29c}$$

$$\beta = \left(KB_{||}/f^2\right)TEC \tag{9.29d}$$

Therefore, the rotation is dispersive in nature. So, similar to the excess range-based estimation of TEC, if signals of two different frequencies are transmitted in the same polarization, TEC may be obtained from the difference in their received polarization angle at the receiver. So if $\Delta\beta$ is the difference in the received polarization of the signals in frequency f_1 and f_2, then:

$$\begin{aligned} TEC &= \Delta\beta/\left[KB_{||}\left(1/f_1^2 - 1/f_2^2\right)\right] \\ &= \left(\Delta\beta/KB_{||}\right)\left(f_1^2 f_2^2\right)/\left(f_1^2 - f_2^2\right) \end{aligned} \tag{9.30}$$

A recent study has been made to use the Kalman filter to estimate in real time the relative ionospheric delay by using the Doppler data or alternatively the range rate (Acharya, 2013). In a navigation receiver, the measured ranges have additional excess equivalent path, over the geometric distance from the satellite, due to the delay in the ionosphere. In addition, it is also corrupted by the receiver clock bias. Therefore, the range rate carries the information of the differential ionospheric delay as well as the clock drift. So for all navigation receivers capable of measuring the range rate, the filter basically segregates the excess path due to ionospheric delay from that occurring due to clock shift.

If R_G is the true geometric range of the user receiver from the satellite, the total measured range R_M is:

$$R_M = R_G + r_i + r_b \tag{9.31}$$

Here, r_i is the path equivalent of delay due to ionospheric TEC and r_b is the path equivalent for the receiver clock bias and is given by $c \times \delta\tau$, where $\delta\tau$ is the clock bias. So differentiating the equation with respect to time, we get:

$$\dot{R}_M = \dot{R}_G + \dot{r}_i + \dot{r}_b \tag{9.32a}$$

Thus, $\dot{r}_b$ becomes $c \times \delta\tau'$, where $\delta\tau'$ is the clock drift. Rearranging the above equation, we get:

$$\dot{r}_i + \dot{r}_b = \dot{R}_M - \dot{R}_G \tag{9.32b}$$

Thus, the difference between the measured range rate and the true geometric range rate provides the sum of the ionospheric delay rate and the rate of change of clock bias, each converted to the rate of equivalent excess path. Utilizing this relation, the ionospheric delay has been derived using two Kalman filters.

Two Kalman filters, KF1 and KF2, are utilized in a closed loop as shown in Figure 9.10. The measured range R_M is a common input to both of the filters.

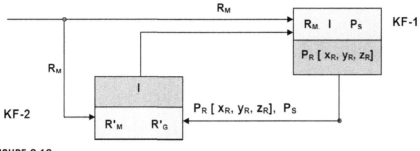

FIGURE 9.10

Kalman filter arrangement for TEC determination.

KF1 is typical position estimator found in a navigation receiver for precise position estimation. Initiated with some approximate ionospheric correction value, it obtains an approximate receiver position. The initiation value may be derived from any alternative information like an empirical model.

The range component due to ionospheric delay r_i and its first derivative $\dot{r}_i$ are two of the state variables of KF2. The other state variable of KF2 is $\dot{r}_b$, which is the excess path rate equivalent of the clock drift. So the augmented state vector for KF2 is given by:

$$S = [r_i, \ \dot{r}_i, \ \dot{r}_b]^T \tag{9.33}$$

The associated error covariance vector P is also required for the purpose. The states are initiated with approximate assumptions of their values and the corresponding error variances. Subsequently, the filter propagates the states forward in time through the filter processes of alternate measurement update and the temporal update (Maybeck, 1982).

This Kalman filter does the measurement update with the measurement equation given by:

$$\dot{R}_M - \dot{R}_G = \dot{r}_i + \dot{r}_b + v$$
$$\text{or,} \quad z = [0 \quad 1 \quad 1]S + v \tag{9.34}$$
$$\text{or,} \quad z = H.S + v$$

where $z = (\dot{R}_M - \dot{R}_G)$ and v is the uncorrelated normal zero mean measurement noise given by N(0, R). This forms the measurement equation and is used for the measurement update. The estimated receiver position along with the satellite position, obtained as an output of KF1, are used to derive the Euclidian geometric range R_G. The measured range R_M and the geometric range R_G thus obtained are numerically differentiated to obtain $\dot{R}_M$ and $\dot{R}_G$, respectively, and their difference $z = (\dot{R}_M - \dot{R}_G)$ is used as input during this update as given in Eqn (9.34). However, this value is noisy due to the obvious presence of errors in the measured range, satellite position errors, errors in the initial position estimation by KF1, and for using numerical methods for differentiation as well. KF2 removes this noise and finds out separately the states in S. It also updates the associated error covariance as per Eqn (9.15).

Then it also estimates the next value of r_i and other states in S during its temporal update phase. The state transition matrix is given by:

$$\Phi = \begin{pmatrix} 1 & dt & 0 \\ 0 & 1 & 0 \\ 0 & 0 & 1 \end{pmatrix} \tag{9.35}$$

And the temporal update is done using the system equation given by:

$$S_{k+1} = \varphi S_k^+ + u \tag{9.36}$$

Where φ is the state transition matrix and u is the process noise given by $N(0, Q)$. The error covariance is also simultaneously updated accordingly. The state updates are done according to Eqns (9.4) and (9.5b).

This updated value of predicted states for the next instant τ_{k+1} is S_{k+1}. Out of these, the value of $r_{i|k+1}$ is again fed to KF1 in the next cycle for ionospheric correction during its position estimation. So, with the continuing iteration in every cycle of the two Kalman filters in turn, the estimations are carried out with their accuracy improving to a limit.

The results are highly sensitive with the initialization value of the state ri in KF2. This is a reason why this method has been found to give better relative than absolute variations. Moreover, it has the limitation that the absolute delay values deteriorate with time. However, better absolute delays may be obtained when the algorithm is initiated with superior values and is calibrated with true values at regular intervals. So when there is an intermittent loss of one of the frequencies in dual-frequency receivers, this method may work as a fair substitute during those gaps when no direct ionospheric estimation information is otherwise available.

Initialization of other state variables and their corresponding error covariance is also important, as they affect the convergence time and accuracy of the filter. These two states may be initialized with zeros with large error covariance values. The technique is insensitive to the errors in position input even up to large errors. The large spatial correlation of the ionosphere is what makes the resulting errors negligible. Moreover, as the Kalman filter derives the first-order differentials out of the geometric range derived from the position estimates, the effects of the position offset errors get almost canceled in successive differencing. So it remains almost unaltered until large deviations occur.

Conceptual questions

1. With a noisy a-priori estimate of a state available, and with the state itself as the measured parameter with added noise, verify, using vectorial geometry, that the expression for the best estimate tallies with that obtained from measurement

update. Assume that the two noises are orthogonal. Also assume that the noise amplitude is the square root of its corresponding variance.

2. What happens when the true noise variance is greater than the assumed one in a Kalman filter?

3. What will be the effect on the estimate x_{k+1} when the measurement z_{k+1} is unavailable? How does the effect propagate in time?

4. State whether the two frequencies used for estimation of ionospheric TEC should be close by or widely separated to have better sensitivity.

5. Express the local production function in terms of normalized height with respect to the local peak.

References

Acharya, R., 2013. Doppler utilized kalman estimation (DUKE) of ionosphere. Advances in Space Research 51 (11), 2171−2180.

Acharya, R., Nagori, N., Jain, N., Sunda, S., Regar, S., Sivaraman, M.R., Bandyopadhyay, K., 2007. Ionospheric studies for the implementation of GAGAN. Indian Journal of Radio and Space Physics 36 (5), 394−404.

AIAA, 1998. Guide to Reference and Standard Ionospheric models. ANSI/AIAA, Washington, DC, USA. G-034.

Appleton, E.V., 1946. Two anomalies in the ionosphere. Nature 157, 691.

Chapman, S., 1931. The absorption and dissociative or ionizing effect of monochromatic radiation in an atmosphere on a rotating earth. Proceedings of the Physical Society of London 43 (26), 484.

Farrell, J.A., 2008. Aided Navigation: With High Rate Sensors. McGraw-Hill Publications, New York, USA.

Godha, S., 2006. Performance Evaluation of Low Cost MEMS-Based IMU Integrated with GPS for Land Vehicle Navigation Application (M.Sc. thesis), UCGE Report No. 20239, Department of Geomatics Engineering, University of Calgary, Canada.

Greenspan, R.L., 1996. GPS and inertial integration. In: Parkinson, B.W., Spilker Jr., J.J. (Eds.), Global Positioning Systems, Theory and Applications, vol. II. AIAA, Washington, DC, USA.

Grewal, M.S., Andrews, A.P., 2001. Kalman Filtering: Theory and Practice Using Matlab. John Wiley and Sons, New York, USA.

Grewal, M.S., Weill, L., Andrews, A.P., 2001. Global Positioning Systems, Inertial Navigation and Integration. John Wiley and Sons, New York, USA.

Klobuchar, J.A., 1987. Ionospheric time-delay algorithm for single-frequency GPS users. IEEE Transactions on Aerospace and Electronic Systems AES-23 (3), 325−331.

Klobuchar, J.A., 1996. Ionospheric effects on GPS. In: Parkinson, B.W., Spilker Jr., J.J. (Eds.), Global Positioning Systems, Theory and Applications, vol. I. AIAA, Washington, DC, USA.

Martyn, D.F., 1947. Atmospheric tides in the ionosphere − I: solar tides in the F2 region. In: Proceedings of Royal Society of London, A189, pp. 241−260.

Maybeck, P., 1982. Stochastic Models, Estimation and Control, vols 1&2. Academic Press, USA.

Mitra, S.K., 1946. Geomagnetic control of region F2 of the ionosphere. Nature 158, 668–669.

Petovello, M., 2003. Real-Time Integration of a Tactical-Grade IMU and GPS for High Accuracy Positioning and Navigation (Ph.D. thesis), UCGE Report No. 20173, Department of Geomatics Engineering, University of Calgary, Canada.

Risbeth, H., Garriot, O.K., 1969. Introduction to Ionospheric Physics. Academic Press, USA.

Strang, G., 1988. Linear Algebra and its Applications, Harcourt, Brace, Jovanovich. Publishers, San Diego, USA.

Applications

10

10.1 Introduction

Navigation services primarily offer position and time to their users. But how do the users utilize this information so that it effectively works to serve their purposes? No service is worthwhile if it is not used for some application. Rather, it is the applications that decide the performance targets for any offered service. It is no different for satellite navigation. Radio signal-based navigation started more than a century ago, in the last decade of the nineteenth century, when the signals from the coasts were used for sending the chronometer information to the ships adjacent to the shore.

Understanding Satellite Navigation. http://dx.doi.org/10.1016/B978-0-12-799949-4.00010-5

However, it is with the advent of the satellite navigation system that the service has taken on a global shape.

The main aim of this chapter is to make the reader aware of the extensive range of applications that are possible with satellite navigation and also to mention the huge volume of work that has been done in this regard. So, in this chapter, instead of concentrating on some specific applications only, we shall first discuss the current scenario of global navigation satellite system (GNSS)-based applications. This will be a roundup of the entire efforts that have gone into its evolution over the time frame. Our aim will be to elucidate the potential applications, showing how they benefit personal, societal, or scientific needs. However, we shall finish the chapter by describing a few selected applications with greater technical detail.

10.1.1 Advantages over other navigation systems

Like any other forms of navigation, the position, velocity, and time estimates obtained from the satellite navigation system can be efficiently used in a variety of applications. But in certain terms, this system has features more advantageous than the other forms of navigation. Here we list those features that are generally exploited to utilize the system, making it more suitable than others.

10.1.1.1 Message-based signal structure

The navigation receiver interacts only with the signal to derive all the possible information it needs to estimate its position. We have learned in Chapter 4 how, by using a three-tiered structure of the signal, all the key purposes like information dissemination, propagation time, and hence range estimation and protection of the signal, are simultaneously served. The messages incorporated into the signal not only provide the data for reference positioning but also aid the receiver in correcting actions, inform on status, or actuate different, related applications. For example, geo-tagged disaster warning information may be disseminated through the system message.

10.1.1.2 Satellite integrity information

In Chapter 2 we learned that the ground system of the satellite navigation system always keeps track of the dynamics of the satellites and also the signals they transmit to the users. So, in such systems there can be provisions to readily transmit any anomaly in the signal or constellation observed, through proper messaging. Thus, message-based navigation allows simultaneous transmission of a variety of useful information, including its own reliability index vis-à-vis disseminating the primary navigation parameters in the signal. This simultaneous availability of integrity information is an added advantage to the users of the satellite navigation system, as it enables them to select or reject a certain satellite based upon its performance.

10.1.1.3 Propagation effects on the signal

The satellite signals, while propagating through the medium between the satellites to the earth, experience the effects that the medium impresses upon it. From these signatures, in the realm of both time and amplitude, certain characteristics of the medium can be efficiently derived. Thus satellite navigation provides a scope for observing the propagating medium through proper derivation of its received signal.

10.2 Applications overview

Considering the global scenario, satellite navigation is becoming more and more a part of the daily life of humankind. It is not only being used for personal navigation purposes, in cars and mobile telephones that give positions, but is also of great use in industrial sectors like energy distribution networks or banking systems, providing time and frequency synchronizations. Applications also range from transport and communication to land survey, agriculture, scientific research, tourism, and whatnot.

The feat of demonstrating the benefits of the satellite navigation technology and then rolling them out on a global basis should be accredited to the US Global Positioning System (GPS). Although the use of navigational products for societal and strategic applications started with its predecessors, such as the Transit and the Timation system, GPS, providing free navigation service across the globe, has seen its applications growing almost since its inception. New applications are still being developed, covering all walks of life and sectors of the world. So, while discussing the applications, the authority and influence of GPS has to be acknowledged. However, with the very recent addition of the other operating and proposed systems (Globalnaya Navigatsionnaya Sputnikovaya Sistema (GLONASS), Galileo, Compass, etc.) into the troupe, the associated applications have received an enormous boost.

In this section, we first briefly review the current applications of the navigational system in a holistic manner. This review will provide an understanding of the different perspectives of utilizing the system, along with their importance and usefulness.

10.2.1 Applications architecture

Before entering further into the main description of the applications, we shall discuss the most general forms of the architecture exploited for their implementation. This will provide a reference and explanation of the application details and will also provide an overview of how the information from different sources is assimilated with position data and furthermore give an idea of the resources and infrastructure that each of these applications demand. There are two main types of architecture used in these applications-standalone and extended architecture-which we shall now describe.

10.2.1.1 Standalone architecture
There are some typical navigational applications that, in general, require only an ordinary navigation receiver with appropriate intelligent utility algorithms. Additional

information needed by the utility may be made available either through preloading of relevant data or by augmenting it with a collocated device. The baseline is that all the resources used for such applications are local and do not require any communication of data to external remote entities.

The position solution obtained by the navigational receiver is intelligently used by the associated utility software with the auxiliary data to produce intelligible information. The value-added information thus created is finally interfaced through a proper graphical user interface (GUI) for convenient and effectual use by the user.

Figure 10.1 shows the schematic for such applications.

10.2.1.2 Extended architecture

There are satellite navigation applications in which the navigation solutions are supported by auxiliary data fetched through a communication link.

Typically, these applications obtain the solution from the receivers and then connect with a centralized information base, typically a server, to exchange data that is used to meet the application goal. The communication may be done through a satellite link or a ground-based radio communication system like GSM (Global System for Mobile Communications), depending upon the application. Thus it needs a synergic use of navigation and communication systems.

The additional information available at the information base, such as geographic information system, meteorological, remote sensing, or any other relevant data, enhances the scope of the application and also may elevate it as a service. The final processing may be done either at this centralized server or at the user's receiver, and the results are then displayed to the user through proper GUI (Figure 10.2).

10.2.2 Applications roundup

Since the 1990s, the navigational applications have become immensely popular across the globe. The popularity of GNSS applications can be gauged by the extent of their usage. Here, we list some important satellite navigation applications that are

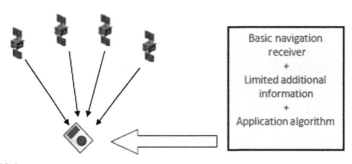

Basic navigation
receiver
+
Limited additional
information
+
Application algorithm

FIGURE 10.1

Schematic for the standalone architecture for global navigation satellite system applications.

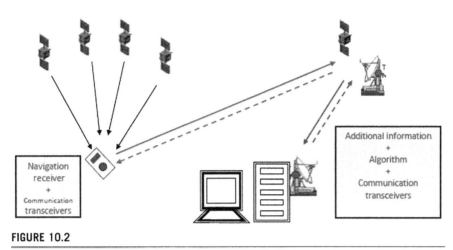

FIGURE 10.2

Basic architecture of extended applications.

being currently used. However, we must keep in mind that these descriptions are not exhaustive in terms of technical intricacies, but are only brief citations of their utilizations.

10.2.2.1 Civil applications

The initial uses of GPS and GLONASS, the respective GNSS implementations by the United States and by Russia, were military. However, the versatility of the system and societal demands available for civilian use. In fact, it is civil applications that have now become the driving force for the further development and advanced research related to this system. Different civil applications are possible with the systems, which we broadly categorize and describe under the following subheads.

10.2.2.1.1 Vehicular applications

Satellite navigation solutions, in conjunction with additional information, generate useful navigation products. These products provide general navigational aids to road, water, and air vehicles and are widely used throughout the world today.

10.2.2.1.1.1 Road navigation Satellite-based navigation has become a popular option for modern road transit systems. Besides showing one's position on digital maps in real time and tracking the locus of one's progress, it can help motorists find the easiest, cheapest or quickest route to their destinations or provide optimal routes for commercial vehicles. This is done by matching the map with the derived positions and is probably the most well known application of satellite navigation (figure 10.3).

When augmented with a communication link, this application may be used for fleet monitoring and management. This is done by transmitting the individual positions of the vehicles in a fleet through a preferred communication channel at regular

FIGURE 10.3

Applications for road navigation.

intervals. Subsequently, all the positions of the whole fleet are received at a central location, collated, and observed through a consolidated display. It also helps in general traffic management by enabling appropriate supervision and decision making for traffic routing and guidance (French, 1991, 1996; Obuhuma and Moturi, 2012).

10.2.2.1.1.2 Rail navigation The basic navigation service may be used in trains to provide passengers with the exact location of the train, its relative position over the entire route, speed, heading, time, next station, expected time of arrival etc. In addition, the tracking of the individual trains can be done and disseminated through a central server in a manner similar to the fleet monitoring described above. This can serve to generate data for the real-time rail information service and also to derive the performance of the rail movement and related statistics, as well. This data can also be used for intelligent rail guidance systems that can automatically recommend varying train speeds and reroute traffic, thus increasing the efficiency of the rail track capacity.

10.2.2.1.1.3 Marine navigation A wide variety of vessels move on water bodies each day. Irrespective of the purpose and size of the vessels, which may vary from the commercial needs of freight movement by giant ships to cruises for entertainment by sea liners or small boats for fishing, the safety and route optimization of marine transportation is a vital need. Therefore, it is important for sea goers to stay on the correct trail to avail the shortest or safest route or to return to the best location identified for entertainment or fishing, etc. This is a part of waypoint planning and management, which includes its risk assessment, as well. It establishes the requirements for position fixing that may be achieved through satellite navigation systems, considering the accuracy, integrity, continuity, and availability the system provides. This may also help in defining international water boundaries and access rights to areas. Resident applications can raise alarms in a receiver whenever the boundaries are crossed. Monitoring of registered vessels may be done when positions derived at these vehicles are transmitted to the monitoring system through a satellite link (Sennott et al. 1996).

10.2.2.1.1.4 Air navigation Reliance on satellite navigation is the future of air navigation and traffic management systems. Improvement of the precision of the position estimation en route is achieved through the integration of inertial navigation or other systems (Eschenbach, 1996) with the satellite navigation system, while an augmentation system provides enough accuracy and confidence for aircraft landing as well as during the en-route phase. Augmentation enables the aircraft to carry out a precision approach for landing with category I precision or better, satisfying the specified accuracy, continuity, and integrity levels (Parkinson et al. 1996). The air traffic controlling scenario is also changing with the proposed advent of systems where the individual aircraft, along with the air traffic controller, can identify and visualize the locations of other aircraft in its vicinity by sharing their precise mutual positions derived through these systems. This provides pilots with added awareness of their surroundings (Braff et al. 1996).

10.2.2.1.1.5 Space navigation Considering the geo-directive nature of the navigation signals, the service may be availed for providing navigation solutions to satellites in low earth orbit (LEO). This may provide high-precision orbit determination with minimum ground control. It may also support attitude determination of the spacecraft, replacing expensive on-board attitude sensors with low-cost multiple antennae placed on it along with specialized algorithms (Lightsey, 1996). Additionally, it may provide the timing solutions to the low earth remote sensing satellites where timing is of importance, replacing expensive spacecraft atomic clocks with low-cost, precise time receivers.

10.2.2.1.2 Personal applications

10.2.2.1.2.1 Child tracking Today's parents can benefit from the advancements of this modern navigation technology to identify and follow the location of their small children. This provides them with the secure knowledge of their child's location should they be out of sight. This application can be implemented by using satellite navigation receivers equipped with a communication channel. The communication may be done through GSM or Wi-Fi to indicate and display a child's positions on the parent's phone or website at regular intervals. Thus a child's movement can be monitored easily (Figure 10.4).

10.2.2.1.2.2 Location-based services (LBS) This is a service that provides specific information explicit to the position where the user is currently located. LBS may be both a push and a pull type of service. In the former, the service provider spontaneously disseminates information to the user, while in the latter, the user fetches information based on his or her query.

The popular form of LBS is a query-based service, in which the user, on estimating his or her positions, raises different queries and asks for a response from a central data server through a communication channel. The type of queries may vary widely. However, some of the most frequently used queries belong to the following classes:

- Amenities: closest hospital, automated teller machine, filling station, restaurant, shopping mall

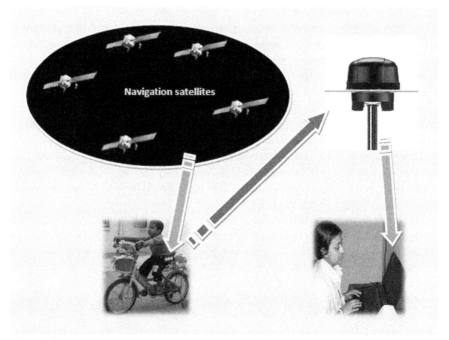

FIGURE 10.4

Applications for child tracking.

- Weather: current weather, temperature, possibility of rain at the location
- Topology: landform, height from sea level, nearest river, lakes, mountains, etc.
- Entertainment: any event near the location on a specific date, movies, theaters, etc.
- Agriculture and industry: soil type, crops, agro-industrial products, etc.

The user query, tagged with its current location, is served by the centralized database. The server, containing the geo-specific and relevant remote sensing database, returns the response to the query to the user through the same channel. Thus a strong convergence and synergy of navigation, communication, and remote sensing is needed here.

10.2.2.1.2.3 Sports and entertainment Satellite navigation is also currently used in sports and entertainment. Primarily, for outdoor exploration, it can eliminate many associated hazards by providing a precise location and timing. Activities like mountain trekking, hiking, etc. carry with them many intrinsic dangers, including getting lost in unfamiliar or unsafe territory. The only form of reliable navigation available here is from satellites. In addition to its all-weather ability to provide position, the digital data may be stored to record and return to waypoints or even to retrace suitable routes. Golfers started using this application to measure precise distances within the course and improve their game. Other applications include motor racing, recreational aviation, yachting and boating. Satellite navigation technology has also generated entirely new sports and outdoor activities. An example of this is 'geocaching,' a sport that rolls a pleasurable day's outing and a treasure hunt into one, and 'geodashing,' a cross-country race to predefined coordinates.

10.2.2.1.3 Societal applications

10.2.2.1.3.1 Disaster management Giving timely warning and providing quick assistance are two major concerns of civilian administrative authorities during and after a disaster. Satellite-based navigation, providing signals from space, can continue services even in case of disasters when other land-based services are likely to fail. These signals can be used efficiently for both pre facto and post facto activities.

10.2.2.1.3.1.1 Disaster warning Several human and animal habitats are located in earthquake-prone areas. Navigation can play a prominent role in generating seismological warnings. This requires a wide network of very accurate receivers with robust algorithms that can help to quickly identify earthquakes and aid in emergency assistance to save lives. The major criticality in the system is the correct identification of the quakes without missed detection or false alarm (Allen and Ziv, 2011). Similarly, occurrences of landslides, river floods and tsunamis can also be identified and warnings raised by monitoring the river and sea surface heights, respectively.

10.2.2.1.3.1.2 Search and rescue Locating and saving stranded people, animals, and assets are very important operations after any disaster. This can be efficiently done through a centralized monitoring and crisis management platform and can be established with the navigation system. On one hand, satellite navigation system users, if additionally enabled with a communication channel like GSM or Wi-Fi, can indicate their specific locations using the preferred channel, if available. These identifications can be observed centrally to make prompt decisions and rush aid. From the rescuers' perspective, once their receivers are registered in the system, they may automatically be tracked for efficient coordination, possibly carrying out exchange of information and instructions with it through any existing communication system. This allows improved planning and optimization of resource allocation from a central location and permits prompt response.

10.2.2.1.3.2 Emergency/medical services More than 1000 million emergency calls are generated worldwide every year. In many cases, emergency vehicles cannot be dispatched on time due to the absence of sufficient location information. A pivotal component of any successful emergency operation is time. The dispatcher, in most cases, either has to ask for an exact location or has to map the calling telephone number, each of which consumes much critical time and lacks accuracy. With satellite navigation, the precise location of the event can be tagged with the call, reducing the approach time of the assistance and thus saving lives.

10.2.2.2 Scientific and technological applications

The navigation may also serve many scientific applications where absolute precision is necessary. In experimental physics or in different applications of technology, position and timing requirements need to be complied with. This also calls for signal integrity in addition to a fair level of accuracy and precision.

10.2.2.2.1 Geodesy and surveying

As technology evolves and expands throughout the world, the surveying and mapping community is resorting more and more to satellite navigation-based techniques. These techniques significantly reduce the amount of equipment and labor hours that are normally required of other conventional surveying techniques. Surveying needs sophisticated receivers with sub-metric accuracy, capable of carrier phase measurements to provide real-time kinematic or post processing, as required for the application.

Geodesy is the scientific study of the shape of the earth. It is very much dependent upon relative positioning of unknown points with respect to a few predetermined points by finding the corresponding baseline. This is facilitated by the relative positioning techniques of satellite navigation systems. One of the major advantages is that each station needs to look skyward toward the visible satellites and hence mutual visibility is not mandatory, allowing large baselines. Another aspect of satellite navigation-based geodesy is the fact that as satellites use geocenters as their natural center of rotation, which is also taken as the origin for the geodesic measurements, there is no relative transformation of the coordinates required and the positions can be easily and conveniently determined (Leick, 1995; Larson, 1996; Torge, 2001).

10.2.2.2.2 Earth and atmospheric studies

Apart from obtaining the position solutions and time, the signals from the navigation satellites may be ingeniously used for earth and atmospheric studies. Propagation features like the delay, bending, and signal fluctuations experienced by the traversing radio waves of the navigation signal may be utilized for deriving the required parameters for the studies. The most common areas involve earth and atmospheric studies, including the ionosphere; some of these studies are described below.

10.2.2.2.2.1 Meteorological studies Meteorologists have started making atmospheric sounding using radio occultation with LEO satellites and GNSS signals. This is done by receiving and measuring the atmospheric bending of beams. Radio occultation is an important application of the navigation signals in which the bending of the rays transmitted by a navigation satellite enables it to be received by a satellite in LEO or by a receiver on the earth surface, which would otherwise be shadowed by the earth. The situation is shown in Figure 10.5. The amount of bending estimated from the positions of the GPS transmitter and the receiver is used for profiling of the tropospheric elements. Sensitive receivers detect the extra path traversed by the signal from the relative phase changes in the radio signals. This indicates the refractivity of the atmosphere from which different meteorological parameters like density, pressure, temperature, and humidity and their distributions can be derived, leading to characterization of the atmosphere.

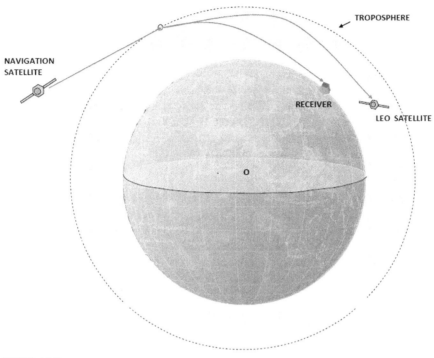

FIGURE 10.5

Applications in radio occultation.

10.2.2.2.2.2 Ocean surface and salinity monitoring Reflectometric measurements of the navigation signals in terms of the delay and power are used to estimate the sea surface height and salinity. In addition to specific algorithms, this needs the receiver to be capable of handling very low power signals as expected from the reflected waves from the sea surface (Camps, 2008; Garrison, 2011) (Figure 10.6).

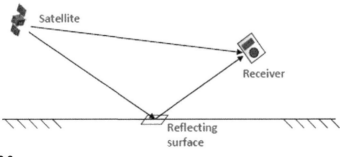

FIGURE 10.6

Reflectometric applications.

10.2.2.2.2.3 Ionospheric studies The dual-frequency receivers can identify the ionospheric total electron content (TEC). Moreover, wherever available, the satellite-based augmentation system message also contains the ionospheric delay data transmitted to the user. Thus the near-real-time information of the vertical TEC can be obtained at a single or multiple locations from the navigation signals. Additionally, from the discrete measurements, the whole spatial profile of the electron density and TEC may be derived using tomography (Hansen, 2002; Ganguly and Brown, 2001; Acharya et al. 2004). This provides great value and volume toward ionospheric research. Simultaneously, the ionospheric scintillation can also be derived from it.

10.2.2.2.3 Environment

Satellite navigation systems can be used to obtain accurate and timely information related to environmental elements. This empowers authorities to make better decisions to sustain the earth's environment while balancing human needs. Coverage estimation of forests and other extensive natural resources like coasts, mines, etc. along with their temporal variation, can be done very easily with satellite-based navigation by delineating the locus of the points about the resource's contour. It can also be used for animal tracking. Endangered species of animals may also be tracked and mapped, helping to preserve and enhance declining populations. The animals can be easily traced or spotted for required medication and other aid, whenever necessary.

10.2.2.3 Reliable timing service

We should also appreciate the time dissemination capability of a navigation system. For most of the typical requirements the timekeepers are in fixed locations and know their position; they only need one satellite to get the offset of their local clock from the navigation system time.

Institutions like industries and scientific laboratories require time synchronization to precisely measure common epochs and events. These require timing sources that can guarantee accuracy at several points. This may be achieved by synchronization of time through time transfer using the common view (CV) technique, in which two stations simultaneously observe the same satellite. We shall learn about these methods in detail in our next subsection. Laboratories across a country may use these techniques to exploit the precision of a common reference source like an atomic clock or to simultaneously establish synchronization between them.

10.3 Specific applications

In this section we shall discuss in detail a couple of important applications that use satellite navigation. The rationale for singling out two applications is in order to become aware of the intricacies involved in harnessing the technology for applications. So, we have chosen applications in two separate categories, one requiring precise position and one that requires precise estimation of time.

10.3.1 **Attitude determination**

An extended massive body has six degrees of freedom. Three of them determine the position while the others define the orientation of the body. The orientation of the body with respect to a fixed reference frame is called the attitude. The relative positioning of some fixed points on any moving body enables us to find the attitude of the body with respect to a fixed frame, typically fixed to the earth. We shall learn in this section only the fundamental ideas of attitude determination.

Attitude determination requires a local frame to be fixed on the moving body. The relative orientation of this body-fixed frame with respect to a standard local or geocentric reference frame is the attitude of the body. For convenience, we shall consider the moving body to be a vehicle and it will be accordingly referred to henceforth.

Relative orientation of a frame with respect to the other is determined by the angles that the coordinate axes of a frame make with that of the other. So, the complete set of direction cosines of the axes of the body-fixed frame of the vehicle in the standard reference frame define its attitude. Now, any vector has three directional cosines in a frame. Directional cosines are the cosines of the angle that the vector makes with the axes of the frame. As the individual axis of the body-fixed frame is a vector, each of these three axes will make three angles with the three axes of the standard frame, making a total of nine.

However, these nine directional cosines are not independent. Let the body-fixed frame defined by three axes be denoted by $F' = [x'\ y'\ z']$ while the axes of the standard reference frame is denoted by $F = [x\ y\ z]$. These are shown in Figure 10.7.

So, to fix the direction of the x' axis, are fixed in the F frame, the three-direction cosine is initially fixed. They maintain the relation $\alpha^2 + \beta^2 + \gamma^2 = 1$. So, we require three independent parameters to fix x' in F.

Once x' is defined in F, the plane perpendicular to it, containing y' and z' also get fixed. However, to define the direction of y' on this plane we need any two of its directional cosines. They will completely define its relative orientation with respect to the axes of F due to its containment on this plane. Once the x' and y' axes are defined in F, the other axis z' is automatically fixed by virtue of the right-handed orthogonal structure of F'. So, in total, we need only five independent relative

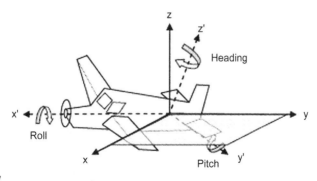

FIGURE 10.7

Attitude determination.

directions that define the orientation of one frame in the other. If we know these five independent elements, we can derive the complete set of nine directional cosines. Then, any vector defined in the F′ frame can be defined in the F frame also and vice versa, and the attitude of the vehicle can be defined.

The matrix that transforms the vectors from one frame to the other is called the *attitude matrix* or the *directional cosine matrix* (Cohen, 1992, 1996). The most convenient way to represent the cosine elements of the matrix is by representing its elements as the dot products of the unit vectors of the two frames. So,

$$A = \begin{pmatrix} x' \cdot x & x' \cdot y & x' \cdot z \\ y' \cdot x & y' \cdot y & y' \cdot z \\ z' \cdot x & z' \cdot y & z' \cdot z \end{pmatrix} \tag{10.1}$$

The reference frame may be chosen as appropriate for a certain application. We have already learned about these frames in Chapter 1. For vehicles like aircraft or land or sea vehicles, local frames are very suitable while the geocentric earth-centered, earth-fixed frame may be set as the reference in certain applications like spacecraft attitude control. Further, nonconventional application-specific frames may also be used for the purpose.

The body frame is also a right-handed Cartesian coordinate that remains fixed with the moving vehicle. Its x' axis is typically along the forward direction of its true motion while the y' axis is normal to it but on the body plane. The z' axis remains normal to both, as shown in Figure 10.7. The x' axis is called the roll axis, as a rotational motion of angle φ about this axis is called a roll. A roll is manifested by a change in the orientation of the $y'z'$ axes with respect to the reference frame. The y' axis is called the pitch axis. A change in the orientation of the $x'z'$ axes with respect to the reference marks a variation in the pitch angle θ of the vehicle. Finally, the z' axis is called the heading axis, as rotation about this axis is termed a heading. The orientation of the $x'y'$ axes changes with respect to the reference when the heading angle ψ changes.

When the body frame is aligned with the reference frame, the heading, pitch, and roll angles are all taken as zero. It makes A an identity matrix. For definite heading followed by pitch, and roll angles of the vehicle, the attitude matrix becomes (Cohen, 1996)

$$A = \begin{pmatrix} \cos\theta\,\cos\psi & \cos\theta\,\sin\psi & \sin\theta \\ -\cos\varphi\,\sin\psi + \sin\varphi\,\sin\theta\,\cos\psi & \cos\varphi\,\cos\psi + \sin\varphi\,\sin\theta\,\sin\psi & \sin\varphi\,\cos\theta \\ \sin\varphi\,\sin\psi + \cos\varphi\,\sin\theta\,\cos\psi & -\sin\varphi\,\cos\psi + \cos\varphi\,\sin\theta\,\sin\psi & \cos\varphi\,\cos\theta \end{pmatrix} \tag{10.2}$$

Here, the element A_{ij} represents the component of the ith coordinate in the body frame on the jth axis of the reference.

10.3.1.1 *Measurements and estimations*
Consider that two navigation receivers, R1 and R2, are placed on the body of the vehicle separated by a distance b', which forms the baseline on the body-fixed frame.

These receivers measure the carrier phase of the signal transmitted by the same satellite. Recall that we learned in Chapter 8 that for short baselines the single-difference phase equation at receiver R1 and R2 from satellite s can be written as

$$\Delta\varphi_{12}^s = \left(-e_1^s \cdot b\right) + N_{12}^s \lambda + c\Delta\delta t_{12} + \varepsilon_{12}^s \tag{10.3}$$

where φ_{12}^s is the difference between the measured phase at the two receivers, b is the baseline in the reference frame F, e_1^s is the unit vectors from the reference receiver 1 to the satellite s, and ε_{12}^s is the differential error between the two receivers considering the receiver noise and all other differential residual errors between the two receivers. For any moving vehicle, the distance between receivers R1 and R2 is small enough to assume all the errors to be identical but the receiver noise. So, ε_{12}^s is mostly contributed by the difference in the receiver noise.

Considering that the two receivers are driven by either the same clock or two clocks that are synchronized, $c\Delta\delta t_{12} = 0$. Since the positional difference of the receivers is negligible compared to the correlation of the ionosphere and troposphere, these errors are assumed to be eliminated on differencing. Hence, the equation turns into

$$\Delta\varphi_{12}^s = \left(-e_1^s \cdot b\right) + N_{12}^s \lambda + \varepsilon_{12}^s \tag{10.4}$$

Now the integer ambiguity is required to be resolved first. On resolving N, we have

$$\Delta\varphi_{12}^s - N_{12}^s \lambda = \left(-e_1^s \cdot b\right) + \varepsilon_{12}^s \tag{10.5}$$

In absence of the clock errors, it can be shown that the left-hand side of Eqn (10.5) is the differential geometric range, $\Delta\rho_{12}^s$. So, expanding b and e in terms of their components along the reference axes, we get

$$\Delta\rho_{12}^s = -\left(b_x e_x + b_y e_y + b_z e_z + \varepsilon_{12}^s\right) \tag{10.6}$$

Now if at any instant A is defined by Eqn (10.1) then the corresponding baseline b in the reference frame may be expressed in terms of the baseline b' in the body frame by the relation:

$$b^T = b'^T A \tag{10.7}$$

So we may write

$$\Delta\rho_{12}^s = -A^T\left(b' \cdot e_1^s\right) + \varepsilon_{12}^s \tag{10.8}$$

In this equation, $\Delta\rho_{12}^s = \Delta\varphi_{12}^s - N_{12}^s \lambda$, out of which $\Delta\varphi_{12}^s$ is measured by the receivers while N_{12}^s is estimated. Unit vector e_1^s can be estimated from the relative position of the satellite and the antenna. b' in the body frame is predefined and hence known a priori. Only the elements of the matrix A are the unknown. This matrix contains the independent elements of the attitude of the vehicle, i.e. orientation of the body-fixed axes, and is independent of the baseline chosen. So these values remain the same irrespective of any baseline chosen on the body-fixed frame for a definite orientation of the vehicle. Therefore, the elements of A can be solved by using similar phase measurements between different pairs of receivers on the vehicle and for different satellites visible.

So, as an estimation process, first the integer ambiguities are resolved. Once this is done, the solution is obtained by minimizing the quadratic attitude determination cost function for the m baselines and n satellites.

$$J(A) = \sum_{i=1}^{m} \sum_{j=1}^{n} \left\{ \Delta\rho_{1i}^{j} \right\} - A^{T} \left(b'_i \cdot e_1^j \right) \Big\}^2 \tag{10.9}$$

where b'_i is the ith baseline defined in the body frame with respect to reference receiver R1 and receiver Ri, and $\Delta\rho_{1i}^{j}$ is the differential geometric range between the same receivers toward the satellite j.

The elements of A may be obtained iteratively. First a trial solution A_0 is obtained, and a better estimate is then derived out of it by linearizing the cost function about the trial solution and solving for the correction matrix δA so that the solution after the correction becomes $A_0 + \delta A$. This process is repeated and the final solution is assumed when the correction δA does negligible alterations to the solution. Intrinsically, we have assumed the baselines to be all coplanar (Cohen, 1992; Cohen and Parkinson, 1992; Cohen, 1996).

10.3.2 Time transfer

When the precise time information obtained and maintained by a clock at a certain location is made available to clocks at other different locations, time transfer is said to have occurred.

The basic architecture for such a time transfer contains two clock systems separated by some distance, in which at least one of them has the capability of steering the clock. There is a calibration channel through which the calibration information is passed between the two clocks. The free-running stability of the clocks and the characteristics of the calibration channel are the parameters that determine the overall performance of the total system in terms of accuracy and precision. If the purpose of the comparison is to calibrate the time of the local device, which we are discussing here, then the delay through the channel must be known or derived, and any uncertainty in this delay will enter into the error budget for the overall calibration procedure.

There are three general methods that are commonly used to transmit time and frequency information: one-way time transfer, two-way time transfer, and Common View (CV). Here, we shall discuss these methods (Miranian and Klepczynski, 1991; Klepczynski, 1996; Levine, 2008) with obvious relation to satellite-based transfer (Figure 10.8). We shall describe the advantages and limitations of the different methods, discuss the uncertainty estimates for systems that are based on these methods.

10.3.2.1 One-way time transfer

One-way time transfer is the time transfer method where the precise clock maintained at one location is made available to any other by transmitting the precise

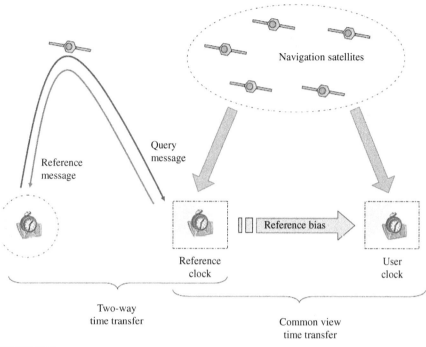

Navigation satellites

Query message

Reference message

Reference bias

Reference clock

User clock

Two-way time transfer

Common view time transfer

FIGURE 10.8

Different time transfer methods.

time from the former to the latter. This time, on reception, is then corrected for the propagation time delay to retrieve the precise current time. The receiving location either measures or models the path delay using ancillary data. We have already discussed one form of one-way time transfer while we learned how the users can derive the time using the signals from the navigation satellites.

10.3.2.2 Two-way time transfer

Two-way time transfer is the method in which both the peer clocks transmit and receive time data sent by the other and the calibration value is derived from the delays observed in two opposite directions.

The precision clock with which another clock at any other location is required to be synchronized, is referred to as the reference and the other clock that needs to be corrected and synchronized with the former is called the user clock.

This method has the inherent assumption that, unlike one-way time transfer, where the receiver is a receive-only terminal, here both the clock locations are able to communicate between them in duplex manner. Moreover, it is assumed that the time for a message to travel between the two stations is the same in both directions. However, the absolute time required to transmit the message from one end to the other is not important but should be considerably small, such that within the round-trip time the characteristics of the clock do not change.

Let the user clock C_u have a positive bias of δt_{ur} with respect to the reference clock C_r. Now, to derive this error. The user clock transmits a message at a time T_{us} toward the reference clock, with a time stamp in it. T_{us} is measured by the clock at the user station. Since this time is δt_{ur} ahead of the reference time, the time at that instant in the reference clock was $T_{us} - \delta t_{ur}$.

Note the suffixes appearing with the time T. The first suffix will be either "u" or "r," representing whether it is the user or the reference that is measuring it. The second represents whether it is during the sending or during the receiving process and will be represented by "s" and "r," respectively.

The message is received at the reference station at time T_{rr}, T_{rr} being measured by the clock at the reference station.

The time for the message movement in terms of the reference clock is

$$\begin{aligned} \Delta_r &= T_{rr} - (T_{us} - \delta t_{ur}) \\ &= T_{rr} - T_{us} + \delta t_{ur} \end{aligned} \tag{10.10}$$

The measured propagation time, as obtained by using the local time by the receiving reference station and the time stamp marked on the message, is

$$\begin{aligned} \Delta_m &= T_{rr} - T_{us} \\ &= \Delta r - \delta t_{ur} \end{aligned} \tag{10.11}$$

At any subsequent time, the reference station sends a message to the user station, also provided with a time stamp. It also sends Δ_m as estimated at the reference along with it. This message is transmitted at time T_{rs}, measured by the clock at the reference station, and is received at the user station at time T_{ur}, measured by the clock at the latter station.

Similarly, the reference time and the measured time of travel are, respectively

$$\nabla_r = T_{ur} - \delta t_{ur} - T_{rs} \tag{10.12}$$

and

$$\begin{aligned} \nabla_m &= T_{ur} - T_{rs} \\ &= \nabla_r + \delta t_{ur} \end{aligned} \tag{10.13}$$

However, the stations have the measured values only and need to derive the bias δt_{ur} from it. So, differencing these two equations and halving it, we obtain

$$\begin{aligned} &\tfrac{1}{2}\left(+\nabla_m - \Delta_m\right) \\ &= \tfrac{1}{2}\left[+(T_{ur} - T_{rs}) - (T_{rr} - T_{us})\right] \\ &= \tfrac{1}{2}\left(\nabla_r + \delta t_{ur} - \Delta_r + \delta t_{ur}\right) \\ &= \tfrac{1}{2}\left(2\delta t_{ur} + \nabla_r - \Delta_r\right) = \delta t_{ur} + \tfrac{1}{2}\left(\nabla_r - \Delta_r\right) \end{aligned} \tag{10.14}$$

So, if the path delays are symmetric, then $(\Delta_r - \nabla_r)$ is zero and thus the user clock obtains its own delay and can correct the same.

However, if the message takes different times for the forward and the reverse journey and if the difference is ε, then the result becomes $\delta t_{ur} + \frac{1}{2}\varepsilon$ and it is necessary to eliminate or to reduce this error to a minimum. In many cases, this error remains proportional to the absolute path difference, and in such cases the effect of any asymmetry can be minimized by keeping the path delay itself small.

The protocol does not depend on any particular relationship among the transmit times at the user and the reference. So, the reference station may simply echo the message from the user station with time stamp or otherwise can reply to the received message at any later time. The delay between the receiving at the reference and its response is not important, provided that it is measured accurately and that both clocks are well behaved during this interval.

Therefore, if we have a satellite navigation receiver at any location with the facility of communicating with the other station, then the accuracy of the clock derived at this station may be disseminated to all other stations, which will thus enjoy almost similar precision even without the provision of a precise clock.

The role of the navigation system here is not in synchronizing the clocks but in providing accuracy and precision to one of them such that it can behave as a reference. So, with this method a reference station located within the service area of a satellite navigation system can provide precise timing to one that is beyond the service area or located at a place where the time dilution of precision is constantly poor to provide a precision timing service. The calibration channel through which the synchronization is obtained may be any other communication channel. This may be implemented through a duplex satellite channel. Here the propagation and the transponder delay remain almost the same for both the forward and reverse path, making it suitable for such implementation. But, this is at the expense of the station being capable of duplex satellite communication with the reference. This mandatory requirement makes the hardware at the user stations more complicated and expensive than the one-way method. Moreover, the messages must maintain a prescribed format assigned a priori and hence no anonymous association is possible. So, two-way time transfer is possible only in closed groups.

The total error in synchronizing the clocks may be divided into components consisting of the measurement errors at the two ends, the path asymmetry error and the correction implementation error.

This can be used in a differential system, where the users correct their position and synchronize their clocks with respect to the reference stations, which are remotely located. When the differential correction channel is established in real time, this can be used for synchronization of the clocks between them. This removes one of the unknown parameters, the relative clock bias at the user end, making the computation less onerous. With one of the unknowns being now alternatively estimated, it eases the estimation process and also reduces the requirement of visibility of satellites.

Two-way time transfer is used for many other different applications, especially for industrial and scientific purposes. It is used where time synchronization between two clocks is required, for purposes like starting two processes at the same instant at two remote locations. It is also used when the precision of an atomic clock located at one place is required to be utilized at some other locations.

10.3.2.3 The CV time transfer

The CV is the time transfer method in which several stations with their respective clocks, including the station with a reference clock, receive data from a common source over paths whose delays are approximately equal and then exchange data mutually to derive the calibration information.

In this method, the time transmission from the navigation satellite is utilized. Further, there exists more than one user who receives this timing signal to obtain their respective bias with that of the satellite. One of the receivers of this timing signal may be a reference atomic clock while the clocks at other locations in the common view are required to be synchronized with the reference. The term common view itself suggests that in such a method the path for the reference and for the users should have common characteristics in all respects.

On receiving the time signals transmitted from a single source, each station measures the time at which a particular signal arrives at its location using its own clock. The time measured at different stations will be

$$T_{uj} = T^s + \delta t^s_{uj} + d^s_{uj} \tag{10.15}$$

where u_j represents the jth user station, T_{uj} is the measured time at the jth station, T^s is the transmit time at the satellite measured with the satellite clock, δt^s_{uj} is the bias of the clock at the jth station with respect to this satellite clock, and d^s_{uj} is the path-traversing time taken by the signal to reach from the satellite to the jth station. The stations then compare these measurements and obtain them to get their mutual timing bias, as described below.

Assume that a reference station is keeping a highly precise reference clock and a user station wants to achieve similar precision with a nominal clock. Both of them receive the timing signals from the satellite navigation system and compare them with their own clock. The reference station measures the receive time as

$$T_r = T^s + \delta t^s_r + d^s_r \tag{10.16}$$

where T_r is the measured time at the reference station in its own clock, T^s is the transmission time with the satellite clock, δt^s_r is the positive bias of the reference station's clock with respect to the satellite clock, and d^s_r is the path delay between the satellite and the reference station. Since T^s can be derived from the time stamp marked on the message, it can be read and the difference can be obtained as

$$\begin{aligned} \Delta^s_r &= T_r - T^s \\ &= \delta t^s_r + d^s_r \end{aligned} \tag{10.17}$$

Similarly, differencing at the user station provides us with

$$\begin{aligned} \Delta^s_{uj} &= T_{uj} - T^s \\ &= \delta t^s_{uj} + d^s_{uj} \end{aligned} \tag{10.18}$$

Note here that the satellite transmission time T^s for the user and the reference need not be the same. However, the satellite clock transmitting the time needs to be stable enough such that it does not drift considerably within this time.

This difference derived at the reference station, i.e. the reference bias, is communicated to the user station, and on further differencing these two derived differences, we get

$$\begin{aligned}
\Delta^s_{r\_uj} &= \Delta^s_r - \Delta^s_{uj} \\
&= \delta t^s_r + d^s_r - \delta t^s_{uj} - d^s_{uj} \\
&= \delta t^s_r - \delta t^s_{uj} + d^s_r - d^s_{uj}
\end{aligned} \tag{10.19}$$

Now, if the one-way path delays from the transmitter to the two stations are equal, then both the path delays cancel out and the difference becomes

$$\Delta^s_{r\_uj} = \delta t^s_r - \delta t^s_{uj} \tag{10.20}$$

So, the time bias between the user and the reference sites can be computed without knowing anything about the source or the path delay. Any change in the path delay that occurs with time gets canceled and does not affect the process to the extent that it remains common to the paths to the two stations. So, if one of the stations has a precise clock, the same precision may be obtained at the other station, exploiting the relative bias to correct the receiver clock.

However, in practice it is difficult to find a situation such that the path delays to both the locations are exactly equal. This problem can be eliminated if the receivers can obtain their precise range from the satellite. The range enables the receivers to estimate the traversing time of the signal. This can be used in Eqns (10.17) and (10.18), for using as known d^s_r and d^s_{uj} values, respectively. Thus the exact clock bias difference at these two locations, both with respect to the satellite clock is obtained even if the distances are different. So, when this information is exchanged, the user clock can derive its relative bias with respect to the reference clock and correct it.

This method may also be extended even to antipodal locations, i.e. even without seeing the same set of satellites, if the satellite sources are all synchronized to a common reference or their relative time skew is precisely known.

References

Acharya, R., Sivaraman, M.R., Bandyopadhyay, K., 2004. Tomographic estimation of ionosphere over Indian region. In: Proceedings of ADCOM. ACCS, pp. 564–567.

Allen, R.M., Ziv, A., 2011. Application of real-time GPS to earthquake early warning. Geophysical Research Letters 38, L16310. http://dx.doi.org/10.1029/2011GL047947.

Braff, R., Powell, J.D., Dorfler, J., 1996. Applications of the GPS to air traffic control. In: Parkinson, B.W., Spilker Jr., J.J. (Eds.), Global Positioning Systems, Theory and Applications, vol. I. AIAA, Washington, DC, USA.

Camps, A., 2008. A hybrid radiometer/GPS reflectometer to improve sea surface salinity estimates from space. In: Microwave Radiometry and Remote Sensing of Environment, MICRORAD 2008. Florence, Italy.

Cohen, C.E., 1992. Attitude Determination Using GPS (Ph.D. thesis). Stanford University, Stanford, USA.

Cohen, C.E., 1996. Attitude determination. In: Parkinson, B.W., Spilker Jr., J.J. (Eds.), Global Positioning Systems, Theory and Applications, vol. II. AIAA, Washington, DC, USA.

Cohen, C.E., Parkinson, B.W., 1992. Aircraft applications of GPS based attitude determination: test flights on a Piper Dakota. In: Proceedings of ION-GPS 92. Institute of Navigation, Washington,DC, USA.

Eschenbach, R., 1996. GPS applications in general aviation. In: Parkinson, B.W., Spilker Jr., J.J. (Eds.), Global Positioning Systems, Theory and Applications, vol. II. AIAA, Washington DC, USA.

French, R.L., 1991. Land vehicle navigation—a worldwide perspective. Journal of Navigation 44 (1), 25—29.

French, R.L., 1996. Land vehicle navigation and tracking. In: Parkinson, B.W., Spilker Jr., J.J. (Eds.), Global Positioning Systems, Theory and Applications, vol. II. AIAA, Washington, DC, USA.

Ganguly, S., Brown, A., 2001. Ionospheric tomography: issues, sensitivities, and uniqueness. Radio Science 36 (4), 745—755.

Garrison, J.L., 2011. Estimation of sea surface roughness effects in microwave radiometric measurements of salinity using reflected GNSS signals. IEEE Geosciences and Remote Sensing Letters 8 (6), 1170—1174.

Hansen, A., 2002. Tomogrpahic Estimation of the Ionosphere Using Terrestrial GPS Sensors (Ph.D. dissertation). Stanford University.

Klepczynski, W.J., 1996. GPS for precise time and time interval measurements. In: Parkinson, B.W., Spilker Jr., J.J. (Eds.), Global Positioning Systems, Theory and Applications, vol. II. AIAA, Washington, DC, USA.

Larson, K.M., 1996. Geodesy. In: Parkinson, B.W., Spilker Jr., J.J. (Eds.), Global Positioning Systems, Theory and Applications, vol. II. AIAA, Washington, DC, USA.

Leick, A., 1995. GPS Satellite Surveying, second ed. John Wiley and Sons, New York, USA.

Levine, J., 2008. A review of time and frequency transfer methods. Metrologia 45, S162—S174. http://dx.doi.org/10.1088/0026-1394/45/6/S22.

Lightsey, E.G., 1996. Spacecraft attitude control using GPS carrier phase. In: Parkinson, B.W., Spilker Jr., J.J. (Eds.), Global Positioning Systems, Theory and Applications, vol. II. AIAA, Washington, DC, USA.

Miranian, M., Klepczynski, W.J., 1991. Time transfer via GPS at USN0. In: Proceedings of the 4th International Technical Meeting of the Satellite Division of the Institute of Navigation (ION GPS 1991), pp. 215—222. Albuquerque, NM, USA.

Obuhuma, J.I., Moturi, C.A., 2012. Use of GPS with road mapping for traffic analysis. International Journal of Scientific and Technology Research 1 (10). ISSN 2277-8616120.

Parkinson, B.W., O'Connor, M.L., Fitzgibbon, K.T., 1996. Aircraft automatic approach and landing using GPS. In: Parkinson, B.W., Spilker Jr., J.J. (Eds.), Global Positioning Systems, Theory and Applications, vol. I. AIAA, Washington, DC, USA.

Sennott, J., Ahn, I.S., Pietraszewski, D., 1996. Marine applications. In: Parkinson, B.W., Spilker Jr., J.J. (Eds.), Global Positioning Systems, Theory and Applications, vol. II. AIAA, Washington, DC, USA.

Torge, W., 2001. Geodesy, third ed. Water de Gruyter, New York, USA.

Appendix 1   Satellite Navigational Systems

There are many major primary satellite navigation systems as on date. Some of them are global while others are regional. A couple of them are fully operational while the rest are either partially operative or about to come up very soon. These systems are namely:

1. Global Positioning System (GPS) of USA;
2. Global Navigation Satellite System (GLONASS) of Russia;
3. GALILEO of the European Union;
4. QZSS of Japan (GPS complementary);
5. COMPASS of China; and
6. IRNSS of India

Out of these, the QZSS and IRNSS are regional systems while other are global. The important features of these systems are mentioned below.

A1.1 Global Positioning System

The GPS was developed by the United States a part of the satellite-based military navigation system. The US Department of Defense (DoD) planned to use the constellation of artificial satellites under its NAVSTAR program and find the position with high accuracy through radio-based ranging. Starting its launch from 1978, the system acquired initial operational capability in 1993 and then the full operational capability in 1995.

GPS, like any typical system, has the architectural components that can be divided into three segments, viz. Ground segment, Space segment, and the User segment, as mentioned in Chapter 2 (El-Rabbani 2006, Parkinson & Spilker 1996, Grewal et al. 2001).

A1.1.1 Ground segment

The GPS Ground segment consists of the ground monitoring stations distributed across the world. There are six monitoring stations at Hawaii and Kwajalein in Pacific Ocean, Diego Garcia in Indian Ocean, Ascension Island and Cape Canaveral, Florida in Atlantic Ocean. One master station is located at Colorado-Springs, Colorado. At these stations, signals from satellites are monitored and their orbits tracked. These are used to predict the near future satellite orbit and correct their clock as described in Chapter 2.

A1.1.2 Space segment

The space segment of GPS consists of the satellite constellation. This constellation, placed in MEO have 24 satellites in almost circular orbits. The satellites are distributed in 6 equispaced orbital planes with four satellites in one plane, which are unevenly phased as shown in Figure A1.1. The semimajor axis of the orbits is about 26,500 km keeping the satellites about 20,000 km above the earth surface. The corresponding period is of 11 h 58 min. The inclination of 55° makes the satellites visible overhead up to 55° on both Northern and Southern Hemispheres of the earth. The constellation is designed in a way that users across the globe can see at least four satellites at all times with few exceptions. Typically six to eight satellites are visible at a time.

A1.1.3 **User segment**

Users employ the receive-only terminals to receive the navigation data transmitted by the satellites and compute its position. Receivers of different capabilities and accuracies are used. The dual-frequency receivers with highly stable clocks are costly whereas the single frequency receivers with less accurate clocks are available at low prices. Receivers with differential capabilities are used for civilian precise applications like surveying.

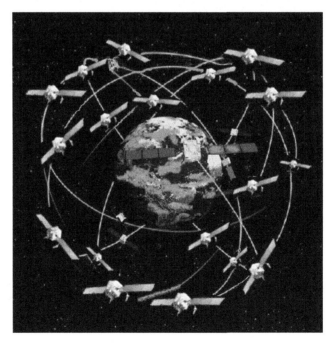

FIGURE A1.1

GPS constellation.

A1.1.4 **GPS services**

GPS provides two categories of services, viz. Precise Positioning Services (PPS) and Standard Positioning Services (SPS). The PPS is the high accuracy, single receiver GPS positioning service available only to US military and other selected agencies. The SPS provides uninterrupted positioning service to any civilian user on a global scale, however it is less accurate than the PPS.

A1.1.5 **GPS signal structure**

A1.1.5.1 Carrier signal

GPS employs sinusoidal signal with frequency 1575.42 and 1227.60 MHz as its two carriers. These signals are designated as L1 and L2 respectively. Each satellite transmits the same navigation signal at these two frequencies. They are coherently selected multiples of a 10.23-MHz master clock, derived from an atomic standard.

GPS uses code division multiplexing that enables its different satellites to use the same carrier frequency simultaneously without interference. Each carrier frequency is BPSK modulated by a "spreading code", which remains multiplied over the navigation data as described in Chapter 3. These ranging codes are different for different satellites and are orthogonal to each other.

A1.1.6 **Ranging code**

GPS uses orthogonal GOLD codes for ranging with suitable correlation characteristics. There are two kinds of such codes. The C/A code of length 1023 chips is issued at 1.023 Mchips/second rate. It is used for the standard positioning service. The other code is called the P code. It is of much larger length and has a rate of 10.23 Mchips/s. L1 is modulated with both P code & C/A code in phase quadrature while L2 is modulated only by the P code. The faster and longer P codes results in better accuracy and reliability. It is accessible only by the PPS service users whereas common SPS users can access only the C/A codes with lesser accuracy.

At times, P code is encrypted into a secure antispoof Y code, also called the P(Y) code. The C/A Code & P code epochs & navigation data are perfectly synchronized. The salient features of C/A and P codes are mentioned below.

A1.1.7 **C/A code**

- The C/A ranging codes are meant for civil users.
- These are short codes with a period of 1023 bits.
- The chip rate is 1.023 MHz, so sequence is of 1-ms duration.
- Short code permits rapid acquisition.
- Gold codes: formed by the products of two equal period 1023 bits PN codes.

A1.1.8 **P code**

- It is meant for use by DoD authorized users only.
- It is a long code.

- Chip rate is 10.23 MHz, i.e., 10 times faster than C/A.
- It is a product of two PN codes, X_1 and X_2.
- X_1 has a sequence of 15,345,000 chips and X_2 has a sequence of 15,345,037 chips.
- The P code has a period of around 38 weeks.
- In GPS, P code is reset every Saturday/Sunday midnight, so that the period of truncated sequence is one week.
- The P code is difficult to acquire without acquisition aids.

A1.1.9 Data

The GPS satellites transmit the Navigation and other data like bias and health data, formatted in a definite manner in a contiguous fashion. A description of the navigation data structure is given below.

- The navigation data is a 50-Hz Bipolar Non Return to Zero (NRZ) encoded data.
- It consists of ephemeris data, clock bias, almanac, and health data.
- It is divided into data frames of 30 s each.
- Each data frame is divided into five subframes of 6 s.
- Subframe 1, 2, 3 are specific to transmitting satellite and 4,5 are common to all satellites.
- Essential satellite ephemeris and clock parameters are repeated in every frame through 1, 2, and 3.
- Twenty-five pages of subframe 4 & 5 are transmitted.
- It takes 12.5 min (750 s) to transmit entire message.

The GPS navigation data is a Bipolar NRZ signal issuing bits at the rate of 50 bits per second. It transmits one data frame of 1500 bits in 30 seconds. Each frame is divided into five subframes of 6 s each. Out of these, subframe 1, 2, and 3 are specific to the transmitting satellite. Each subframe is of length 300 bits and starts with a definite preamble followed by data and parity bits.

Word 1 and 2 have the same format in every subframe. Word 1 is the telemetry word. Its first 8 bits constitute the preamble 10,001,011 ($8B_H$) and the rest is the telemetry data. Word 2 is the handover word and contains the truncated Z count that indicates the time of end of the subframe in quantum of 1.5 s.

The data section of subframe 1 contains, 2nd order polynomial coefficients to calculate satellite clock offset in addition to its time-of-applicability. Subframes 2 and 3 contain the ephemeris, consisting of the orbital parameters. The ephemeris time-of-applicability is also transmitted along with. The same data content continues for consecutive frames until new update. New data sets of subframes 1, 2 and 3 usually begin to be transmitted precisely on the start of the hour.

Subframes 4 and 5 contain the almanac data and some related health and configuration data. Due to the large volume of the data, it is subcommutated in a contiguous manner as the last two subframes in every frame. It takes 25 such frames to complete one set of this data. So, the data repeats after 25 consecutive frames, which take

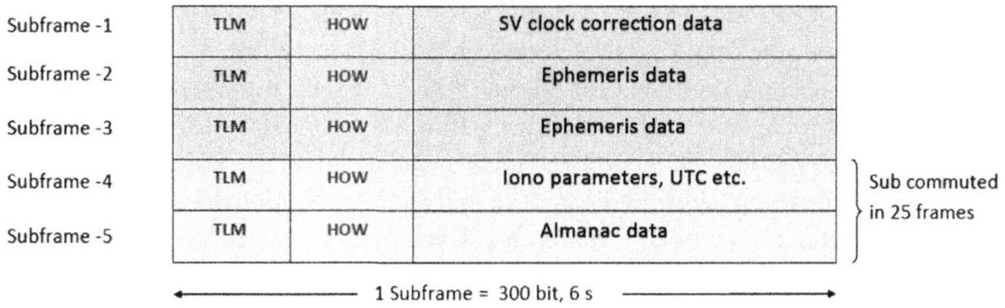

Subframe -1	TLM	HOW	SV clock correction data
Subframe -2	TLM	HOW	Ephemeris data
Subframe -3	TLM	HOW	Ephemeris data
Subframe -4	TLM	HOW	Iono parameters, UTC etc.
Subframe -5	TLM	HOW	Almanac data

Sub commuted in 25 frames

← 1 Subframe = 300 bit, 6 s →

FIGURE A1.2

GPS navigation data format.

12.5 min for the same. The same data is transmitted until next upload (Global Positioning System Directorate 2012(a), Global Positioning System Directorate 2012(b), Kaplan and Hegarty 2006, Misra and Enge 2001). The structure of the GPS data is shown in Figure A1.2.

Elements of the Navigation Data

- **TLM:** Telemetry data
- **HOW:** Handover word containing Z count
- **CC:** Clock correction data
- **EPM:** Ephemeris parameters
- **IONO:** Ionospheric correction data (Klobuchar coefficients)
- **UTC:** Universal coordinated Time
- **ALM:** Almanac data

Navigation data parameters like the ephemeris, almanacs, clock-offset coefficients, etc. are updated in every 2 h. The beginning of transmitting a new data is called a cutover, which occurs at hour boundaries. Coefficients for ionospheric delay change at a slower rate. These data, however, are uploaded by the ground stations, in advance, once in every 24 h. The coordinate datum used by the GPS is WGS-84 (Parkinson & Spilker 1996).

A1.1.10 Modernization of GPS (www.gps.gov)

In order to meet the growing demands of the users in both the segment of the services, major decisions have been taken to improve the GPS service quality. The efforts include upgrading of the GPS space and control segments with new features to improve GPS performance. These features include new civilian and military signals in addition to introduction of modern technology (www.gps.gov: Official U.S. Government information about the Global Positioning System (GPS) and related topics).

One important step forward for modernization of the GPS is to switch from the fixed frame-based data format to the flexible message type that will enable it to transmit different navigation parameters in the data on priority basis at variable intervals (Kovach et al. 2013). The new civilian signals planned for the purpose of modernization include the L2C, L5, and L1C. The legacy civil signal L1 C/A will also be continued, making a total of four signals for the civilian use.

L2C is the second civilian GPS signal which, when combined with L1 C/A enables ionospheric correction, while for existing dual-frequency users, will deliver faster signal acquisition, enhanced reliability, and greater operating range. It has a modern signal design with dedicated channel for codeless tracking. The higher effective transition power of L2C makes it easier to receive under foliage and indoor conditions.

L5 is the third civilian GPS signal, primarily conceived for the safety-of-life transportation and other high-performance applications. In addition to higher power, it offers greater bandwidth for improved immunity to jamming and an advanced signal design, including multiple message types and forward error correction.

L1C is the fourth civilian GPS signal, designed to enable interoperability between GPS and other systems. The design including forward error correction and multiplexed binary offset carrier modulation will improve mobile GPS reception in cities and other challenging environments.

As an endeavor of improving the performance, the satellite slots in the space segment has been expanded. The GPS constellation is a mix of new and legacy satellites. There are still satellites from block II-A with advanced technology compared to the Block II satellites. Some of them, whose life has expired, have been replaced by the Block II-R. Even few Block II-R(M) satellites has been put in place, which replace the older ones and also have modernized signal L2C. The follow on series of satellites in Block II-F carry both L2C and L5 signal with longer life expectancy. The newest of the satellites, comprising the Block III are soon to be put in space and consist of all three civilian signals including L1C.

A1.2 GLONASS (www.glonass-iac.ru)

The GLONASS (Global'naya Navigatsionnaya Sputnikovaya Sistema) was also developed as the military navigation system by the then USSR in early part of the 1970s.

The GLONASS space segment is comprised of 28 satellites out of which 24 are in operational phase, 3 are spares, and 1 in in-flight test phase.

The satellites are distributed in three planes separated by 120° with eight satellites in each plane. Different frequency channels are allocated to each of these satellites with appropriate frequency reuse plan. The orbits of these satellites have radius of about 25500 Km. This is slightly lesser than that of the GPS and consequently the period of rotation is 11 h 15 min which is also obviously lesser than GPS. The inclination of the orbital planes is 64°. This increased inclination helps in obtaining the satellites evenly distributed across the sky for even places at higher latitudes.

The GLONASS system was formally declared operational in 1993 while in December 1995, the full constellation was achieved. But, due to no further launches, the system was left with only six operational satellites in 2001. However, GLONASS recovered from this stage very quickly and in October 2011, full constellation was again restored. Apart from the first generation satellites, the active satellites now have representations from both second and third generations. 2G satellites are of type GLONASS-M with increased lifetime. The current version of the satellites is of type GLONASS-K, which has reduced mass and hence further improved lifetime.

The ground stations for monitoring the GLONASS satellites are almost entirely located within former USSR. The System Control Centre (SCC) is at Krasnoznamensk and the Central Clock is located at Schelkovo, near Moscow. The network of five Telemetry, Tracking and Command (TT&C) stations are at Schelkovo, Komsomoisk, St-Peteburg, Ussuriysk and Yenisseisk. Some of these stations also have the Laser Ranging capability. The monitoring stations are all within the Russian territory with the only station outside this region located at Brasilia, in Brazil (GPS World, 2013).

GLONASS offers both standard precision and high precision services for the civilians and military respectively. So, the GLONASS satellites transmit two types of signal, viz. the Standard precision (S) signal and the High precision (P) signal.

S signals use Direct Sequence Spread Spectrum (DSSS) technique with a single ranging code at a rate of 0.511 Mcps. Each satellite transmits on a different frequency using a 15-channel Frequency Division Multiple Access (FDMA). The carrier frequencies span either side from 1602.0 MHz (L1) and are given by (Grewal et al. 2001).

$$f1(n) = 1602 \text{ MHz} + n \times 0.5625 \text{ MHz}.$$

where n represents the index and $n = -7, -6, -5, \ldots 0, \ldots, 6, 7$. These signals are then BPSK modulation and transmitted using Right Handed Circular Polarization (RHCP).

P signal uses same DSSS with ranging codes 5.11 Mcps. It is transmitted in FDMA in the L1 band signals with same center frequencies but with phase orthogonality. It is also broadcasted in L2, with the center frequency f2, where,

$$f2(n) = 1246 \text{ MHz} + n \times 0.4375 \text{ MHz}.$$

GLONASS uses a coordinate datum named "PZ-90" in contrast to the GPS's WGS-84.

A1.3 GALILEO

GALILEO is the global primary satellite navigation system by the European Union, a project managed by European commission with Support from European Space Agency. In addition, this system is also compatible and interoperable with GPS.

With the endeavor for such a navigation system starting in around 1999, the Galileo project was supposed to be completed in three phases. However, unlike GPS and GLONASS, this was fully controlled by the civilian authority from its beginning. The first phase for characterizing the critical technologies for this system was completed with the GIOVE-A and GIOVE-B experimental satellites. The full operational capability with complete deployment of the ground and space infrastructure will be following.

In addition to the typical navigation services, namely the Open service (OS) and Public regulated service (PRS), GALILEO will also offer commercial and search and rescue services.

The Galileo is designed for 30 satellites in three orbital planes with the radius of about 30000 km, i.e., at about 24000 km above the earth surface. The corresponding period of revolution is 14 h 22 min with the orbital inclination of 56°. It will use the L1 (1559–1591 MHz) and the L5 (1164–1300 MHz) frequency for the purpose in Code Division Multiplexing Access (CDMA) mode and with both BPSK and BOC modulation (Margaria et al. 2007, Shivaramaiah and Dempster 2009).

The free OS provides positioning and time information intended mainly for high-volume satellite navigation applications. The Commercial Service (CS) for the development of applications for professional or commercial use by means of improved performance and data with greater added value than those obtained through the OS.

The PRS restricted to government-authorized users, for sensitive applications that require a high level of service continuity. The PRS uses strong, encrypted signals.

The Search and Rescue Support Service (SAR) works by detecting distress signals transmitted by beacons, locating these beacons and relaying messages to them.

References

El-Rabbani, A. (2006). *Introduction to GPS* (second ed.) Boston, MA, USA: Artech House.

Global Positioning System Directorate. (2012a). *Navstar GPS Space Segment/Navigation User Interfaces: IS-GPS-200G*. USA: Global Positioning System Directorate.

Global Positioning System Directorate. (2012b). *Navstar GPS Space Segment/User Segment L1C Interfaces: IS-GPS-800C*. USA: Global Positioning System Directorate.

Grewal, M. S., Weill, L., & Andrews, A. P. (2001). *Global Positioning Systems, Inertial Navigation and Integration*. New York, USA: John Wiley and Sons.

Kaplan, E. D., & Hegarty, C. J. (Eds.). (2006). *Understanding GPS Principles and Applications* (second ed). Boston, MA, USA: Artech House.

Kovach, K., Haddad, R., & Chaudhri, G. (2013). *LNAV Vs. CNAV: More than Just NICE Improvements, ION-gnss⁺—2013*. Nashville, USA: Nashville Convention Centre.

Margaria, D., Dovis, F., & Mulassano, P. (2007). An innovative data demodulation technique for Galileo AltBOC receivers. *Journal of Global Positioning Systems, 6*(1), 89–96.

Misra, P., & Enge, P. (2001). *Global Positioning System: Signals, Measurements and Performance*. Ganga Jamuna Press.

Parkinson, B. W., & Spilker, J. J., Jr. (Eds.). (1996). *Global Positioning Systems, Theory and Applications, vol. I*. Washington DC, USA: AIAA.

Shivaramaiah, N. C., & Dempster, A. G. (2009). In: *The Galileo E5 AltBOC: Understanding the Signal Structure, IGNSS Symposium 2009—Australia*. International Global Navigation Satellite Systems Society.

First Glonass Station Outside Russia Opens in Brazil. February 20, 2013. GPS World.

Index

Note: Page numbers with "f" and "t" denote figures and tables, respectively.

A

Absolute differential methods
  ephemeris error, 291
  ionospheric delay, 292
  IPP, 294
  reference station clock, 291—292
  tropospheric delay, 292—293
Accelerometer, 24—25, 25f
Active hydrogen maser (AHM), 45
Air navigation, 357
Analog to digital converter (ADC), 173, 174f
Antenna, 167—168
Asymmetric/two-key cryptosystems, 132
Atmospheric noise/sky noise, 167
Attitude determination
  body frame, 364
  directional cosines, 363
  measurements and estimations, 364—366
  relative orientation, 363

B

Bancroft's method, 233—236
Bandpass sampling (BPS), 173
Baseband signal processor
  closed loop arrangement, 181, 181f
  code and carrier wipe off section, 179,
    180f
  double estimation, 200—201
  FDMA and CDMA system, 181
  local oscillator, 180, 180f
  signal acquisition
    description, 182
    Doppler frequency, 182
    parallel search method, 185—186
    serial search method, 182—185
  signal tracking
    carrier tracking, 186—192, 188f, 191f
    code tracking, 188f, 193—196, 196f
    coherent code tracking, 196—198,
      197f
    noise performance, 192—193
    noncoherent code tracking, 198—199
    numerically controlled oscillator, 186
  tracking imperfections
    correlation loss, 201, 201f
    jitter and phase noise, 201—202, 203f
  VEVL, 199—200, 200f

Binary offset carrier (BOC)
  Alt-BOC, 146—151, 147f—150f
  autocorrelation, 145, 145f
  definition, 84—85
  double estimation, 200—201
  mathematical representation, 139—141
  reference frequency, 141
  spectrum, 143, 144f
  time variation, 141, 142f
  VEVL correlator, 199—200, 200f
Binary phase shift keying (BPSK)
  power spectrum, 140f
  spectrum/bandwidth efficiency, 138
  time domain representation, 137
Bit error rate (BER), 93
Bureau International de Poids et de Mesures
  (BIPM), 44

C

Carrier phase-based methods
  double differencing technique, 305—306
  single differencing technique, 303—305
  triple difference method, 306—307
Carrier phase-based ranging
  cycle slip, 210
  Doppler, 210
  integer ambiguity, 209
  receivers, 159
  schematic representation, 207—208, 208f
Carrier tracking, 186—192, 188f, 191f
Channel coding, 93
Child tracking, 357, 358f
Cipher message, 130
Civil applications
  disaster warning, 359
  emergency/medical services, 359
  personal applications
    child tracking, 357, 358f
    location-based services, 357—358
    sports and entertainment, 358—359
  search and rescue, 359
  vehicular applications
    air navigation, 357
    marine navigation, 356
    rail navigation, 356
    road navigation, 355—356, 356f
    space navigation, 357

Printed in the United States
By Bookmasters